Basic
Photographic
Materials
and
Processes

Leslie Stroebel

John Compton

Ira Current

Richard Zakia

South Essex College
Further & Higher Education, Southend Campus
Luker Road Southend-on-Sea Essex SS1 1ND
Tel: 01702 220400 Fax: 01702 432320
Minicom: 01702 220642

Focal Press
Boston   London

Focal Press is an imprint of Butterworth Publishers.

Cover design: Instructional Media Services/Rochester Institute of
Technology
Cover photograph: Barbara Morgan. *Cadenza.* 1940. Courtesy of the
Willard and Barbara Morgan Archives.
Illustrations: J. Wesley Morningstar

Library of Congress Cataloging-in-Publication Data
  Basic photographic materials and processes / Leslie Stroebel . . .
[et al.].
    p.   cm.
  ISBN 0-240-80026-5
  1. Photography. I. Stroebel, Leslie D.
TR145.B25 1989
771—dc20   89-36389

British Library Cataloguing in Publication Data
Basic photographic materials and processes.
  1. Photography
  I. Stroebel, Leslie
  770

  ISBN 0-240-80026-5

Butterworth Publishers
80 Montvale Avenue
Stoneham, MA 02180

10 9 8 7 6 5 4 3 2 1

Printed in the United States of America

Contents

Zeke Berman. Untitled. 1984.

Preface

Nathan Lyons.

Since the invention of the Daguerreotype process in 1839, the procedure for making photographs has become progressively easier. With the automated photographic equipment now available, there are some who claim it is no longer necessary to know anything about the technology of photography to make good photographs. This thought is not new. Over a hundred years ago, George Eastman used the slogan "You press the button, we do the rest." Serious photographers, however, have always known that knowledge of how photography works enables them to better achieve the results they want. This book was designed to be used as a textbook by college level students, but it also is appropriate for anyone who wants to learn more about how the photographic process works.

Evaluation of the esthetic attributes of photographs is considered to be subjective in nature since feelings play a more important role than logical reasoning, and different viewers can rightfully reach different conclusions. Technical aspects of photographs, on the other hand, are considered to be objective in nature since they can be measured, and different persons should obtain approximately the same measurements.

Although this book does not deal with photography as an art form, it does deal with esthetics in two important ways. First, by providing the knowledge needed to control the various components of the photographic process, technology enables the photographer to achieve desired image effects. For example, graininess of a photographic image can be affected by every step of the process from choice of subject to choice of printing paper. Information on how to control graininess is presented without value judgment, leaving it to the photographer to vary the graininess as seems appropriate.

Second, by providing the photographer with data that have previously been correlated with viewer responses to photographs representing systematic variations of a factor, technology reduces the number of photographs that are unacceptable due to related esthetic considerations. For example, published film speeds enable photographers to obtain a high proportion of correctly exposed photographs. Exposure failure occurs due to deviation from one or more of the conditions under which the tests were conducted. When picture-making conditions differ from the test conditions, technological procedures can be used to enable the photographer to arrive at an adjusted or personalized exposure index.

Photographic technology has evolved from traditional scientific disciplines that include mathematics, physics, chemistry, psychology, and physiology. Most of the technical material in this book is related to the process of making photographs, as distinct from the study of photographic theory for its own sake. It should be noted, however, that many students who earn a college degree in photography are attracted to photographic careers that do not have a major emphasis on working with a camera. Many of these positions require the same understanding of the photographic process as is expected for professional picture-making photographers.

The authors' objectives are that this book will enable the reader to

1. understand and apply published information about photographic equipment and materials (such as manufacturers' data publications),

2. learn how to obtain reliable information about the subject, equipment, materials, and processes involved in making photographs (such as how to determine the accuracy of an exposure meter),

3. learn to solve technical problems involved in the production of photographs (such as how to use controlled fogging to change image contrast),

4. build a technical foundation for the more advanced formal or informal study of photography and related imaging systems (such as video, computer imaging, and photomechanical reproduction), and

5. learn the basics of how the process of visual perception operates, and learn to anticipate and compensate for variations in visual perception (such as those associated with adaptation and defective color vision.

Concise and practical marginal notes have been used throughout the book to provide the reader with highlights of important concepts and to serve as a preview of each section.

Sample questions are included at the end of each chapter so that readers may check their understanding of the material.

1 Light and Photometry

Andrew Davidhazy.

THE NATURE OF LIGHT

Light is fundamental to photography. The function of photographic materials is to record patterns of light. The function of photographic equipment used for taking pictures is to produce light (lamps), to measure light (exposure meters and color-temperature meters), and to control light (lenses, shutters, apertures, and filters). It is therefore desirable for the photographer to possess some understanding of the nature of light.

Light is defined as the form of radiant energy that our eyes are sensitive to and depend upon for the sensation of vision. The obvious importance of light has resulted in it being the object of an enormous amount of experimentation and study over many centuries. One of the first persons to make significant headway in understanding the nature of light was Isaac Newton. In the seventeenth century he performed a series of experiments and proposed that light is emitted from a source in straight lines as a stream of particles. This theory was called the "corpuscular theory."

However, the facts that light bends when it passes from one medium to another, and that light passing through a very small aperture tends to spread out, are not easily explained by the corpuscular theory. As a result, Christian Huygens proposed another theory called the "wave theory." This theory holds that light and similar forms of electromagnetic radiation are transmitted as a waveform in some media. (This theory was elaborated considerably by Thomas Young in the nineteenth century after he performed a number of experiments.) The wave theory satisfactorily explained many of the phenomena associated with light that the corpuscular theory did not, but it still did not explain all of them.

One of the more notable of these unexplained effects was the behavior of "blackbody radiation." (Blackbody radiation is radiation produced by a body that absorbs all the radiation that strikes it, and emits radiation by incandescence, depending on its temperature.) In 1900 Max Planck suggested the hypothesis of the "quantization of energy" to explain the behavior of blackbody radiation. This theory states that the only possible energies that can be possessed by a ray of light are integral multiples of a quantum of energy.

Light travels about 25,000 miles in 1/10 of a second. Sound would require about 33 hours to travel the same distance.

In 1905, Einstein proposed a return to the corpuscular theory of light with light consisting of photons, each photon containing a quantum of energy. These suggestions, along with others, gradually developed into what is known today as "Quantum Theory" or "Quantum Electrodynamics." This theory combines aspects of the corpuscular and wave theories, and satisfactorily explains all of the known behavior of light. Unfortunately, this theory is difficult to conceptualize, and can be rigorously explained only by the use of sophisticated mathematics. As a result, the corpuscular and wave theories are still used to some extent where simple explanations of the behavior of light are required.

LIGHT WAVES

If the idea that light moves as a wave function is accepted, it becomes necessary to determine the nature of its waves and the relationship of light to other forms of radiation. Actually, light is but a fractional part of a wide range of radiant energy that exists in the universe, all of which can be thought of as traveling in waves. These forms of energy travel at the same tremendous speed of approximately 186,000 miles (3×10^8 meters) per second, differing only in wavelength and frequency of vibration. These waves have been shown to vibrate at right angles to their path of travel. The distance from the crest of one wave to the crest of the next is termed the wavelength, with the Greek letter λ (lambda) used as the symbol. Figure 1–1 illustrates this concept. The number of waves passing a given point in a second is called the frequency of vibration; the symbol f is used to specify it. The wavelength multiplied by the frequency of vibration equals the speed or velocity (symbol v) of the radiation. Thus, $\lambda \times f = v$.

Since the wavelength of radiant energy can be determined with far greater accuracy than the frequency, it has become common practice to specify a particular type of radiation by its wavelength. Because of the extreme shortness of the wavelengths encountered with light sources, the most frequently employed unit of measure is the nanometer (nm) which is equal to one-billionth of a meter. A somewhat less commonly used measure is

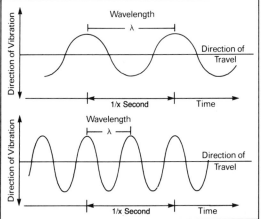

Figure 1–1 A simple form of a light wave, illustrated in a longitudinal cross section. In reality, the wave is vibrating in all possible right angles to the direction of travel. The second wave has a wavelength one half that of the first and, therefore, a frequency twice as great.

the Angstrom (Å), which is equal in distance to 1/10 of a nanometer (e.g., 400 nm equals 4,000 Å). Table 1–1 summarizes these measurement concepts.

THE ELECTROMAGNETIC SPECTRUM

When the various forms of radiant energy are placed along a scale of wavelengths, the resulting continuum is called the *electromagnetic spectrum*. Although each form of radiant energy differs from its neighbors by an extremely small amount, it is useful to divide this spectrum into the generalized categories shown in Table 1–2. All radiations are believed to be the result of electromagnetic oscillations. In the case of *radio waves*, the wavelengths are extremely long, being on the order of 10^{10} nm, and are the result of long

Table 1–1 Units of length

Unit	Symbol	Length
Meter	m	3.218 ft. (38.6 in.)
Centimeter	cm	0.01m (10^{-2}m)
Millimeter	mm	0.001m (10^{-3}m)
Micrometer	μ (mu)	0.000001m (10^{-6}m)
Micron	"	"
Nanometer	nm	0.000000001m (10^{-9}m)
Millimicron	mμ	"
Angstrom	Å	0.0000000001m (10^{-10}m)

Table 1–2 The electromagnetic spectrum

Frequency Hertz (cycles per second)	Wavelength nm		Type of Radiation
	10^{15}		
10^{4}	10^{14}		Maritime radio
	10^{13}		communications
10^{6}	10^{12}	1 km	AM radio
	10^{11}		
10^{8}	10^{10}		
	10^{9}	1 m	FM radio
10^{10}	10^{8}		Radar
	10^{7}	1 cm	TV
10^{12}	10^{6}		Microwave
	10^{5}		
10^{14}	10^{4}		Infrared
	10^{3}		Light
10^{16}	10^{2}		
	10^{1}		Ultraviolet
10^{18}	10^{0}	1 nm	
	10^{-1}	1 Å	
10^{20}	10^{-2}		X-rays
	10^{-3}		
10^{22}	10^{-4}		
	10^{-5}		
	10^{-6}		Gamma rays

electrical oscillations. The fact that such energy permeates our environment can easily be substantiated by turning on a radio or television receiver in any part of the technologically developed world. This form of radiant energy is not believed to have any direct effect upon the human body. Radio waves are customarily characterized by their frequency, expressed in hertz (cycles per second).

The portion of the electromagnetic spectrum that is sensed as heat is called the *infrared* region. The origin of this form of radiant energy, which is shorter in wavelength than radio waves, is believed to be the excitation of electrons by thermal disturbance. When these electrons absorb energy from without, they are placed in an elevated state of activity. When they suddenly return to their normal state, electromagnetic radiation is given off. It has been shown that all objects at a temperature of greater than −273°C give off this type of radiation. The temperature of −273°C is referred to as absolute zero or 0° Kelvin (K), after Lord Kelvin, who first proposed such a scale. Thus, all the objects we come into contact with give off some infrared energy. In general, the hotter an object, the more total energy it produces and the shorter the peak wavelength.

If the object is heated to a high enough temperature, the wavelength of the energy emitted will become short enough to stimulate the retina of the human eye and cause the sensation of vision. It is this region of the electromagnetic spectrum that is termed *light*. Notice that it occupies only the narrow section between approximately 380 nm and 720 nm. Because the sensitivity of the human visual system is so low at these limits, 400 and 700 nm are generally considered to be more realistic values. Objects with very high temperatures produce *ultraviolet* energy, which is shorter than 400 nm in wavelength.

To produce radiant energy shorter in wavelength than about 10 nm requires that an object be bombarded by fast-moving electrons. When these rapidly moving electrons strike the object, the sudden stopping produces extremely short wave energy called *X-radiation*, or more commonly, *X-rays*. Still shorter wavelengths can be produced if the electron bombardment intensifies, as occurs in a cyclotron. In addition, when radioactive material decomposes, it emits energy shorter in wavelength than X-rays. In these two cases the energy is referred to as gamma rays, which are usually 0.000001 nm (10^{-6} nm) in wavelength and shorter. These forms of electromagnetic energy are the most energetic, penetrating radiation known.

Thus it can be seen that the wave theory of radiant energy provides a most useful system for classifying all the known forms of radiation.

THE VISIBLE SPECTRUM

Although the terms "UV light" and "black light" are commonly seen in print, "UV radiation" is a more appropriate term.

Light occupies a very small part of the electromagnetic spectrum.

Located near the middle of the electromagnetic spectrum are the wavelengths of energy referred to as light. It is important to note that the location of this region is solely dictated by the response characteristics of the human eye. In fact, the international standard definition states: "Light is the aspect of radiant energy of which a human observer is aware through the visual sensations which arrive from the stimulation of the retina of the eye." Stated more simply, light is that energy which permits us to see. By definition, all light is visible, and for this reason the word *visible* is an unnecessary (and perhaps confusing) adjective in the common

expression *visible light*. This definition also may be interpreted to mean that energy which is not visible cannot (or should not) be called light. Thus it is proper to speak of ultraviolet radiation and infrared radiation, but not ultraviolet light and infrared light. The popular use of such phrases as *black light* and *invisible light* to describe such radiation makes it impossible to determine what type of energy is being described, and they should be avoided.

To understand more about light, it is necessary to become familiar with the way in which the human eye responds to it. The graph in Figure 1–2 represents the results of testing 100 observers with normal color vision for the perception of brightness in respect to different wavelengths of energy. The plot comes from the average of all observers and shows the sensitivity of the eye to different wavelengths (or colors) of light. In this case, sensitivity is similar to a film speed. These data indicate that the sensitivity of the eye drops to near zero at 400 nm and at 700 nm, thus specifying the limits within which radiant energy may be referred to as light. It also shows that the response of the eye is by no means uniform throughout the visible spectrum. In fact, if equal physical amounts of different colors of light are presented to an observer, the curve shows that the middle (green) portion of the spectrum would appear the brightest, and the extreme (blue and red) parts would look very dim. It is for this reason that a green "safelight" is used when processing panchromatic film. Since the eye

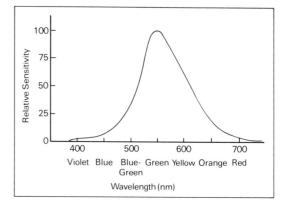

Figure 1–2 The sensitivity curve of the human eye. The plot indicates the relative brightness of the energy at each wavelength. Note that the curve is asymmetrical.

is more sensitive to green light, it takes less of it to illuminate the darkroom than any other color of light.

The plot shown in Figure 1–2 has been accepted as an international standard response function for the measurement of light. Therefore, any meter intended for the measurement of light must possess a sensitivity function identical to it. Most photoelectric meters used in photography have response functions significantly different from the standard and are not properly called light meters, although the international standard curve can be approximated by the use of appropriate filters with some meters. (It should be noted that the determination of the proper f-number and shutter speed for a given photographic situation does not require a meter with this response function.)

When all of the wavelengths between 400 nm and 700 nm are presented to the eye in nearly equal amounts, the light is perceived as white. There can be no absolute standard for white light because the human visual system easily adapts to changing conditions in order to obtain the perception of whiteness. For example, the amounts of red, green, and blue light in daylight are significantly different from those of tungsten light; however, both can be perceived as white due to physiological adaptation and the psychological phenomenon known as color constancy. Thus our eyes readily adapt to any reasonably uniform amount of red, green, and blue light in the prevailing illumination. This means that our eyes are not to be relied upon when judging the color quality of the prevailing illumination for the purposes of color photography.

If a beam of white light is allowed to pass through a glass prism as illustrated in Figure 1–3, the light is dispersed into a series of colors termed the *visible spectrum*. This separation of the colors occurs as a result of the light of various wavelengths being bent by varying amounts. The shorter-wavelength blue light is bent to a greater extent than the longer-wavelength green and red light. The result is a rainbow of colors that range from a deep violet to a deep red. Experiments indicate that human observers can distinguish nearly 100 different spectrum colors. However, the visible spectrum is often arbitrarily divided into the seven colors listed in Figure

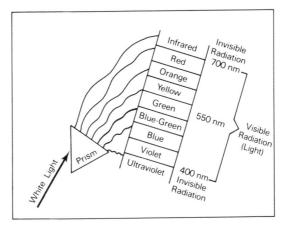

Figure 1–3 The dispersion of white light into the visible spectrum.

1–3. For the purpose of describing the properties of color photographic systems in simple terms, the spectrum is divided into just three regions: red, green, and blue. The color of the light may be specified at a given wavelength, thereby defining a spectral color. Such colors are the purest possible because they are unaffected by mixture with light of other wavelengths. It is also possible to specify a certain region or color of the spectrum by the bandwidth of the wavelengths. For example, the red portion of the spectrum could be specified as the region from 600 nm to 700 nm.

SOLID-OBJECT SOURCES AND THE HEAT-LIGHT RELATIONSHIP

Light is a form of energy that can only be produced from some other form of energy. The simplest and perhaps most common method is from heat energy, a process called *incandescence*. Whether the source is a filament in a tungsten lamp, a candle flame, or anything that has been heated until it glows, incandescence is always associated with heat. In fact, the amount and color of light produced by an incandescent source is directly related to the temperature to which it is heated. Consider, for example, an iron poker, one end of which is placed in a fire. Initially the poker feels somewhat cold; but as it is left in the fire, its temperature rises and it begins to feel warm. By increasing the temperature of the poker, we can become aware

Leonardo da Vinci discovered that white light contains different colors 200 years before Newton, but he identified only five colors.

of a change in its radiating properties through our sense of touch, although it looks the same. Soon the poker will become too hot to touch and we will be able to sense its radiation as heat at a short distance. As the temperature is raised even higher, the poker reaches its point of incandescence and begins to emit a deep red glow. If the poker is allowed to get hotter still, the color of the light it produces will become more yellowish, and the light will become brighter. At extremely high temperatures, the end of the poker will look white, and ultimately blue, in addition to becoming still brighter. All of this illustrates the fact that a solid object, when heated to its point of incandescence and higher, produces light that varies in color as a function of its temperature. When describing the temperature of such sources, it is common practice to employ the absolute or Kelvin scale for reasons discussed earlier.

The best solid-object radiator is an absolutely black body, as it absorbs all the energy that strikes it. All light sources are radiators, some of which are more efficient than others. Thus, since a perfectly black object would absorb and emit energy but not reflect it when heated, it would be the most efficient source. A blackbody source is achieved in practice by the design shown in Figure 1–4. An enclosure surrounded by a boiling material becomes the source. Since the interior surface of the object is darkened and everywhere

All objects emit radiation at room temperature, but objects must be heated to a temperature of approximately 1200° Fahrenheit before visible radiation is emitted.

concave, any light that enters will be absorbed, either immediately or after one or more reflections. Consequently the hole will appear perfectly black. As the walls of the oven are heated, they will emit radiant energy in all directions; that which escapes through the hole is called blackbody radiation. When such an experiment is performed and the blackbody is heated to a variety of temperatures, the characteristics of the radiant energy it produces change systematically, as shown in Figure 1–5.

As discussed previously, every object emits energy from its surface in the form of a spectrum of differing wavelengths and intensities when heated to temperatures greater than −273°C (absolute zero). The exact spectrum emitted by the object is dependent upon its absolute temperature and its *emissivity* (the flow of radiation leaving a small area divided by the area). Since the blackbody has perfect emissivity, the temperature in degrees Kelvin becomes the only important factor. It can be seen in Figure 1–5 that as the temperature increases, the curves are everywhere higher, indicating that the amount of energy at each wavelength increases. Additionally, as the temperature increases, the wavelength at which the peak output occurs becomes shorter. The peak position shifts from the long-wavelength end of the infrared region

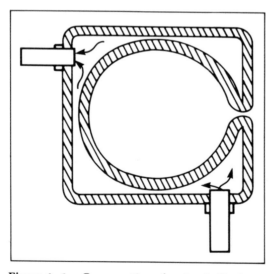

Figure 1–4 Cross section of a simple blackbody radiator consisting of an enclosure surrounded by boiling or molten material.

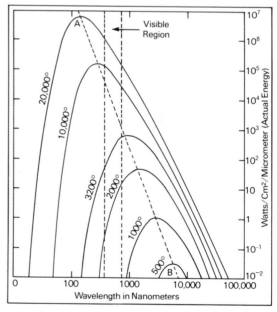

Figure 1–5 Spectral-energy curves for a blackbody heated to various temperatures.

toward the short-wavelength end of the ultraviolet region as the temperature increases. The portion of emitted energy that would be perceived as light is contained within the narrow band between the broken lines.

Some observations are now in order. First, since all of the objects with which we come into contact are at temperatures greater than −273°C, they are emitting some form of radiant energy, principally in the long-wavelength infrared region. For example, the human body at a temperature of 98.6 F (40 C, 313 K) emits infrared energy from 4000 nm to 20,000 nm, with the peak output at approximately 9600 nm. Second, most objects must be heated to a temperature of greater than 1000 K (727 C, 1340 F) in order to give off energy short enough in wavelength to be sensed by the human eye. For example, iron begins to glow red when it is heated to a temperature of about 1200 K, and a typical household tungsten lamp operates at a filament temperature of nearly 3000 K. Both sources will emit large amounts of infrared energy in addition to the visible energy. Third, in order for it to emit ultraviolet energy, a solid-object source must be heated to extremely high temperatures. Since tungsten steel melts at 3650 K, incandescent sources are not typically used when large amounts of ultraviolet energy are needed.

VAPOR SOURCES AND THE EFFECT OF ELECTRICAL DISCHARGE

A fundamentally different method for producing light makes use of radiation emitted from gases when an electrical current is passed through them. Such sources are called discharge lamps and generally consist of a glass tube containing an inert gas, with an electrode at each end. An electrical current is passed through the gas to produce light and ultraviolet energy. This energy may be used directly or to excite phosphors coated on the inside of the glass tube, as in a fluorescent lamp.

The basic process involved in the production of light is the same for all vapor lamps. The light emission from the vapor is caused by the transition of electrons from one energy state to another. When the electrical current is applied to the lamp, a free electron leaves one of the electrodes at high speed and collides with one of the valence electrons of the vapor atom. The electron from the vapor atom is bumped from its normal energy level to a higher one and exists for a short time in an excited state. After the collision, the free electron is deflected and moves in a new direction at reduced speed. However, it will excite several more electrons before it completes its path through the lamp. The excited electron eventually drops back to its former energy level and, while doing so, emits some electromagnetic radiation.

The radiation emitted may be at any of several wavelengths, depending primarily upon the properties of the vapor in the tube. Each type of vapor atom, when excited, gives off energy at wavelengths determined by its structure. Some gases emit radiation only at a few wavelengths, while others emit energy at many different wavelengths. These sources are said to show a discontinuous or line spectrum. For example, the spectrum of sodium vapor shows a bright yellow line near 600 nm, as shown in Figure 1–6, while mercury vapor produces energy at many different wavelengths, both in the ultraviolet region and in the visible region, as illustrated in Figure 1–7.

The pressure under which the vapor is dispersed in the tube has a significant effect upon the amount of energy that will be emitted. The terms *low pressure* and *high pressure* are often used to describe such lamps; low pressure indicates some small fraction of atmospheric pressure, while high pressure is

Solids emit continuous spectrums. Gases and vapors emit discontinuous spectrums.

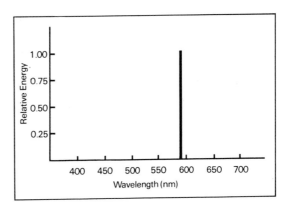

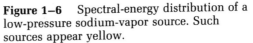

Figure 1–6 Spectral-energy distribution of a low-pressure sodium-vapor source. Such sources appear yellow.

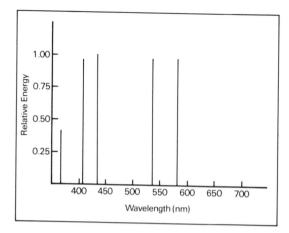

Figure 1–7 Spectral-energy distribution of a low-pressure mercury-vapor source. This source would appear violet.

applied to sources working above atmospheric pressure. High-pressure sodium-vapor lamps are often used for illuminating streets at night. Low-pressure sodium-vapor lamps are often used as safelights in photographic darkrooms when working with orthochromatic materials because they produce light in regions of the spectrum where the emulsion shows low sensitivity. High-pressure mercury-vapor sources are often used when making blueprints, while low-pressure mercury-vapor sources are used in greenhouses as plant lights because the ultraviolet energy they emit is beneficial to plant growth. It is important to note that the spectral characteristics of the radiation emitted by these sources is dependent primarily upon the properties of the vapor in the tube.

Perhaps the most commonly encountered vapor source is the fluorescent lamp. These are typically low-pressure mercury-vapor tubes that have phosphors coated on the inside of the glass envelope. When bombarded by the large amount of ultraviolet radiation emitted by the mercury vapor, these phosphors begin to glow and give off visible energy at all wavelengths in the visible spectrum. Thus, the light emitted by a fluorescent lamp is the result of both the discontinuous energy emitted by the vapor and the continuous energy emitted by the fluorescing phosphors.

There are many classes of phosphors that can be used for this purpose, with each phosphor emitting its own color of light. Figure

1–8 illustrates the spectral energy distribution for a typical cool white fluorescent lamp. The light that corresponds to the discontinuous line spectrum produced by the mercury vapor may not be apparent to a human observer because of the ability of the visual system to adapt to variations in the color balance of "white" light. Photographic films do not have this capability. Therefore, color transparencies made on daylight type color film with "daylight" fluorescent illumination tend to have a greenish cast unless an appropriate filter is used.

LUMINESCENCE

Luminescence can be described as the emission of light due to causes other than incandescence. It is generally caused by the emission of photons from atoms by exciting them with energy of relatively short wavelengths. As the electrons return to a lower energy level, energy of longer wavelengths is released. The rate at which this occurs can be affected by the presence of an *activator*, which is made up of ionized atoms that trap, then release the electrons slowly for recombination. The exciting energy is usually in the ultraviolet region, but can be caused by energy in the visible and infrared regions. There are many forms of luminescence. *Bioluminescence* is the result of biochemical reactions, and is typically seen in fireflies and glowworms. *Chemiluminescence* occurs as the result of some chemical reactions. *Gal-*

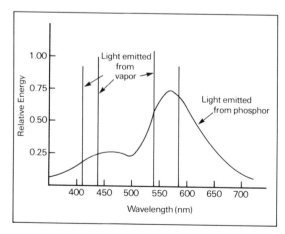

Figure 1–8 Spectral-energy distribution of a cool white fluorescent lamp.

vanoluminescence occurs as the result of galvanic action (the result of the flow of an electrical current) and, in the days when vacuum-tube rectifiers were common, could be seen as a "glow" when they were operating. *Triboluminescence* occurs when a material is ground; an example would be the grinding of sugar with a mortar and pestle. Solid substances that luminesce are called *phosphors*. Photographers are most concerned with the luminescence that occurs as a result of excitation with ultraviolet energy and light.

Fluorescence is luminescent emission of electromagnetic radiation in the visible region that occurs during the time of excitation. Thus, phosphors that are radiated with ultraviolet energy fluoresce. When the excitation is stopped, the fluorescence ceases within about 10^{-8} seconds; but it sometimes continues for as long as 10^{-1} seconds, depending on the activator.

Phosphorescence is the emission occurring after the excitation has been stopped, and which continues for somewhat more than 10^{-8} seconds up to several hours. The duration is strongly dependent on temperature. Phosphorescence is similar to fluorescence, but has a slower rate of decay.

Some dyes fluoresce, including fluorescein, eosin, rhodamine, and a series of dyes (and other materials) that are used as brighteners in substances such as paper and cloth. Photographic papers may contain brighteners to give them "cleaner" and more brilliant "whites." Most modern papers are treated with brighteners, a distinguishing characteristic when comparing modern prints to those made 30 or 40 years ago, without brighteners. The occurrence of brighteners in fabrics and similar materials can present problems in color photography when the ultraviolet component of electronic flash energy (or other light sources) causes them to fluoresce, most often in the blue region of the spectrum. This has a strong effect on the reproduced color of the material, often rendering it "blue" or "cyan" when the other materials in the scene have been reproduced satisfactorily in the photograph. The effect is minimized or eliminated by the use of an ultraviolet-absorbing filter over the electronic flash, or other source, to prevent it from exciting the fluorescing dyes or pigments in the material. An ultraviolet filter over the camera lens in this case would not correct for the fluorescence.

Fluorescent lamps excite gas molecules within the tube by means of electrons, to produce energy of discrete wavelengths, or lines, largely in the blue and ultraviolet regions (but dependent to a great extent on the gas elements in the tube). Some of this energy is absorbed by phosphors coated on the inside of the tube and is converted to longer-wavelength (visible) radiation. The color of the fluorescent emission is highly dependent on the phosphor material, and the activators incorporated in it.

Fluorescing screens are extensively used in radiography. In this application, they are called *intensifying screens*. They fluoresce when activated by the X-rays used for exposure, and the visible radiation from the screens is considerably more effective in exposing photographic emulsions than the X-rays themselves. The screens are placed in contact with both sides of the film, in the film holder, during exposure.

THE USE OF COLOR TEMPERATURE

As discussed previously, the amount and color of radiation emitted by a solid-object source is very much temperature dependent. In fact, the color of light being emitted from such a source can be completely specified by the Kelvin temperature (K) at which it is operating. Such a rating is referred to as the *color temperature* of the source. A color temperature may be assigned to any light source by matching it visually to a blackbody radiator. The temperature of the blackbody radiator is raised until the color of its light visually matches that from the lamp, and the Kelvin temperature of the blackbody is then assigned as the color temperature of the lamp. Thus, the use of color-temperature ratings in photography presumes that the lamp being described adheres to the same heat-to-light relationship as does the blackbody radiator. For incandescent lamps, this presumption is correct. However, the color of light emitted by some sources, such as fluorescent lamps, has no relationship whatever to the operating temperature. In these cases, the term *correlated color temperature* is used to indicate

Brighteners in photographic papers fluoresce to produce whiter whites.

Brighteners in clothing, paper products, hair rinses, etc., can adversely affect the color of the objects in color photographs.

As color temperature increases, the color balance of the light shifts from reddish to bluish.

that the color of light emitted by a blackbody radiator of that temperature produces the closest visual match that can be made with the source in question. It is important to note here that a visual match of two sources does not ensure a photographic match. In fact, in most cases, the photographic results will be different.

The color temperatures of a variety of light sources are given in Table 1–3. It is apparent that photographers are faced with a tremendous range of color temperatures, from the yellowish color of a candle at about 1800 K to the bluish appearance of north sky light, rated at 15,000 K. Notice that as the color temperature increases, the color of light emitted from the source shifts from red to white to blue in appearance. In order for photographers to produce excellent photographs under widely varying illuminant conditions, color films are designed for use with three different color temperatures. Type A color films are designed to produce images having optimum color balance with light rated at 3400 K, while Type B films are intended for a source rated at 3200 K. Daylight-balanced color films are designed to be used with 5500 K light, which is often referred to as photographic daylight. The actual color temperature encountered from a lamp in practice can vary significantly as a result of reflector and diffusor characteristics, changes in the power supply and the age of the bulb. Consequently, it is often necessary to measure the color temperature of the source with a color-temperature meter when using color films. Most

Tungsten color film can be used in daylight by using a filter that converts daylight to tungsten-quality light.

Light Conversions. An orange 85B filter converts daylight to tungsten quality. A bluish 80A filter converts tungsten light to daylight quality.

color-temperature meters employ photoelectric cells used in conjunction with color filters to sample specific regions of the spectrum being produced. These instruments measure the blue energy compared to the red energy output and give a reading directly in color-temperature values, or indirectly through the use of calibration charts. Such meters can be successfully used in both daylight and tungsten lighting conditions, where the source is producing a continuous spectrum. However, when used with vapor sources that produce a discontinuous spectrum, the results may be very misleading because of significant differences in this response compared to the visual response.

When the color temperature of the light from a solid-object source does not match the response of the film, light-balancing filters must be used over the camera lens. If the color temperature is too high, a yellow filter over the camera lens will lower the color temperature of the light that the film receives. If the color temperature of the light is too low, a blue filter can be used to raise it. If the filters are desaturated in color, then the changes to color temperature will be relatively small, while saturated filters will give large changes in color temperature.

The filters necessary to properly correct the color of light from vapor sources, such as fluorescent lamps, must either be determined from manufacturer's literature or from tests performed with the color film itself.

THE MIRED SCALE

Although color temperature provides a useful scale for classifying the light from continuous-spectrum sources, it does have some limitations. For example, a 500 K change at the 3000 K level does not produce the same visual or photographic effect as a 500 K change at the 7000 K level. This is because there is a nonlinear relationship between changes in a source's color temperature and the changes in the color of light it produces, which is illustrated in Figure 1–9.

This awkwardness in the numbers can be eliminated if the reciprocal of the color temperature is used, because of the more nearly linear relationship existing between this value and the effect. Since this number would be

Table 1–3   The color temperatures of some common light sources

Source	Color Temperature (°K)
Candle	1800
60-watt tungsten lamp	2800
100-watt tungsten lamp	2900
250-watt photographic lamp	3200
250-watt studio flood lamp	3400
Flashbulb (uncoated)	4000
Direct sunlight	4500
Photographic daylight (sunlight plus sky light)	5500
Electronic flash	6000
Sky light (overcast sky)	8000
North sky light	15,000 (approximately)

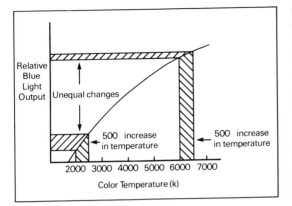

Figure 1–9 The relationship between color temperature and the relative amount of blue light being given off by a solid-object source.

quite small, it is commonly multiplied by 1 million to obtain a number of appropriate size. The resulting value is called the *mired* and is an acronym for the term "micro-reciprocal-degrees." Therefore, to convert color temperatures to the mired scale, the following formula is used:

$$\text{Mired Value} = \frac{1}{\text{Color temperature}} \times 10^6$$

Some publications also refer to this value as "reciprocal megakelvins."

Figure 1–10 illustrates the relationship between changes on the mired scale and changes in the resulting effect. It shows that equal changes in the mired scale produce equal changes in the effect. The relationship between color temperature and mireds is illus-

trated in Figure 1–11, which indicates that the higher the color temperature, the lower the mired value. Consequently, bluish sources with very high color temperatures will have very low mired values, while reddish-appearing sources with low color temperatures will have high mired values.

The mired scale is most frequently used for color-compensating filters that are intended to correct the color of light from a source to match the response of the color film. For example, if it is desired to expose a daylight color film balanced for 5500 K to a tungsten lamp operating at 3000 K, a color correction filter will be necessary to obtain the proper color balance. The color temperatures are converted to mired values, as shown below.

$$\text{Mired value of film} = \frac{1}{5500} \times 10^6 = 182$$

$$\text{Mired value of source} = \frac{1}{3000} \times 10^6 = 333$$

The mired value of the source, 333, is subtracted from the mired value of the light for which the film is balanced, 182; the resulting mired shift is −151. Thus, to correct this light source, a filter providing a mired shift of −151 would have to be used over the camera lens. The filter manufacturer's data sheet will provide the information necessary to determine the appropriate filter for this mired shift value. Negative mired values require bluish filters to increase the color temperature, while positive mired values require yellowish filters to decrease the color temperature. A major benefit of using the mired scale for designating filters is that a given mired change will have the same effect on the color of the light from solid-object sources at any color temperature.

Daylight has a higher color temperature than tungsten light— and a lower mired value.

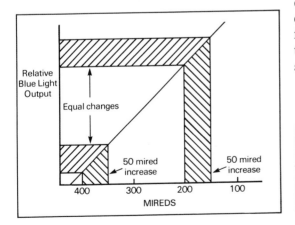

Figure 1–10 The relationship between mireds and the relative amount of blue light given off by a solid-object source.

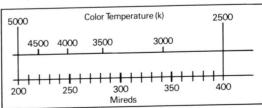

Figure 1–11 Comparison of color temperature and mired scales.

The color-temperature meters discussed previously often read out directly in mired values, and the mired shift can be calculated from that.

COLOR-RENDERING INDEX

Correlated color-temperature specifications are quite adequate for describing the visual appearance of vapor and fluorescent sources. However, when considering the way object colors are represented, such data are often misleading. The color-rendering ability of a light source is a measure of the degree to which the perceived color of objects illuminated by the source will match the perceived colors of the same objects when illuminated by a standard light source of high color-rendering ability, under specified viewing conditions. Thus, the color-rendering properties of a light source relate to color perception of the objects it illuminates.

Since fluorescent lamps do not follow the energy-emission properties of a blackbody radiator, correlated color temperatures do not indicate the amount of energy being produced at each wavelength. For example, the spectral-energy distributions for two fluorescent lamps of equal correlated color temperature (4200 K) are shown in Figure 1–12. Note that the lamp labeled Cool White—Deluxe is producing more red light than the lamp labeled Cool White, which would result in widely different color rendering for some objects. For example, if a person's face were illuminated by these sources, the Cool White—Deluxe lamp would provide more red light, resulting in a healthier, more natural skin-complexion appearance than that given by the Cool White lamp.

In an effort to address this problem, the concept of a color-rendering index (R) was established by the CIE (Commission Internationale de l'Eclairage or International Commission on Illumination). Determining the color rendering index, R, for any given light source requires that the spectral energy distribution and the correlated color temperature are known, so that an appropriate reference source can be selected. In this method, eight arbitrary Munsell color samples are used, and the CIE chromaticities under the given source are calculated. A similar

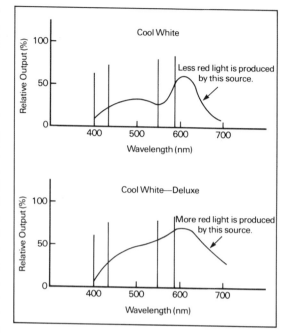

Figure 1–12   Spectral-energy distributions for two fluorescent lamps with same correlated color temperature (4200 K).

set of calculations under the selected reference source will give eight different CIE chromaticities. The differences between the two sets of data indicate the color shift of the given light source in relation to the reference source. The eight differences are then averaged to arrive at the color-rendering index, R. The R value is based upon an arbitrary scale that places a specific warm white fluorescent lamp with a correlated color temperature of 3000 K at R = 50 and the reference source at R = 100. The reference source always has an R equal to 100. Thus, the higher the R value for a given source, the more closely object colors will appear to their counterparts under the reference source, and, therefore, the better the color rendition. The color rendering indexes for a variety of fluorescent lamps are given in Table 1–4.

Some limitations to this concept should be noted. First, since the calculated R value is obtained in relation to a reference source, two given sources can only be compared to each other if their reference sources are similar (within approximately 200 K). Second, since the R value is an average taken from eight different colors, it does not specify the performance of the source on a specific color. Third, the R value is based upon the visual

Table 1–4 The correlated color temperatures and color rendering indexes for a variety of fluorescent lamps

Lamp Name	Correlated Color Temperature	Color-Rendering Index
Warm white	3000 K	R = 53
Warm white—deluxe	2900 K	R = 75
White	3500 K	R = 60
Cool white	4200 K	R = 66
Cool white—deluxe	4200 K	R = 90
Daylight	7000 K	R = 80

appearance of the object colors and not their photographic appearance. Thus, a source can have a high R value (90 or greater) and not give desirable color reproduction in a color photograph.

THE POLARIZATION OF LIGHT

As discussed previously, light, like other forms of radiation in the electromagnetic spectrum, is believed to move in a wave motion. The vibrations of these waves occur in all directions at right angles to the path of travel, as illustrated by the light beam to the left of the filter in Figure 1–13. This is described as an unpolarized light beam. However, the wave can be made to vibrate in one direction only, and when this occurs, the light is said to be polarized. The light beam on the right of the filter in Figure 1–13 illustrates this condition. Such polarization of light occurs naturally in a number of ways described below.

1. Light emitted from a portion of clear blue sky, at right angles to a line connecting the viewer (or camera) and the sun, is

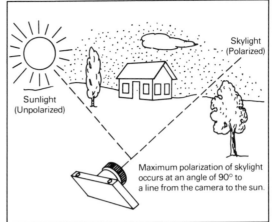

Figure 1–14   The polarization of sky light.

highly polarized. This is caused by the scattering of the light rays by very small particles in the atmosphere such as dust and molecules of water vapor and other gases. As the angle decreases, the effect also decreases, as illustrated in Figure 1–14. Thus, a portion of the light blue sky encountered in typical outdoor scenes is polarized.

2. Light becomes polarized when it is reflected from a flat, glossy, non-metallic surface, such as glass, water, and shiny plastics. This effect is maximized at an angle whose tangent is equal to the refractive index of the reflecting material; this is known as the Brewster angle. For common surfaces, such as glass and water, this angle is approximately 55° from the perpendicular (35° from the surface) and is illustrated in Figure 1–15.

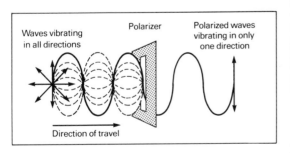

Figure 1–13   The polarization of light.

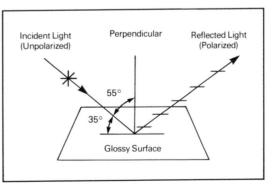

Figure 1–15   The polarization of light reflected from a glossy surface.

3. Light becomes polarized when it is transmitted through certain natural crystals or commercially manufactured polarizing filters. Among the substances having the ability to polarize light are the dichroic crystals, particularly the mineral tourmaline. When a beam of unpolarized light passes through a thin slab of tourmaline, it becomes polarized. This property can be demonstrated by rotating a second slab of tourmaline across the direction of polarization, as illustrated in Figure 1–16. When the second slab is rotated to a position of 90° from that of the first, no light will be transmitted through the second filter. Commercially available polarizing filters are made from dicrylic substances whose chemical composition calls for the grouping of parallel molecular chains within the filter. This is the nature of the filters illustrated in Figures 1–13 and 1–17.

Polarized light will become depolarized when it strikes a scattering medium. This most frequently occurs when the light is reflected from a mat surface, such as a sheet of white blotting paper. It can also be depolarized upon transmission through a translucent material, such as a sheet of mat acetate.

The phenomenon of polarized light provides the photographer with a very useful tool for controlling light. For example, a polarizing filter will absorb light that is in one plane of polarization. If the plane of polarization of the filter and the plane of polarization of the light are the same, the maximum amount of light will be transmitted. However, if they are at right angles to each other, no light will pass, as shown in Figure 1–17. At angles between these two, varying amounts of light will be transmitted, allowing for the useful control of light. Thus, polarizing filters can be thought of as variable-neutral density filters. (Also see Chapter 13.)

PRACTICAL SOURCES OF LIGHT AND THEIR PROPERTIES

Thus far, the discussion of light has centered on methods of production, properties, and measurement. At this point it is appropriate to consider some practical sources of light commonly encountered by the photographer.

Daylight

Daylight is usually composed of at least two different sources of light: (1) direct sunlight, which is modulated by the earth's atmosphere, and (2) blue sky light, which is the light reflected from the atmosphere. Additional light may also reach the subject by reflection from objects. The nature of daylight at any given time is dependent upon the geographical location, the season of the year, the time of day, the prevailing weather conditions, and the surroundings. When the sky is clear, the amount of direct sunlight is at a maximum, which is about 88% of the total

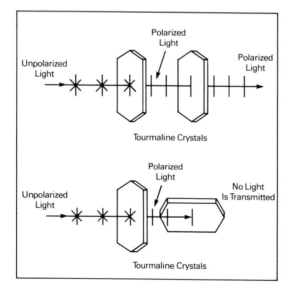

Figure 1–16 The polarizing effect of tourmaline crystals.

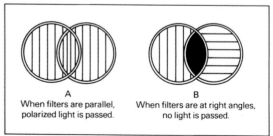

Figure 1–17 The use of polarizing filters to control light.

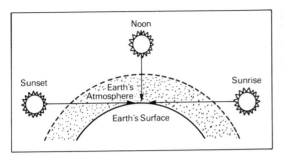

Figure 1–18 The relationship between position of the sun and the length of the sunlight's path of travel through the earth's atmosphere.

light on the ground, producing a lighting ratio of approximately 8:1. The ratio decreases as the amount of overcast increases.

Sunlight is produced by the high temperature of the sun, and its energy distribution outside the earth's atmosphere closely approximates that of a blackbody source at 6500 K. In passing through the earth's atmosphere, however, a considerable amount of this light is lost, particularly in the blue region, and the color temperature of the sunlight plus the skylight varies between 5000 and 6000 K on the ground during the middle of the day. Daylight type color films are balanced for a color temperature of 5500 K. Variations in color temperature with the time of day are illustrated in Figures 1–18 and 1–19. Color compensating filters can be used on the camera lens when the color temperature of the daylight does not match that for which color film was manufactured. The use of color temperature to describe the appearance of daylight

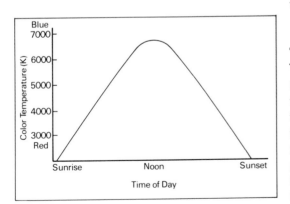

Figure 1–19 The variation in color temperature of light with time of day at the surface of the earth (no cloud cover).

is not completely appropriate, however, since the energy-output characteristics of daylight only approximate those of a blackbody radiator.

Skylight is blue because the small particles of the atmosphere selectively scatter light, especially the short wavelengths, away from the direction of travel of the sunlight and toward the observers below. This effect is known as Rayleigh scattering, and varies inversely with the fourth power of the wavelength. Thus the scattering at 400 nm is 9.4 times that at 700 nm. The color temperature of blue skylight ranges between 10,000 and 20,000 K.

Tungsten-Filament Lamps

Incandescent lamps emit light because of the high temperature reached by the tungsten filament wire as a result of its resistance to the passage of electricity—thus, electricity is converted to heat which is converted to light. The upper temperature limit for this type of source is 3650 K, the melting temperature of tungsten. Color temperature is an appropriate measure of the color quality of the light emitted by tungsten-filament lamps, and it ranges from approximately 2700 K for 15-watt household light bulbs to 3400 K for photoflood lamps, which have a relatively short life of 10 hours or less. Photographic studio lamps rated at 3200 K have approximately 10 times the life expectancy since lamp life depends primarily upon the rate at which tungsten evaporates from the filament, which in turn varies with the temperature. (See Figures 1–20, 1–21, and 1–22.)

Tungsten-Halogen Lamps

Sometimes referred to as quartz-iodine lamps, tungsten-halogen lamps differ from conventional tungsten lamps in that iodine is added to the gas in the tube and the envelope is made from quartz or other high-temperature glass. The iodine combines with the tungsten that evaporates from the filament, but the high temperature of the filament produces a decomposition that redeposits tungsten on the filament, thereby increasing the life of the filament and preventing the deposit of tung-

The color temperature of daylight is considered to be 5500 K.

Heating a tungsten filament to the temperature required to produce daylight-quality light would cause the filament to melt.

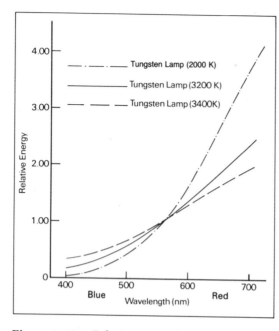

Figure 1–20   Relative spectral-energy distributions for three incandescent sources.

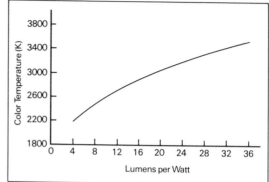

Figure 1–22   The average relationship between luminous efficiency and color temperature of tungsten filament lamps.

sten on the envelope. Figure 1–23 illustrates the loss in total light output for conventional tungsten lamps and tungsten-halogen lamps during their operating lives.

Fluorescent Lamps

Fluorescent lamps produce light by establishing an arc between two electrodes in an atmosphere of very low pressure mercury vapor contained in a glass tube. This low-pressure discharge produces ultraviolet radiation at specific wavelengths that excite crystals of phosphors lining the wall of the tube. Phosphors such as calcium tungstate have the ability to absorb ultraviolet energy and to re-radiate this energy as light. The color of light emitted by a fluorescent tube depends largely on the mixture of fluorescent materials used in the phosphor coating.

The light that reaches the eye (and the camera) from such a lamp, therefore, consists of the light given off by these fluorescent compounds plus such part of the light from the mercury vapor that gets through them without being absorbed. The result is a continuous spectrum produced by the fluorescent material, superimposed upon the line spectrum of energy produced through the electrical discharge of the mercury vapor. The spectral-energy distributions for three of the

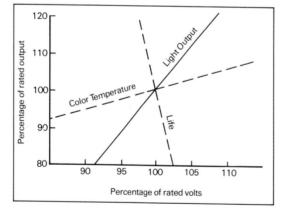

Figure 1–21   The variation in light output, color temperature, and life of a tungsten lamp at different voltages.

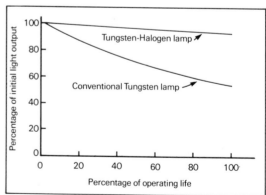

Figure 1–23   Typical lamp depreciation for a conventional tungsten lamp and a tungsten-halogen lamp.

more commonly encountered fluorescent lamps are illustrated in Figure 1–24.

Fluorescent lamps with a correlated color temperature of 5000 K (such as the General Electric Chroma 50®) are believed to give a better match to daylight and are now specified in ANSI standards for transparency illuminators and viewers. Fluorescent lamps find widespread use because they generate very little heat and are less expensive to operate (i.e., have higher luminous efficiency) than tungsten lamps. Additionally, they are low luminance sources, because they possess larger surface areas than do tungsten lamps. The result is that these lamps have less glare and produce softer illumination. Fluorescent lamps are nearly always used for commercial lighting and, consequently, photographers working under such light conditions should be familiar with their characteristics.

The luminous efficiency of fluorescent lamps is generally higher than that of tungsten sources, ranging between 40 and 60 lm/W, making them more economical to operate. Additionally, the average life of a fluorescent lamp is approximately 5,000 hours, which is nearly five times as long as that of a conventional tungsten lamp.

5000 K fluorescent lamps are recommended for transparency illuminators and print viewers.

High-Intensity Discharge Lamps

Similar in operation to fluorescent lamps, high-intensity discharge lamps produce light by passing an arc between two electrodes that are only a few inches apart. The electrodes are located in opposite ends of a small sealed transparent or translucent tube. Also contained within the tube is a chemical atmosphere of sodium and/or mercury. The arc of electricity spanning the gap between the electrodes generates heat and pressure much greater than in fluorescent lamps, and for this reason these lamps are also referred to as high pressure discharge sources. The heat and pressure thus generated are great enough to vaporize the atoms of the various metallic elements contained in the tube. This vaporization causes the atoms to emit electromagnetic energy in the visible region. Since the physical size of the tube is small, it allows for the construction of optical sources that have excellent beam control. Such sources are frequently employed for nighttime illumination of sports stadiums, highways, exteriors of buildings, and the interiors of large industrial facilities.

As in the operation of low-pressure discharge sources (i.e., fluorescent lamps), high-intensity discharge lamps produce spikes of energy at specific wavelengths. These peaks are the result of specific jumps of the electrons within the atomic structure of the metallic elements. Energy is emitted in peaks located in different positions of the visible spectrum for each element. Thus these lamps do not have a true color temperature, since they are not temperature dependent for the color of light emitted. The three most commonly encountered high-intensity discharge sources are mercury, metal halide, and sodium. (See Figures 1–25, 1–26, and 1–27.)

Since their introduction in the 1950s, high-

Camera filtration is recommended when making color photographs with discharge-lamp light sources.

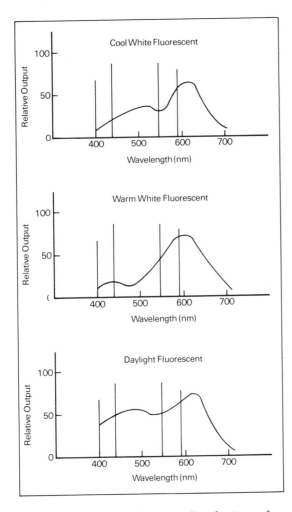

Figure 1–24 Spectral-energy distributions of three different fluorescent lamps.

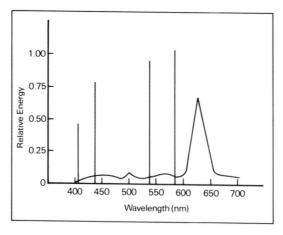

Figure 1–25   Spectral-energy distribution of a high-intensity mercury discharge source with the name DeLuxe White.

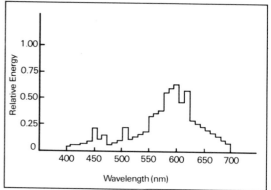

Figure 1–27   Spectral-energy distribution of a high-intensity sodium-vapor discharge source, the General Electric Lucalox®.

intensity discharge sources have steadily increased in use. Mercury vapor lamps were the first lamps of this type available. Although their efficiency was high (17 to 46 lm/W) compared to tungsten, their color-rendering ability was poor. Today, the metal halide (54 to 92 lm/W) and sodium lamps (59 to 106 lm/W) are preferred for both efficiency and improved color rendering. However, these sources pose the same problems photographically as do fluorescent lamps. The use of color temperature and the mired scale is entirely inappropriate for these sources. The results of testing specific sources with specific color films to obtain acceptable color balance are summarized in Table 1–5. The filtration conditions only represent starting

points for individual tests, since lamps and films are affected by problems of variability. Additionally, alternating current causes fluorescent and high-intensity discharge lamps to flicker. This effect is typically not visible due to the persistence of vision (see Chapter 12), but it can lead to unevenness of exposure across the film plane at shutter speeds faster than 1/60 second with focal-plane shutters. The same phenomenon can lead to underexposure when using a between-the-lens shutter at higher speeds.

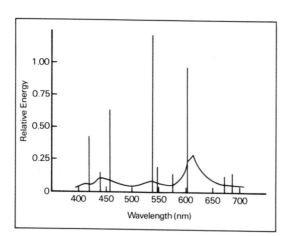

Figure 1–26   Spectral-energy distribution of a high-intensity metal halide discharge source, the General Electric Multi-Vapor® lamp.

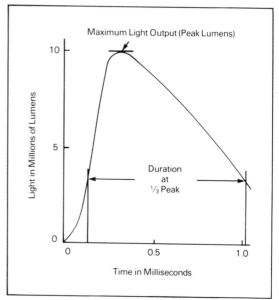

Figure 1–28   Light output curve of an electronic flash unit.

Table 1–5 Guides for initial tests when exposing color films with fluorescent and high intensity discharge lamps

Type of Lamp	Kodak Color Film Type[a]			
	Group 1 Daylight Balance	Group 2 Daylight Balance	Group 3 Tungsten Balance	Group 4 Tungsten Balance
Fluorescent				
Daylight	40M + 40Y	50M + 50Y	85B[b] + 40M + 30Y	85[d] + 40M + 40Y
	+ 1 stop	+ 1⅓ stops	+ 1⅔ stops	+ 1⅔ stops
White	20C + 30M	40M	60M + 50Y	40M + 30Y
	+ 1 stop	+ ⅔ stop	+ 1⅔ stops	+ 1 stop
Warm white	40C + 40M	20C + 40M	50M + 40Y	30M + 20Y
	+ 1⅓ stops	+ 1 stop	+ 1 stop	+ 1 stop
Warm white deluxe	60C + 30M	60C + 30M	10M + 10Y	No filter
	+ 2 stops	+ 2 stops	+ ⅔ stop	None
Cool white	30M	40M + 10Y	10R + 50M + 50Y	50M + 50Y
	+ ⅔ stop	+ 1 stop	+ 1⅔ stops	+ 1⅓ stops
Cool white deluxe	20C + 10M	20C + 10M	20M + 40Y	10M + 30Y
	+ ⅔ stop	+ ⅔ stop	+ ⅔ stop	+ ⅔ stop
Average fluorescent[c]	10B + 10M	30M	50R	40R
	+ ⅔ stop	+ ⅔ stop	+ 1 stop	+ ⅔ stop
High-Intensity Discharge Lamps				
General Electric Lucalox®[d]	70B + 50C	80B + 20C	50M + 20C	55M + 50C
	+ 3 stops	+ 2⅓ stops	+ 1 stop	+ 2 stops
General Electric Multi-Vapor®	30M + 10Y	40M + 20Y	60R + 20Y	50R + 10Y
	+ 1 stop	+ 1 stop	+ 1⅔ stops	+ 1⅓ stops
Deluxe white mercury	40M + 20Y	60M + 30Y	70R + 10Y	50R + 10Y
	+ 1 stop	+ 1⅓ stops	+ 1⅔ stops	+ 1⅓ stops
Clear mercury	50R + 30M + 30Y	50R + 20M + 20Y[e]	90R + 40Y	90R + 40Y
	+ 1⅔ stops	+ 1⅓ stops	+ 2 stops	+ 2 stops
Sodium vapor	NR	NR	NR	NR

Note: Red or blue filters have been included to limit the number of filters to three.

[a]Group 1 includes KODAK EKTACHROME 200; KODACHROME 25; KODAK VERICOLOR II, Type S; KODACOLOR II and KODACOLOR 400 Films.
   Group 2 includes KODACHROME 64, KODAK EKTACHROME 64, and KODAK EKTACHROME 400 Films.
   Group 3 includes KODAK EKTACHROME 50; KODAK EKTACHROME 160; and VERICOLOR II, Type I, Films.
   Group 4 includes KODACHROME 40 Film.
[b]KODAK WRATTEN Gelatin Filter No. 85B. ‡KODAK WRATTEN Gelatin Filter No. 85.
[c]Use of these filters will yield results of less than optimum correction. They are intended for emergency use only, when it is impossible to determine the type of fluorescent lamp in use.
[d]This is a high-pressure sodium vapor lamp; however, the *Guides for Initial Tests When Exposing Color Films with Fluorescent and High Intensity-Discharge Lamps* may not apply to other makes of high-pressure sodium vapor lamps due to differences in spectral characteristics.
[e]For KODAK EKTACHROME 400 Film, use 25M + 40Y and increase the exposure 1 stop.

Electronic Flash

An electronic flash lamp consists of a glass or quartz tube that is filled with an inert gas such as xenon and has electrodes placed at either end. When a high-voltage current from the discharge of a capacitor passes between the electrodes, the gases glow, producing a brilliant flash of light. The total time of operation is exceedingly short, with the longest times being on the order of 1/500 second and the shortest times approaching 1/100,000 second. The time-light curve for a typical electronic flash unit is shown in Figure 1–28. The effective flash duration is typically measured between one-third peak power points, and the area contained under the curve between these limits represents nearly 90% of the total light produced. If this number is measured in lumen seconds, the light being emitted in all directions is considered. A more revealing measurement is the number of effective beam-candle-power seconds, which is a measurement of the light output at the beam position of the lamp.

The spectral-energy distribution of these sources shows a line spectrum, the exact nature of which is determined by the type of gas dispersed in the tube. Although the gas gives a line spectrum, there are so many lines

The color temperature of most electronic-flash units is close to that of photographic daylight.

and they are so well distributed throughout the visible spectrum that no serious error is involved in considering the spectrum to be continuous, as shown in Figure 1–29. The spectrum from the gas in these tubes approaches incandescence due to the high-current density at the time of discharge. The resulting light has a correlated color temperature of approximately 6000 K, which is conveniently close to the 5500 K level at which daylight color films are designed to operate. The small difference in color temperature is often compensated for by using a color-correction filter built into the lens of the flash unit. If no such filter exists in the source, a color-correction filter can be used over the camera lens to avoid a bluish cast in the color images.

The total light output for an electronic-flash unit depends upon the design of the flash-lamp and reflector, the voltage, and the capacity of the capacitor. In the early days of electronic flash photography, the exposure was controlled by setting the f-number at the number obtained by dividing a guide number by the flash-to-subject distance. Electronic flash meters are now commonly used, and many portable electronic flash units have built-in sensors that monitor the light reflected from the subject and quench the flash when sufficient light falls on the subject to produce the desired exposure effect.

Electronic flash units have proven to be a most useful source of light for photographers. Their consistency in color output, quantity of light output from flash to flash, and the extremely short duration of the flash are all

LASER is an acronym for Light Amplification by Stimulated Emission of Radiation.

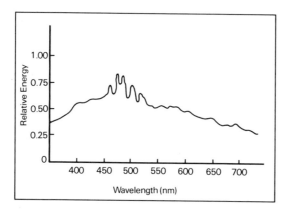

Figure 1–29   Relative spectral-energy distribution of a typical xenon-filled electronic flash unit.

reasons for the widespread use of these sources.

LASERS

The word *laser* is an acronym for "light amplification by stimulated emission of radiation." Stimulated emission is a process by which a photon of light stimulates an excited atom to emit a second photon that is at the same frequency and moves in the same direction as the initiating wave. This concept is sometimes referred to as optical pumping, which is the process whereby matter is raised from a lower to a higher energy state.

For example, the potential energy of water is raised when it is pumped from an underground well to a storage tank on top of a tower. There are many forms of matter that can be elevated or pumped, from one energy level to a higher one, by optical methods. It has been discovered that the atoms of some substances will absorb photons of light and when doing so increase their energy level. It has also been discovered that the excited atoms do not remain in the higher energy state but fall randomly and spontaneously to their lower or ground state. When this occurs, the stimulated atom emits a photon of light. Therefore, the laser medium is a substance such as a gas, liquid, or solid that amplifies the light waves by means of the stimulated emission process. Energy must be pumped into the laser medium to make it active. One way to supply this pumped energy is with light from an external source such as a flash tube.

Figure 1–30 illustrates the basic properties of a ruby laser. A ruby crystal rod has parallel polished ends which are mirrored surfaces. One end is only partially silvered and acts as a window for the light to escape. Energy is supplied to the ruby crystal by a powerful electronic flash tube, which serves to pump the atoms of the crystal to a higher energy state. They exist at this level for a few millionths of a second before dropping to their ground level, resulting in the emission of a photon of light as shown in Figure 1–30A. Although many of these photons will pass out of the crystal walls and be lost, eventually one photon will move directly along the rod and be reflected from the polished ends,

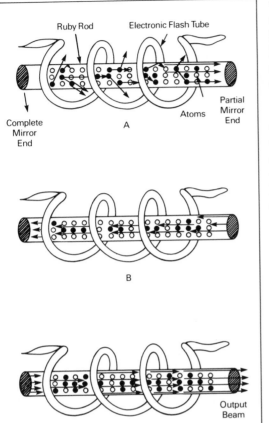

Figure 1–30   Basic operation of a ruby laser (the principle of "optical pumping").

varying wavelengths in and out of phase with each other, which is described as noncoherent light. This concept is illustrated in Figure 1–31. It is this high degree of coherence that makes laser light different from that of all other sources.

THE LANGUAGE OF LIGHT

The intelligent selection and application of light sources require a familiarity with the units of light measurement. Photometry is the branch of physics dealing with the measurement of the strength of light emitted by a source or the light falling on, transmitted by, or reflected from a surface. From the early 1900s, when the science of light measurement was first seriously undertaken, candles were used as the standard sources of light. Consequently, many of the definitions and units of measurement are based on the candle. However, since 1940 the international standard unit for light sources is based on the light emitted by one square centimeter of a blackbody radiator heated to the temperature of solidification of platinum, providing a standard that is more exactly reproducible. This standard unit of light is called the candela. The difference in light emitted by this new standard and the old standard candle is less than 2%.

Candela is a unit of measurement of the intensity of a light source.

passing back and forth along the rod until it encounters an atom in the excited elevated state, as shown in Figure 1–30B. As it strikes this excited mirrored atom, the atom radiates its photon in exact phase with the photon that struck it. This second photon will in turn stimulate another atom and, in this cascade process, continue to fill the rod with in-phase radiation that is oscillating back and forth between the mirrored ends of the rod. A portion of this radiation is emitted through the partially silvered end of the rod and becomes the laser beam (see Figure 30C).

The entire process occurs within a few thousandths of a second, and as the flash tube fires again, the process repeats itself. The result is an intense, monochromatic beam of light that can be focused to a tiny spot. Lasers emit light waves that are in phase with each other, and are described as being coherent. Conventional light sources radiate light of

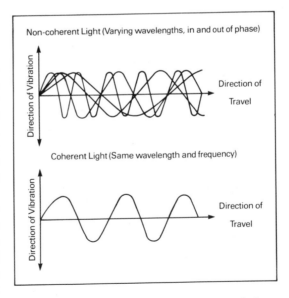

Figure 1–31   Noncoherent and coherent light waves.

THE MEASUREMENT OF INTENSITY

Intensity (sometimes called luminous intensity) is a measure of the rate at which the source emits light in a given direction. Initially, the intensity of a source was determined by comparison to the old standard candle, with the result expressed in terms of candlepower. Thus a 50-candlepower source was one that emitted light equivalent to that which would come from 50 of the standard candles. The correct unit now is the candela. One candela is defined as one sixtieth the light from one square centimeter blackbody heated to the freezing temperature of platinum. However, the term *candlepower* is still often used to describe intensity. Practical sources of light such as tungsten lamps and fluorescent tubes always vary in their intensity with direction, and therefore no single measurement of intensity can completely describe such sources. Perhaps the easiest way to understand the concept of intensity is to think of a ball-shaped lawn sprinkler with many holes through which the water can flow. The intensity of such a source would be similar to the rate at which water was being emitted through one of those holes in a specific direction. Such information would be of limited value since it would not describe the variation in intensities around the ball nor the total amount of water being emitted by the sprinkler.

Since the intensity of a real source changes with direction, it is desirable to learn about the distribution of intensities. Such information is usually provided by the lamp manufacturer in the form of a two-dimensional graph based on polar coordinates, an example of which is shown in Figure 1–32. In this graph, intensity is plotted against the angle on special paper, called polar-coordinate graph paper, that is constructed like a protractor with the angles marked around the margins and each angle having a marked value in candelas. The zero angle is head-on to the lamp, with the intensity in this direction known as the beam intensity (candlepower). From such a graph the intensity at any desired angle can be found.

For the lamp-reflector combination illustrated in Figure 1–32, the beam intensity is approximately 800 candelas. This intensity is nearly uniform, within 25° either side of

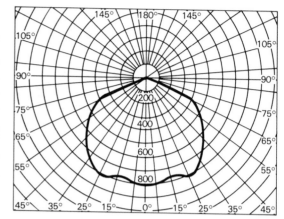

Figure 1–32   Polar-coordinate plot of intensities for a lamp-reflector combination.

the beam position, indicating that this lamp-reflector combination provides nearly uniform illumination over a 50° angle of projection. At 65° off the beam position the intensity drops to nearly 400 candelas. The same lamp equipped with a more narrowly curved reflector would produce a distribution of intensities much narrower than what is shown in Figure 1–32. Therefore, such polar-coordinate plots give fundamental information about the nature of the light that will be emitted.

When the intensity of a source is reported as a single value, there are three ways in which this value can be obtained, as illustrated in Figure 1–33. In the simplest case, the intensity in only one direction is measured and the value reported as a single candela value. When a number of readings are taken at uniform intervals on a horizontal plane around the source and then averaged, the result is the mean horizontal intensity (candlepower) of the light source. Instead of taking a large number of individual readings, this result is obtained in ordinary practice by rotating the source rapidly upon its vertical axis while a single reading is made. The intensity of light in all directions can be determined by measuring intensities at uniform intervals around the light source. An average of these readings would give the mean spherical intensity (candlepower) of the illuminant. It should be noted that this value is related to the total light output of the lamp. In each of these cases, the intensity is determined through comparison to a standard lamp at a variety of distances.

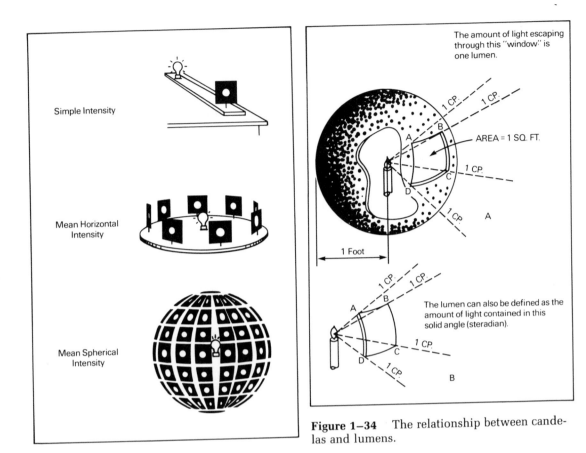

Figure 1–33   The measurement of intensity.

Figure 1–34   The relationship between candelas and lumens.

THE MEASUREMENT OF FLUX

Flux (sometimes called luminous flux) is the rate at which a source emits light in all directions. The flux of a source is usually calculated from measurements of intensity and is closely related to the measurement of mean spherical intensity previously discussed. The unit of measurement is the lumen. Since flux involves the output of light in all possible directions, the lumen therefore involves a three-dimensional concept.

The lumen may be defined as the amount of light falling on a surface one square foot in area, every point of which is one foot from a uniform source of one candela (candlepower). The relationship between the candela and the lumen is illustrated in Figure 1–34. If the opening indicated by A,B,C,D is one square foot of the surface area of a sphere of one-foot radius, the light escaping will be one lumen. If the area of this opening is doubled, the light escaping will be 2 lumens.

Since the total surface area of a sphere with a one-foot radius is 12.57 square feet (i.e., $4\pi r$), a uniform one-candela source of light emits a total of 12.57 lumens. Thus a source of 10 mean spherical candelas emits 125.7 lumens. Since an area of one square foot on the surface of a sphere of one-foot radius subtends a unit solid angle (i.e., one steradian) at the center of the sphere, the lumen may also be defined as the amount of light emitted in a unit solid angle by a source having an average intensity of one candela throughout the solid angle. Therefore, when considering a point source that emits light equally in all directions, there will be 12.57 (4π) lumens of flux for every candela of intensity.

THE MEASUREMENT OF ILLUMINANCE

Light sources may be termed the cause, and illumination the effect or result. Since candelas and lumens are both a measure of the cause, they apply only to the light source and not the effect obtained. Illuminance is defined as the light incident upon a surface. For the measurement of illumination, a unit known as the footcandle is often used.

A footcandle represents the illumination at a point on a surface that is one foot distant from, and perpendicular to, the light rays from a one candela source. For example, if the light source in Figure 1–35 has an intensity of one candela, the illuminance at point A, which is one foot distant from the source, will be equal to one footcandle. The illuminance at points B and C will be somewhat less because they will be at greater distances than one foot. Therefore, an illuminance reading applies only to the particular point where the measurement is made. By averaging the number of footcandles at a number of points, the average illumination of any surface can be obtained. This is often done when evaluating the evenness of illumination on an enlarger easel.

The footcandle is the unit of measure most closely associated with the everyday use of light. A working idea of this unit may be obtained by holding a lighted candle one foot from a newspaper in an otherwise darkened room. The result will be approximately one footcandle of illumination. A full moon on a clear night gives approximately 0.02 footcandle; a well-lighted street gives approximately 5 footcandles; a well-lighted classroom has nearly 50 footcandles of illumination; in daylight in open shade there are approximately 1,500 footcandles of illumination, and in direct sunlight approximately 12,000 footcandles. To photograph a moonlit scene on ISO 100 speed film would require an exposure of about 10 seconds at f/2, plus any additional exposure required to compensate for the reciprocity effect resulting from the long exposure time.

Referring again to Figure 1–34, it is evident that the surface A,B,C,D fulfills the conditions for a surface illuminated to a level of one footcandle. Every point on this square foot of surface is perpendicular to the rays of a one-candela source that is one foot distant. This illustrates an important relationship between lumens and footcandles. A lumen is the light flux spread over one square foot of area that will illuminate that area to a level of one footcandle. Therefore, one footcandle is equal to one lumen per square foot. This relation forms the basis of a simplified method of lighting design known as the lumen method. When the number of square feet to be lighted is known and the desired level of illumination determined, it is simple to determine the number of lumens that must be provided on the working plane.

With a one-candela source, as shown in Figure 1–36, the level of illumination on point A, which is one foot distant, is one footcandle. However, if plane A is removed and the same beam of light is allowed to pass on to plane B, which is two feet away, this same beam of light would now cover an area four times that of plane A. The average illumination on plane B would be one quarter as great as that on plane A, which would be equal to one-quarter of a footcandle. In the same fashion, if the beam of light is allowed to fall upon plane C, which is three feet away from the source, it will be spread over an area nine times as great as plane A, and so on.

Thus, illumination falls off (decreases) not in proportion to the distance but in propor-

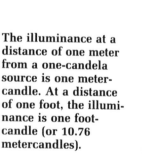

Illuminance meters are calibrated in footcandles and/or metercandles (lux).

One footcandle equals 10.76 metercandles.

The illuminance at a distance of one meter from a one-candela source is one metercandle. At a distance of one foot, the illuminance is one footcandle (or 10.76 metercandles).

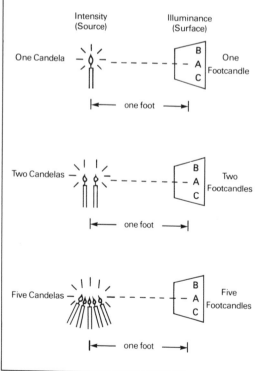

Figure 1–35 The relationship between intensity and illuminance for a constant distance of one foot.

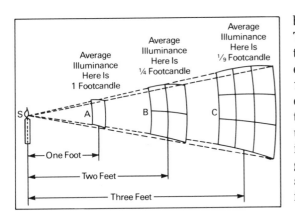

Figure 1–36 The relationship between intensity and illuminance for a constant intensity and varying source-to-surface distances (inverse-square law).

tion to the square of the distance. This relationship is known as the inverse-square law. It should be emphasized that this law is based on a point source of light from which the light rays diverge as shown in Figure 1–36. In practice, it applies with close approximation when the diameter of the light source is not greater than approximately one tenth the distance to the illuminated surface. The formula for the inverse square law is as follows:

$$E = \frac{I}{d^2}$$

where E is the illuminance in footcandles, I is the intensity of the source in candelas, and d is the distance of the surface from the source in feet. Likewise, illuminance measurements can be used to determine the intensity (expressed in candelas) of a source using the following relationship:

$$I = E \times d^2$$

An alternative unit commonly used in photography for expressing the measure of illuminance is the metercandle. The definition of a metercandle is similar to that of a footcandle, illustrated in Figure 1–35, except that the distance from the source to point A would be increased to one meter. Consequently, one metercandle is equal to the amount of light falling on a surface at a point one meter from a one candela source. Since the meter is a greater measure of distance than a foot, the inverse-square law governs the relationship

between a metercandle and a footcandle. There are approximately 3.28 feet in one meter, and therefore one metercandle would be equal to 1 divided by 3.28², which equals 1/10.76, or 0.0929 footcandle. Therefore, 1 footcandle is equal to 10.76 metercandles. The term metercandle is becoming less frequently used; the preferred name is now *lux*. One lux is equal to one metercandle. In the photographic literature the lux (metercandle) measure of illuminance is more common than the footcandle.

When exhibiting reflection prints, the level of illumination can have a significant effect upon their appearance. Consequently, a number of standard conditions have been specified for viewing purposes and are summarized in Table 1–6. The variety of levels suggested is due in part to the differing visual tasks being performed. For example, the viewing of prints for comparison purposes requires a higher level of discrimination and, therefore, a higher illuminance level than does the viewing of a pictorial display. Further, since print judging and display involve subjective opinions, it should not be surprising that a variety of standards exist.

THE MEASUREMENT OF LUMINANCE

Luminance can be defined as the rate at which the unit area of a source emits light in a specific direction. If a source is not a point source but has an appreciable size (as all real sources do), it is less useful to describe the intensity of such a source than to specify its luminance. Luminance is derived from intensity measurements, which are then related to the projected surface of the source.

Luminance is expressed in candelas per unit area (candelas per square foot, candelas per square inch, candelas per square centi-

Some galleries display color prints at a low light level of only 12 footcandles (about 130 lux) to minimize fading.

The inverse-square law does *not* apply to luminance, which does not change with viewing or measurement distance.

Table 1–6 Standard illuminance levels for viewing reflection prints and luminance levels for viewing transparencies

ANSI PH2.30–1988	
Comparison viewing/ critical appraisal	2200 +/− 470 lux
Display, judging, & routine inspection	800 +/− 200 lux
Transparency viewing	1400 +/− 300 cd/m²

meter, depending upon the size of the surface area being considered). For example, Figure 1–37 shows a frosted tungsten lamp with an intensity of 100 candelas in the direction of point A. This lamp has a projected area in that direction of 5 square inches. The luminance in that direction would then be 100 candelas divided by 5 square inches, equal to 20 candelas per square inch. Luminance is the photometric quantity that relates closely to the perceptual concept of brightness. The term *brightness* is used exclusively to describe the appearance of a source and, therefore, cannot be directly measured.

Generally, as the luminance of a source increases, so does the brightness of that source. If two 60-watt tungsten lamps are placed side by side, and one lamp is frosted while the other is clear, the clear lamp looks much brighter than the frosted lamp. Since they are both consuming the same amount of electrical energy and they both are using a tungsten filament, the intensities of the two lamps would be the same. However, the luminance of the clear bulb will be much greater since the projected area of the filament is much smaller than the projected area of the glass envelope on the frosted lamp. From this example it should be evident that knowledge only of the intensities of the sources would be very misleading, while knowledge of the luminance of the sources relates more directly to the perception of brightness of the sources. Consequently, luminance data for real sources are always preferred over intensity data when the visual appearance of the source is desired.

The concept of luminance applies with equal validity to reflecting and transmitting surfaces as well as to light sources, since it makes no difference whether the surface being

The concept of luminance applies to light sources, reflecting surfaces, and transmitting surfaces.

considered is originating the light or merely reflecting or transmitting the light. In this respect, if all of the light falling on a perfectly diffusing surface were reradiated by the surface, the luminance would be numerically equal to the illuminance. This does not happen, since real surfaces never reflect 100% of the light that strikes them. For this reason, it is necessary to determine the reflection factor of the surface, which is the ratio of the reflected light to the incident light. The following formula may be used:

$$\text{Reflection factor} = \frac{\text{Reflected light}}{\text{Incident light}}$$

Thus, it can be seen that for perfectly diffusing surfaces, luminance equals illuminance multiplied by the reflection factor.

The most commonly used unit of luminance when considering reflecting surfaces is candelas per square foot, and the formula is:

$$L = \frac{K \times E}{\pi}$$

where L is the surface luminance in candelas per square foot, E is footcandles incident on the surface, and K is the reflection factor of the surface. As shown in Figure 1–38, the product of the reflectance and the illuminance must be divided by π (3.14), since the light is actually being emitted into a hemisphere of unit (one-foot) radius, and π is the ratio of the radius to the surface area of the hemisphere ($\pi = A/(2r)^2$). For example, if a perfectly diffusing surface with 18% reflec-

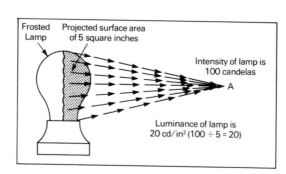

Figure 1–37 The concept of luminance.

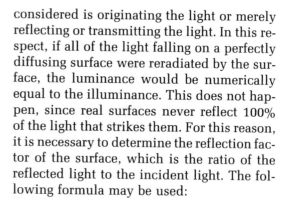

Figure 1–38 A perfectly diffusing surface; sometimes referred to as a lambertian surface.

tance is being illuminated with 100 footcandles of light, the luminance of that surface is found by multiplying 0.18 by 100 divided by 3.14, which equals 5.73 candelas per square foot.

To avoid the necessity of dividing by π all the time, the concept of a footlambert was invented. The footlambert is defined as 1 divided by π candelas per square foot. Thus the above relationship reduces to:

$$L = K \times E$$

where L is now expressed in footlamberts. The footlambert is actually defined as the luminance of a 100%-reflecting surface illuminated by one footcandle of light. Therefore, the luminance of the graycard described previously would be found by multiplying 0.18 × 100 footcandles, which equals 18 footlamberts. Although the calculations are simpler for footlamberts, many modern photoelectric meters read the luminance directly in candelas per square foot, and therefore this is the more commonly used unit of measure in photography.

(These examples assume that the illuminated surfaces are perfectly diffusing [i.e., reflect equally in all directions], which is approximately true for mat surfaces. However, shiny surfaces give highly directional reflections and do not follow these formulas.)

The concept of luminance is particularly useful in photography since it provides a way of describing the light reflected from the surfaces of the subject being photographed. Whenever a reflected-light meter reading is made, it is a luminance measurement. Such data have the unique characteristic of being independent of the distance over which the measurement is made. For an example, if a hand-held meter with a 50° angle of view is used to measure the reflected light from a given surface, the luminance obtained will be identical to that which will be given by a spot-meter reading on the same area at the camera position. The reason for this independence is the fact that as the amount of light being measured from the original area decreases with increasing distance, the projected surface area included in the angle of view increases in direct proportion. Thus, the number of candelas per square foot will remain the same. This assumes, of course, a clear atmosphere exists between the meter and the surface area being measured, which is usually the case.

The various conversions of illuminance data to luminance data are summarized in Table 1–7.

The photometric concepts discussed above are often employed to describe the properties and characteristics of all types of light sources. Table 1–8 summarizes the basic light terms and presents typical values for comparison purposes. It should be obvious from this table that light sources can be described in a wide variety of ways. Thus it is important to understand the differences in order to obtain appropriate information.

Table 1–7 Illuminance-luminance conversions (Examples based on 18% reflectance neutral test card)

Illuminance to Luminance
Luminance = illuminance × reflectance
1. Footlambert = footcandle × reflectance
   Luminance = 1.0 footcandle × 0.18 = 0.18 footlambert
2. Apostilb = metercandle × reflectance
   Luminance = 1.0 metercandle × 0.18 = 0.18 apostilb
3. Candela per square foot = footcandle × reflectance/π
   Luminance = 1.0 footcandle × 0.18/3.1416 = 0.057 C/ft.²
4. Candela per square meter = metercandle × reflectance/π
   Luminance = 1.0 metercandle × 0.18/3.1416 = 0.057 C/m²

Luminance to Illuminance
Illuminance = luminance/reflectance
1. Footcandle = footlambert/reflectance
   Illuminance = 1.0 footlambert/0.18 = 5.56 footcandles
2. Metercandle = apostilb/reflectance
   Illuminance = 1.0 apostilb/0.18 = 5.56 metercandles
3. Footcandle = candela per square foot × π/reflectance
   Illuminance = 1.0 candela/ft.² × 3.1416/0.18 = 17.45 footcandles
4. Metercandle = candela per square meter × π/reflectance
   Illuminance = 1.0 candela/m² × 3.1416/0.18 = 17.45 metercandles

A perfect diffusely reflecting surface (100%) illuminated by 1 footcandle (1 lumen per square foot) will reflect 1 footlambert (1 lumen per square foot or 1/π candela per square foot).
Metric: A perfect diffusely reflecting surface (100%) illuminated by 1 metercandle (1 lumen per square meter or 1 lux) will reflect 1 apostilb (1 lumen per square meter of 1/π candela per square meter).

Reflected-light exposure meters measure luminance. Incident-light exposure meters measure illuminance.

Conversion tables make it easier to compare different units of measurement of light.

Table 1–8   Ten basic light terms with typical examples of each

Popular Concept	Technical Term	Symbol	Unit	Abbreviation	Measurement (Practical)
1. Strength	Luminous intensity	1	Candela	c	Compute from illuminance and distribution
2. Strength	Luminous flux	F	Lumen	lm	1 (Mfr. data) or estimate: watt × luminous efficiency
3. Strength/watt	Luminous efficiency	μ	Lumens/watt	lm/W	2 (Mfr. data)
4. Total light	Luminous energy	Q	Lumen-second	lm-sec	3 (Mfr. data) or integrate area under curve
5. Color	Wavelength	λ	Nanometer	nm	Spectrometer (electronic) Spectroscope (visual) Spectrograph (photographic)
6. Color balance	Color temperature	T	Degree Kelvin	κ°K	4 (Mfr. data) or color-temperature meter
7. Source conversion	Mired shift		Mired	μrd	5 (Mfr. data) or compute: $\mu rd = \dfrac{10^6}{K_1} - \dfrac{10^6}{K_2}$
8. Brightness	Luminance	L	Candela/sq. foot	c/ft²	Reflected-light meter
9. Illumination	Illuminance	E	Footcandle Metercandle Lux	ftc (fc) mc	Incident-light meter with flat surface
10. Exposure	Photographic exposure	H	Footcandle-second Metercandle-second	ftc-sec mc-sec	Compute: $H = $ illuminance × time

Note: Concepts 1–8 apply to light sources. Concept 8, Luminance, also applies to reflected and transmitted light. Concepts 9–10, Illuminance and Photographic exposure, apply to light falling on a surface.

Typical Values:

	1—Luminous Flux	2—Luminous Efficiency
40W tungsten lamp	472 lm	11.8 lm/W
75W tungsten lamp	1,155 lm	15.4 lm/W
100W tungsten lamp	1,750 lm	17.5 lm/W
200W tungsten lamp	4,000 lm	20 lmW
500W tungsten lamp (3200 K)	13,500 lm	27 lmW
500W tungsten lamp (Photoflood)	17,000 lm	34 lmW
100W fluorescent	4,600 lm	46 lmW
200W high pressure mercury	19,000 lm	45 lm/W
200W high pressure mercury	22,000 lm	75 lm/W

3—Luminous Energy

AG-1	7,000 lm-sec
No.22	70,000 lm-sec

4—Color Temperature

Photographic lamp	3200 K
Photoflood lamp	3400 K
Photographic daylight	5500 K

5—Mired Shift

85B Filter	+ 131 μrd
80A Filter	− 131 μrd

REVIEW QUESTIONS

1. Diffraction of light that passes through a small opening is more easily explained with the . . . (p. 2)
   A. wave theory
   B. quantum theory
2. The maximum sensitivity of the human eye, with moderate to high levels of luminance, is at a wavelength of approximately . . . (p. 4)
   A. 500 nm
   B. 550 nm
   C. 600 nm
   D. 650 nm
   E. 700 nm
3. The seven hues that can be identified in the spectrum of white light by most people with normal color vision are . . . (p. 5)
   A. blue, blue-green, green, yellow-green, yellow, orange, red
   B. violet, blue, green, yellow, yellow-orange, orange, red
   C. violet, blue, blue-green, green, yellow, orange, red
   D. blue, green, red, cyan, magenta, yellow, white
4. As the temperature of a blackbody increases, the wavelength at which the peak output of radiation occurs . . . (p. 6)
   A. becomes smaller

B. remains constant

C. becomes larger

5. A color temperature rating of 3200 K for a studio lamp signifies that . . . *(p. 9)*
   A. the temperature of the tungsten filament is 3200 degrees kelvin
   B. the spectral energy of the light emitted matches that of a blackbody at a temperature of 3200 degrees kelvin
   C. the light emitted visually matches that emitted by a blackbody at a temperature of 3200 degrees kelvin

6. The higher the color temperature of a light source, the . . . *(p. 10)*
   A. bluer the light
   B. redder the light

7. The maximum amount of polarized light in light reflected from nonmetallic surfaces occurs at an angle to the surface of approximately . . . *(p. 13)*
   A. 25 degrees
   B. 35 degrees
   C. 45 degrees
   D. 55 degrees

8. An advantage tungsten-halogen lamps have over conventional tungsten lamps is that the tungsten-halogen lamps . . . *(p. 15)*
   A. remain more constant in light output with use
   B. operate at a lower temperature
   C. are less expensive to purchase
   D. produce light of daylight quality

9. The correlated color temperature of unfiltered electronic flash lamps is approximately . . . *(p. 20)*
   A. 3200 K
   B. 5000 K
   C. 5500 K
   D. 6000 K

10. A distinctive feature of laser light is that it . . . *(p. 21)*
    A. is plane polarized
    B. contains an equal amount of all wavelengths in the visible part of the spectrum

C. consists of waves that are all in phase

D. travels at twice the speed of ordinary light

E. cannot penetrate ordinary window glass

11. The candela, as a unit of measurement, is based on the light emitted by . . . *(p. 21)*
    A. a candle
    B. a candela
    C. melted platinum
    D. a heated blackbody

12. The illuminance at a distance of one foot from a light source with a luminous intensity of 10 candelas is . . . *(p. 24)*
    A. 1 footcandle
    B. 10 footcandles
    C. 100 footcandles
    D. 126 footcandles
    E. 200 footcandles

13. The term lux is a synonym for . . . *(p. 25)*
    A. footcandle
    B. metercandle

14. The concept of luminance applies to . . . *(p. 26)*
    A. light sources only
    B. reflecting surfaces only
    C. both light sources and reflecting surfaces

15. When the luminance of a surface of uniform tone is measured with a spot exposure meter, the luminance will vary . . . *(p. 27)*
    A. directly with the meter to surface distance
    B. inversely with the meter to surface distance
    C. directly with the meter to surface distance squared
    D. inversely with the meter to surface distance squared
    E. None of the above.

2 | Camera and Printing Exposure

Willie Osterman. Copyright © by Willie Osterman.

CAMERA EXPOSURE VS. PHOTOGRAPHIC EXPOSURE

Relative aperture is the same as f-number, but camera exposure is not the same as photographic exposure.

The term *camera exposure* refers to the combination of shutter speed and f-number used to expose film in a camera. Although the terms *camera exposure* and *photographic exposure* are related, they are not synonyms. *Photographic exposure* is defined as the quantity of light per unit area received by film or other photosensitive material, and it is calculated by multiplying the illuminance and the time. The relationship is commonly expressed as $H = E \times t$. Assuming that a camera shutter is accurate, the shutter setting is a measure of exposure time, but the f-number setting is not a measure of the illuminance. In a typical picture-making situation, the film receives many different photographic exposures in different areas. Opening the diaphragm by one stop doubles all the different illuminances that constitute the light image on the film. Thus the f-number and shutter settings on a camera enable the photographer to control the photographic exposures received by the film, even though the quantitative values of those exposures are not known. The function of an exposure meter is to indicate a combination of f-number and shutter speed settings that will produce a correctly exposed photograph, taking into account the amount of light falling on or reflected by the subject and the speed of the photographic material being exposed.

F-NUMBERS

F-number = focal length/effective aperture.

Relative aperture is another name for f-number. The word *relative* suggests that the value of the relative aperture or f-number depends upon two things—these are the *focal length* of the lens and the *effective aperture*. The effective aperture is defined as the diameter of the entering beam of light that will just fill the opening in the diaphragm of a camera lens or other optical system. (The diameter of the opening in the diaphragm is known as the *aperture*.) When the diaphragm is located in front of a lens, the effective aperture is the same as the aperture, as illustrated in Figure 2–1. Rarely is the diaphragm located in front of a photographic lens, which makes it necessary to take into account any change in di-

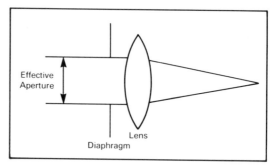

Figure 2–1   When a diaphragm is located in front of a lens, the effective aperture is the same as the aperture or diaphragm opening.

rection of the light rays between the time they enter the lens and when they pass through the diaphragm opening. Since a lens compresses the entering beam of light into a converging cone, a diaphragm with a fixed aperture will transmit more light when it is positioned behind the lens than in front (see Figure 2–2).

CALCULATING F-NUMBERS

F-numbers are calculated by dividing the lens focal length by the effective aperture, or $F\text{-}N = f/D$. Since the effective aperture is rarely the same size as the aperture (the physical opening in the diaphragm), the diameter of the entering beam of light that just fills the diaphragm opening must be measured. One procedure is to place a point source of light one focal length behind the lens so that the beam of light emerging at the front of the lens is restricted by the diaphragm to the same size as an entering beam that just fills the

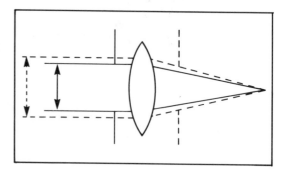

Figure 2–2   A diaphragm will transmit more light when located behind the lens than in front.

diaphragm opening (see Figure 2–3). The diameter of the beam of light can be measured with a ruler in front of the lens, or a permanent record can be made by exposing a piece of photographic paper to the beam of light by placing the paper in a lens cap on the front of the lens.

The focal length of a simple lens can be determined by focusing the lens on a distant object, such as the sun, and measuring the distance from the middle of the lens to the sharp image. (A more precise procedure is described in Chapter 5.) Thus if the focal length of an uncalibrated lens is found to be eight inches, and the diameter of the effective aperture at the maximum opening is one inch, the f-number is the focal length divided by the effective aperture—8/1, or f/8. Note that when the lens is stopped down so that the effective aperture becomes *smaller*, 1/2 inch, for example, the f-number becomes *larger* since $\frac{8}{1/2}$ is f/16. Not only do f-numbers become larger as the amount of light transmitted by the lens becomes smaller, but the relationship is not a simple inverse ratio. The reason for this unfortunate situation is that the f-number is based on the diameter of the entering beam of light, whereas the relative amount of light transmitted is based on the area of a cross section of the beam of light.

The relationship between the area and the diameter of a circle is $A = \frac{\pi \times D^2}{4}$. Thus the amount of light transmitted by a lens varies directly with the diameter squared and inversely with the f-number squared.

WHOLE STOPS

Photographers find it useful to know the series of f-numbers that represents whole stops: f/0.7, 1.0, 1.4, 2.0, 2.8, 4, 5.6, 8, 11, 16, 22, 32, 45, 64. The series can be extended in either direction by noting that the factor for adjacent numbers is the square root of 2, or approximately 1.4. Also, by remembering any two adjacent numbers the series can be extended by noting that alternate numbers vary by a factor of 2, with a small adjustment between 5.6 and 11, and between 22 and 45, to compensate for the cumulative effect of fractional units.

Each time the diaphragm is stopped down one stop (e.g., from f/8 to f/11), the amount of light transmitted is divided by two, and the exposure time required to obtain the same photographic exposure on the film is multiplied by two.

MAXIMUM DIAPHRAGM OPENINGS

The f-number range on most camera lenses is approximately seven stops. A typical 35 mm camera lens may have, for example, a range from f/1.4 to f/16 (seven stops), and a typical view camera lens may have a range from f/5.6 to f/45 (six stops). There has been a constant demand for lens designers over the years to make faster and faster lenses. In the 1940s f/2 lenses were considered fast for 35 mm cameras. Now f/0.7 lenses are mass-produced, a gain of three stops. The fastest one-element lens that can be made with conventional optical design is f/0.5. Such a lens would be half a glass sphere with the film in contact with the flat surface, as illustrated in Figure 2–4.

The need to make photographs under very low light levels, such as moonlight and even starlight for surveillance and other purposes, has led photographic engineers to explore al-

F/2 is two stops faster than f/4.

The fastest single-element glass lens that can be made is f/0.5.

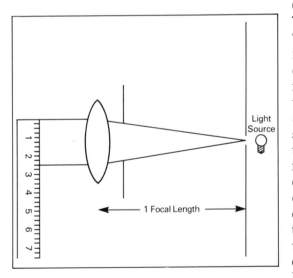

Figure 2–3 A procedure for determining the effective aperture is to place a point source of light one focal length behind the lens and measure the diameter of the beam of light that emerges from the front of the lens.

Stopping a lens down can either increase or decrease image definition.

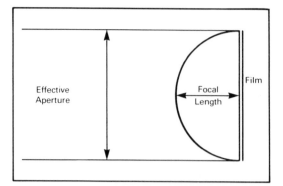

Figure 2–4   An f/0.5 lens, where the lens is half a sphere. The focal length is the radius of the sphere, and the effective aperture is the diameter.

ternatives to the difficult task of further increasing the speed of lenses and films. The most successful are "image-intensifiers," which can electronically amplify the light image formed by the camera lens by as much as 30,000×. To achieve the same effect by designing a lens faster than f/0.7 would require an f-number of f/0.004, which represents an increase in speed of approximately 15 stops. To achieve the same effect by making film faster than ISO 400/27° would require the film to have a speed of ISO 12,000,000/72°. With image intensifiers that are now available, it is necessary to trade off some resolution of fine detail for the increase in speed.

MINIMUM DIAPHRAGM OPENINGS

Since depth of field increases as a lens is stopped down, it would seem desirable to mark f-numbers down to very small openings on all lenses. In 1932 a group of photographers including Ansel Adams, Imogen Cunningham, and Edward Weston formed an organization called Group f/64, with the name implying that they favored making photographs in which everything was sharp.[1] Later, Ralph Steiner used a pinhole with his camera lenses and began his own "East Coast f/180 School" (see Figure 2–5).[2]

[1]Newhall, B. *The History of Photography.* New York: Museum of Modern Art, 1984, p. 128.

[2]Steiner, R. *Ralph Steiner: A Point of View.* Middletown, CT: Wesleyan University Press, 1978, p. 9.

Although stopping down increases depth of field, it also increases the diffraction of light, which tends to reduce image sharpness overall. With a lens stopped down to f/64, the maximum resolving power that can be obtained—no matter how good the lens or accurate the focusing—is approximately 28 lines/mm. For an 8×10-inch camera this is quite good because 10 lines/mm is considered adequate in a print viewed at a distance of 10 inches. Stopping a 35 mm camera down to f/64 would produce the same resolving power of 28 lines/mm in the negative, but since the negative must be enlarged eight times to obtain an 8×10-inch print, the resolving power in the print would be only 28/8, or 3.5 lines/mm. (The calculation of diffraction-limited resolving power is discussed in Chapter 5.) Lens manufacturers normally do not calibrate lenses for 35 mm cameras for openings smaller than f/22 because at that opening the diffraction-limited resolving power is approximately 80 lines/mm in the negative or 10 lines/mm in an 8×10-inch print (see Figure 2–6).

INTERMEDIATE F-NUMBERS

There are situations where it is necessary to know the f-number of divisions smaller than whole stops, such as 1/3, 1/2, or 2/3 stop larger or smaller than a given f-number. (Some lenses have click-stops to obtain 1/2-stop settings, and most exposure meters have 1/3-stop markings on their scales and calculator dials.) Intermediate values can be determined easily on instruments having interval scales, such as thermometers and balances for weighing chemicals, by measuring or estimating between markings or numbers. The f-number series, on the other hand, is a ratio scale in which each number is determined by multiplying or dividing the preceding number by a constant factor—the square root of 2 (approximately 1.4) for whole stops. For 1/2 stops the factor is the square root of 1.5 (1.22). For 1/3 stops the factor is the square root of 1.33 (1.15), and for 2/3 stops the factor is the square root of 1.67 (1.29). Multiplying f/2, for example, by these factors to determine f-numbers that represent stopping down by 1/3, 1/2, 2/3, and 1 stop, the f-numbers are f/2.30, f/2.45, f/2.58, and f/2.8.

Figure 2–5 Photograph made in 1921 by Ralph Steiner, originator of the East Coast f/180 School, using a pinhole in combination with a camera lens. (Photograph by Ralph Steiner.)

LIMITATIONS OF THE F-NUMBER SYSTEM

F-numbers quite accurately indicate changes in the amount of light transmitted by a lens and, therefore, changes in the exposure received by film in a camera. However, they cannot be relied upon to provide the correct exposure based on exposure meter readings of the light falling on a subject or reflected by a subject. The most obvious shortcoming of the f-number system is that it is based on the focal length of the lens. The only time the film is located one focal length from the

lens is when the camera is focused on infinity, and cameras are rarely focused on infinity.

As a result, it is necessary to make an exposure adjustment when the image distance is larger than one focal length—that is, when the camera is focused on objects that are closer than infinity. In practice, it is usually not necessary to make any adjustment until the camera is focused on a distance equal to 10 times the lens focal length or closer. At an object distance of 10 focal lengths, the exposure error would be 23% or about one-quarter stop if no adjustment were made. With short focal length lenses 10 focal lengths is

The f-numbers marked on lenses are accurate only when the camera is focused on infinity.

A 1-to-1 closeup photograph requires a four-times increase in the exposure.

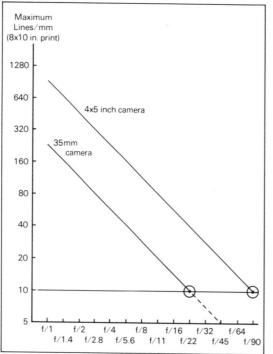

Figure 2–6   The sloping lines show that the diffraction-limited resolving power decreases as a lens is stopped down. A just-acceptable resolving power of 10 lines/mm is reached at f/22 with a 35 mm camera and at f/90 with a 4 × 5-inch camera. Resolving-power values are based on 8 × 10-inch prints, which represent a magnification of 2 for 4 × 5-inch negatives and 8 for 35 mm negatives.

a small distance, only 20 inches, for example, with a normal 50 mm or 2-inch focal length lens on a 35 mm camera. With a normal 12-inch focal length lens on an 8 × 10-inch view camera, on the other hand, 10 focal lengths amount to an object distance of 10 feet. Thus it is dangerous to think that an exposure correction is necessary only when the camera is focused on a very close object.

There are various methods for making the adjustment when photographing objects within the 10-focal-length range. One method is to calculate the *effective f-number*, and then determine the exposure time for that number rather than for the f-number marked on the lens. The effective f-number is found by multiplying the marked f-number by the ratio of the image distance to the focal length, or

Effective f-number

$$= \text{f-number} \times \frac{\text{Image distance}}{\text{Focal length}}$$

With lenses of normal design, the image distance is approximately the distance from the center of the lens to the film plane. (The procedure for determining the image distance with telephoto and other special lenses is covered in Chapter 5.) Thus if the image distance is 16 inches (406 mm) for a closeup photograph with a 4 × 5-inch view camera equipped with an 8-inch (203 mm) focal length lens, the effective f-number when the lens is set at f/11 is f/11 × 16/8 = f/22. When an exposure meter reading is made, the exposure time for f/22 would be used even though the lens is set at f/11. Since f/22 is two stops from f/11, the exposure meter will indicate four times the uncorrected exposure time at f/11 (see Figure 2–7).

An alternative method of determining the exposure correction is to divide the image distance by the focal length and square the result. In the example above, the exposure factor equals $(16/8)^2$ or 4. In this case the correction can be applied either by using the indicated exposure time and opening the diaphragm two stops from f/11 to f/5.6, or by multiplying the exposure time indicated for f/11 by 4.

LENS TRANSMITTANCE

A second shortcoming of f-numbers is that they do not take into account differences between lenses with respect to the amount of

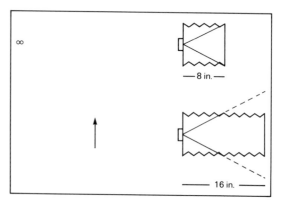

Figure 2–7   F-numbers are based on an image distance of one focal length (above). Doubling the image distance to photograph a close-up object produces an effective f-number that is double the marked f-number; that is, a lens set at f/8 acts as though it is set at f/16.

light lost due to absorption and reflection by the lens elements. The introduction of anti-reflection coating of lenses has reduced variability between lenses due to this factor, but in situations where accurate exposures are essential an adjustment is appropriate. An early zoom lens for 35 mm cameras was approximately three-quarters of a stop slower than indicated by the f-number due to the large number of elements, even though they were coated.

The *T-number* system of calibrating lenses was devised as an alternative to the f-number system in order to use transmittance as a basis, so that any two lenses set at the same T-number would transmit the same amount of light under identical conditions. As a simple example, if a lens marked f/2 in the f-number system transmits only one-half the light that falls on the lens, it would be recalibrated as T/2.8. When an exposure meter reading is made, the exposure time is then selected for f/2.8 rather than for f/2, and the longer exposure time would exactly compensate for the loss of light due to absorption and reflection. T-number is defined as the f-number of an ideal lens of 100% transmittance that would produce the same image illuminance on axis as the lens under test at the given aperture (see Figure 2–8).[3]

EXPOSURE TIME

Photographers can vary either of two factors to alter the exposure received by the film or other light-sensitive material, namely illuminance or time. In many picture-making situations the photographer has a choice of any of a number of combinations of f-numbers and shutter speeds, all of which will produce the same level of photographic exposure but with different depth-of-field and action-stopping characteristics.

Photographers cannot always select a shutter speed solely on the basis of the exposure time that will produce the correct exposure. When photographing a rapidly moving object it may be necessary to use a very short ex-

[3]Stroebel, L., and Todd, H. *Dictionary of Contemporary Photography*. Dobbs Ferry, NY: Morgan & Morgan, 1974, p. 200.

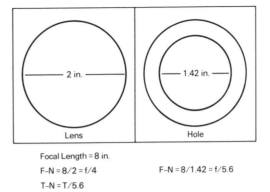

Focal Length = 8 in.

F–N = 8/2 = f/4   F–N = 8/1.42 = f/5.6

T–N = T/5.6

Figure 2–8   The f-number of the lens (left) is calculated by dividing the focal length (8 inches) by the effective aperture (2 inches). The T-number is found by dividing the focal length by the diameter of the opening in the opaque card on the right that transmits the same amount of light (1.42 inches).

posure time to prevent objectionable blurring of the image, and occasionally a photographer will select a slow shutter speed to obtain a blurred image (see Figure 2–9).

Unwanted lack of sharpness frequently results from hand-holding cameras at the slower shutter speeds, especially with long focal length lenses. The rule of thumb is that it is unsafe to hand-hold a camera with an exposure time longer than the reciprocal of the focal length of the lens in millimeters—for example, 1/50 second with a 50 mm lens and 1/500 second with a 500 mm lens. A Mamiya camera introduced in 1980 features a buzzer that warns the user when the shutter is set for a longer exposure time than is considered safe for hand-holding the camera, with an automatic adjustment for different focal length lenses.

LENS SHUTTERS

Most shutters fall into one of two categories: *lens* (or *front*) shutters and *focal-plane* (or *back*) shutters. The ideal position for a lens shutter is between the elements close to the optical center, which is also the ideal position for the diaphragm; such shutters are commonly referred to as *between-the-lens* shutters.

Rule of thumb: The slowest safe shutter speed for handheld cameras is 1/the focal length of the lens in millimeters; 1/50th of a second for a 50 mm focal length lens.

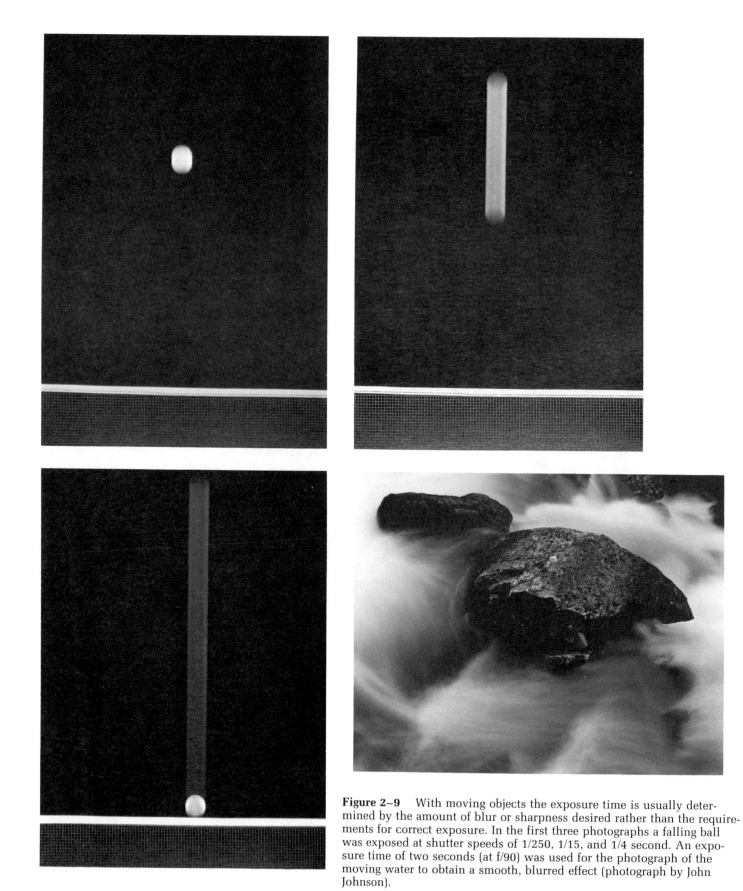

Figure 2–9   With moving objects the exposure time is usually determined by the amount of blur or sharpness desired rather than the requirements for correct exposure. In the first three photographs a falling ball was exposed at shutter speeds of 1/250, 1/15, and 1/4 second. An exposure time of two seconds (at f/90) was used for the photograph of the moving water to obtain a smooth, blurred effect (photograph by John Johnson).

EFFECTIVE EXPOSURE TIME

If one could construct a perfect shutter, the blades would uncover the entire lens, or the entire diaphragm opening, simultaneously. A high-speed motion picture of the operation of a lens shutter shows the blades uncovering the center of the lens first and then gradually uncovering more and more of the lens. At the higher shutter speeds, the shutter blades no sooner uncover the outer edges of the lens than they again cover them at the beginning of the closing operation. Thus the center of the lens is uncovered for a longer time than are the edges, and the total amount of light that is supposed to be transmitted by a lens is actually transmitted only during the time that the diaphragm opening is completely uncovered by the shutter blades (see Figure 2–10). If we used a light meter to measure the change in the amount of light transmitted by a lens during the operation of the shutter, we would find that the reading would increase during the opening stage, remain constant while the shutter blades remained fully open, and decrease during the closing stage. Figure 2–11 shows the change of illuminance with time in the form of a graph.

If lens manufacturers calibrated shutters from the time they start to open until they are completely closed again, the loss of light during the opening and closing parts of the cycle would lead to underexposure—in other words, the effective exposure time is shorter than the total exposure time. To avoid this problem, shutters are calibrated from the half-open position to the half-closed position, as illustrated in Figure 2–12. Thus the loss of light during the opening and closing operations is compensated for by the fact that the total exposure time is longer than the time marked on the shutter. Shutters are normally calibrated from the half-open to the half-closed positions of the shutter blades with the diaphragm wide open.

Unfortunately, when the diaphragm is stopped down, the smaller opening is uncovered sooner and remains completely uncovered longer. As a result, the half-open to half-closed time (which is the effective exposure time) increases as the diaphragm is stopped down, even though the shutter blades are opening and closing exactly the same as

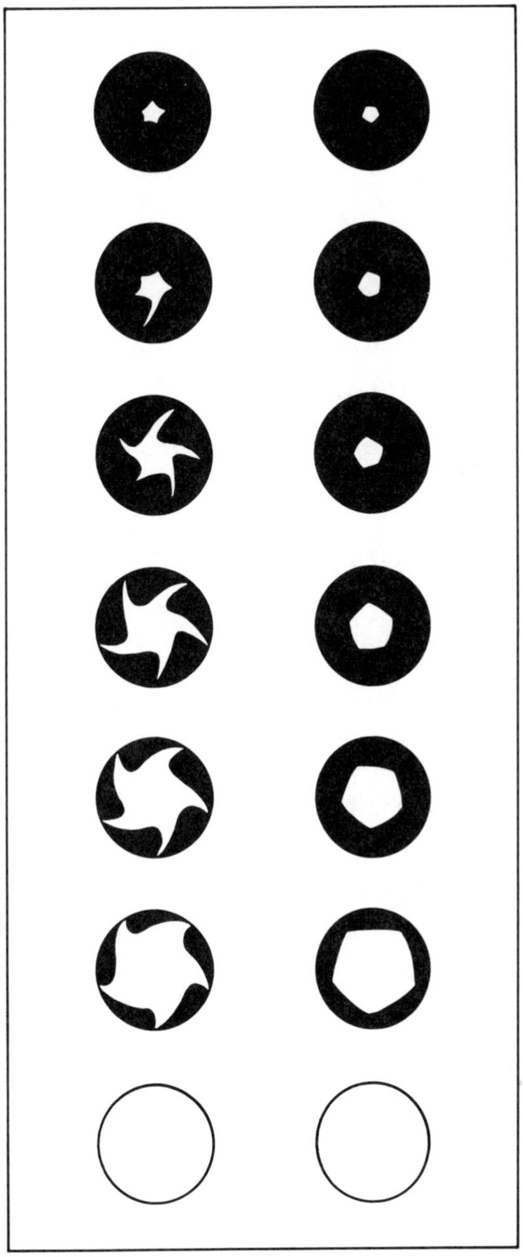

Figure 2–10 On the right are the diaphragm openings for the whole stops on a typical lens. On the left are the positions of the shutter blades at which the diaphragm openings are just completely uncovered. Since the smaller openings are uncovered sooner and stay uncovered longer, the effective exposure time is longer than with a larger opening and the same shutter setting.

when the diaphragm was wide open (see Figure 2–13).

Shutters that are calibrated with the diaphragm wide open, as is the practice, tend to overexpose film at small diaphragm open-

The effective exposure time of a between-the-lens shutter increases as the diaphragm is stopped down.

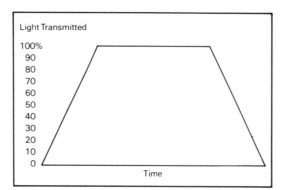

Figure 2–11   The amount of light transmitted by a lens increases as the shutter blades open, remains constant as long as the diaphragm opening is completely uncovered, and decreases as the shutter blades close.

ings due to the increase in the effective exposure time. The error is small at slow shutter speeds but approaches double the indicated exposure, or the equivalent of a one-stop error, at combinations of high shutter speeds and small diaphragm openings. Compensation for this error should be made by stopping down farther than an exposure meter reading indicates for the selected shutter speed, or by selecting the f-number specified for the effective exposure time rather than the marked time. Table 2–1 indicates the correction needed for different combinations of shutter speeds and diaphragm openings.

FOCAL-PLANE SHUTTERS

The basic design of the focal-plane shutter is a slit in opaque material placed close to the

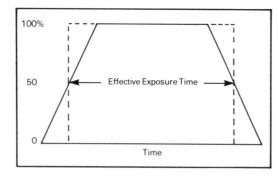

Figure 2–12   Shutters are calibrated from the half-open position to the half-closed position. The total exposure time is longer than the effective exposure time to compensate for the loss of light during the opening and closing stages of the cycle.

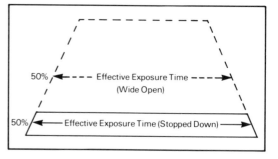

Figure 2–13   When a lens is stopped down, the smaller opening is uncovered sooner and remains uncovered longer, increasing the effective exposure time from half open to half closed.

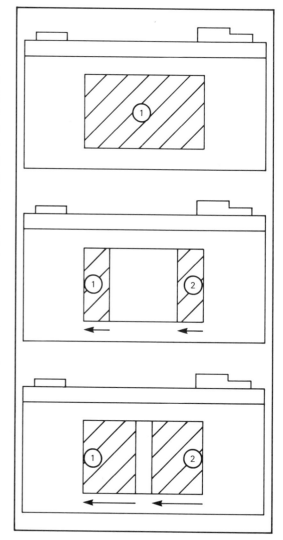

Figure 2–14   At slower shutter speeds, curtain 1, which protects the film from light before the shutter is tripped, completely uncovers the film before curtain 2 begins to cover it again. At higher shutter speeds, curtain 2 begins to cover the film before it is completely uncovered. The space between the two curtains decreases as the shutter speed increases.

Table 2–1   Top: Chart for determining the number of stops between a given f-number and the f-number at the maximum diaphragm opening. Bottom: The reduction in exposure (in stops) required to compensate for the increase in the effective exposure time with various combinations of f-numbers and exposure times

| | Stopped Down | | | | | | | | |
Maximum Aperture	1	2	3	4	5	6	7	8	Stops
f/1.4	f/2	f/2.8	f/4	f/5.6	f/8	f/11	f/16	f/22	
f/2	f/2.8	f/4	f/5.6	f/8	f/11	f/16	f/22	f/32	
f/2.8	f/4	f/5.6	f/8	f/11	f/16	f/22	f/32	f/45	
f/4	f/5.6	f/8	f/11	f/16	f/22	f/32	f/45	f/64	
f/5.6	f/8	f/11	f/16	f/22	f/32	f/45	f/64	f/90	
f/8	f/11	f/16	f/22	f/32	f/45	f/64	f/90	f/128	
Exposure Time[a]									
1/60	0	0	0	0	0	0–¼	0–¼	0–¼	
1/125	0–¼	¼	¼	¼	¼	¼	¼	¼	
1/250	¼	¼	½	½	½	½	½	½	
1/500	½	¾	1	1	1	1	1	1	

[a]Additional stopping down required to compensate for changes in shutter efficiency with between-the-lens shutters.

film that moves across the film from edge to edge. Exposure could be altered with early focal-plane shutters by selecting one of several slit widths in the shutter curtain and selecting one of a number of spring tensions that controlled the speed of the slit's movement. Rewinding the curtain before inserting the dark slide in the film holder resulted in fogged film. Modern focal-plane shutters have an adjustable opening between two curtains (or blades) with a self-capping feature to eliminate the slit when the curtain is rewound (see Figure 2–14).

A basic advantage of focal-plane shutters over between-the-lens shutters is that the shutter can be built into the camera and used with interchangeable lenses rather than incorporating a shutter in each lens. A second advantage is that stopping a lens down has little effect on the effective exposure time so that compensation is not necessary as with between-the-lens shutters.

Disadvantages of focal-plane shutters are: Changes in speed of the slit across the film result in uneven exposure (see Figure 2–15), images of rapidly moving objects are dis-

A distorted image may result when photographing rapidly moving objects with a focal-plane shutter camera.

Figure 2–15   Cold weather caused this focal-plane shutter to malfunction, producing variations of exposure from side to side.

torted in shape, and it is more difficult to synchronize flash and electronic flash at higher shutter speeds (see Figure 2–16).

ELECTRONIC SHUTTERS

For a typical mechanical between-the-lens shutter, the power to open and close the shutter is provided by a spring that is placed under tension, and the timing is controlled by a watch-type gear train. Early focal-plane shutters on sheet-film cameras provided the photographer with two controls over the exposure time: variable tension on the spring that controlled the movement rate of the focal-plane curtain, and a choice of various slit widths in the curtain.

Electronic shutters can be divided into two categories: those using electronics to control the timing with a spring-activated shutter (more properly called electromechanical shutters) and those using electronics for both the timing and the power to activate the shutter. Electronically controlled focal-plane shut-

Electronic flash cannot be used with the highest speed settings of focal-plane shutters because the shutter never completely uncovers the film.

It is possible for a shutter to be accurate at one speed setting, to overexpose at another setting, and to underexpose at a third setting.

ters and between-the-lens shutters (on the mass-produced Polaroid Auto-100 camera) were introduced in 1963.

The timing of electronic shutters is normally controlled by a capacitor charged by a current from a battery, which can be altered with a variable resistor, causing the shutter to close when the capacitor is filled. By adding a camera exposure meter to the circuit, the exposure time is controlled automatically by the amount of light received by the meter photocell. One way electricity is used to open and close a shutter is with solenoids, where a magnetic field produced by current flowing through a coil moves a plunger and the shutter. The liquid Kerr cell, used for scientific photography, is an electronic shutter that is capable of providing exposure times as short as 1 billionth of a second (see Figure 2–17).

SHUTTER TESTING

The discussion above concerning effective exposure times with between-the-lens shutters is based on a shutter that has been accurately calibrated (at the maximum diaphragm opening) and is in perfect working condition. Some shutter manufacturers do not guarantee their shutters to be any more accurate than ±40% of the marked speed. In past years it has been possible to test 100 shutters selected at random representing various brands and periods of use without finding more than one or two that were accurate within ±5% at every setting. Shutters with electronic timing control tend to be more accurate than mechanical shutters, where the timing depends upon gear trains, springs and cams, but the accuracy of any shutter should not be taken for granted. Shutter variability for 140 35 mm single-lens reflex cameras is illustrated by the frequency histogram in Figure 2–18.

The most satisfactory method of testing a shutter is with an electronic shutter tester that displays the shutter's opening and closing as a line on an oscilloscope where deviations from 100% accuracy can be determined from a calibrated grid, or where the effective exposure time is presented as a digital readout (Figure 2–19). Such instruments are capable of being very accurate since they can be calibrated with the 60-cycle al-

Figure 2–16   Electronic-flash illumination used with a camera having a focal-plane shutter at shutter speeds of 1/60 second (top), 1/125 second (center), and 1/250 second (bottom). At speeds higher than 1/60 second the second curtain begins to cover the film before the first curtain has completely uncovered it. The maximum shutter speed that can be used with electronic flash has gradually increased with newer model cameras. At the present time, 1/250 second is the maximum speed.

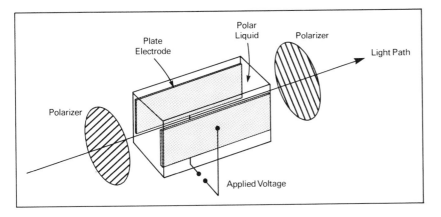

Figure 2–17 A Kerr cell electronic shutter.

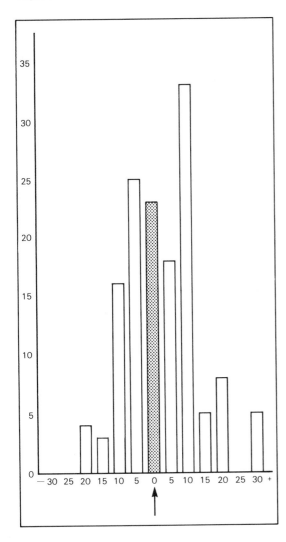

Figure 2–18 The shutters on 140 35 mm cameras were tested at a setting of 1/125 second. The range of variability is from 20% underexposure to 30% overexposure. The conversion data in Table 2–2 indicate that −20% and +30% are both equivalent to a 1/3-stop change in exposure, producing a range of 2/3 stop.

ternating current with which they are normally used. Such testers can also be used to determine the change in the effective exposure time of between-the-lens shutters as the diaphragm is stopped down, and to check the consistency of focal-plane shutters across the film (see Figures 2–20 and 2–21). Shutter-speed settings that are found to be inaccurate still can be used by recording the test results and then using the appropriate f-number for the actual shutter speed rather than for the marked speed.

In the absence of a shutter tester, a practical test can be conducted by exposing film in the camera at each setting provided certain precautions are taken. It is best to use reversal color film since it has a small exposure lat-

Between-the-lens shutters are calibrated by the manufacturer at the maximum diaphragm opening.

Figure 2–19 An electronic shutter tester that displays the shutter's operation as a trace on an oscilloscope.

Table 2–2 Conversion of percentage over- and underexposure to stops. This conversion table (see Figure 2–18) indicates that −20% and +30% are both equivalent to a ⅓-stop change in exposure

Over	Ratio Actual/Marked	Log Ratio	Stops (Log/0.3)
0%	1.0	0.00	0.00
10%	1.1	0.04	0.13
20%	1.2	0.08	0.27
30%	1.3	0.11	0.34 (⅓)
40%	1.4	0.15	0.50 (½)
50%	1.5	0.17	0.58
60%	1.6	0.20	0.67 (⅔)
70%	1.7	0.23	0.77
80%	1.8	0.25	0.85
90%	1.9	0.28	0.93
100%	2.0	0.30	1.00 (1)
300%	4.0	0.60	2.00 (2)
700%	8.0	0.90	3.00 (3)

Under	Ratio Actual/Marked	Log (Bar) Ratio	Log (Neg) Ratio	Stops (Log/0.3)
0%	1.0	0.00	−0.00	0.00
10%	0.9	$\bar{1}$.954	−0.046	0.15
20%	0.8	$\bar{1}$.903	−0.097	0.32 (⅓)
30%	0.7	$\bar{1}$.845	−0.155	0.52 (½)
40%	0.6	$\bar{1}$.778	−0.222	0.74 (¾)
50%	0.5	$\bar{1}$.699	−0.301	1.00 (1)
60%	0.4	$\bar{1}$.602	−0.398	1.33
70%	0.3	$\bar{1}$.477	−0.523	1.74
80%	0.2	$\bar{1}$.301	−0.699	2.33
90%	0.1	$\bar{1}$.000	−1.000	3.33
100%	0.0	—	—	—

itude that permits small changes in exposure to be detected. A subject having a normal range of tones should be used with front illumination. The exposure meter used should be tested against a standard light source or against a number of other meters or one meter of known accuracy, and care must be taken to make an accurate meter reading. With between-the-lens shutters the diaphragm must be wide open, but different shutter speeds can be tested by using neutral-density filters, adding 0.3 in density each time the exposure time is doubled. With focal-plane shutters it is safe to stop down the diaphragm one stop for each change of shutter speed. Finally, it is important to be sure the film is processed normally, and the transparencies should be viewed under standardized viewing conditions.

CAMERA EXPOSURE LATITUDE

Assuming that in each picture-making situation there is a combination of f-number and shutter speed that produces the optimum level of exposure, photographers should know what will happen to the image quality if the level of exposure is either increased or decreased. It is optimistic to think that all the various factors calibrated to determine the "correct" exposure are 100% accurate—film-speed rating, exposure meter, shutter, and f-number—and that there is no human error involved in the process, but for now the assumption will be made that there are no such inaccuracies. The range over which the exposure can be increased and decreased from the "correct" exposure and still produce acceptable results is known as the *exposure latitude*.

As exposure is decreased from the optimum level, the darker areas or shadows of the scene will first lose contrast and then detail. As the exposure continues to be reduced, the detail will be lost in progressively lighter areas of the scene, and eventually there will be no image, even in the lightest areas or highlights of the scene (see Figure 2–22).

On the overexposure side, the first noticeable change is commonly an increase in

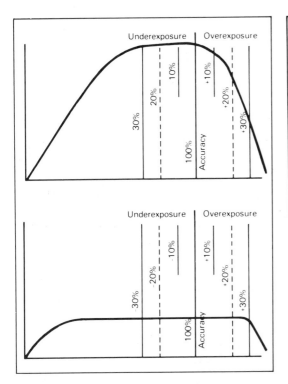

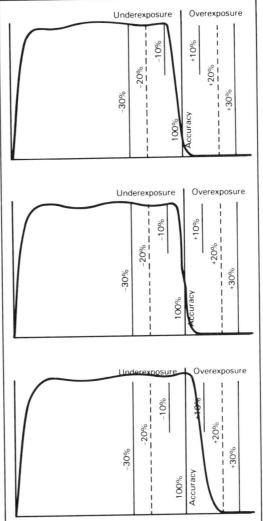

Figure 2—20 The two shutter-tester traces reveal the change in the effective speed of a between-the-lens shutter at a shutter setting of 1/250 second with the diaphragm at the maximum aperture of f/4 (top) and stopped down to f/11.

Figure 2—21 The three traces reveal the change in the effective speed of a focal-plane shutter at (from top) the right edge, center, and left edge of the film aperture. The shutter was set at 1/60 second, and the shutter traveled from right to left.

shadow contrast (and therefore an increase in overall contrast). Overexposure also produces a decrease in image definition or, more specifically, an increase in graininess, a decrease in sharpness, and a decrease in detail. These changes may be apparent only with small-format images magnified considerably in printing. As the exposure continues to be increased, the image contrast will decrease, first in the highlight areas and then toward progressively darker areas of the scene. Seldom is all detail lost with overexposure of negative-type films, although extreme overexposure can result in a partial or complete reversal of tones known as solarization.

Overall changes in contrast due to underexposure or overexposure can be compensated for at the printing stage with black-and-white negatives, but local loss of contrast and detail are not correctable. The penalties for underexposure (loss of shadow contrast and detail) occur more quickly and are more severe than the penalties for overexposure (modest change in contrast and decrease of

definition), leading to the long-standing admonition "It is safer to overexpose than to underexpose" (see Figure 2—23). Stated in terms of exposure latitude, there is more exposure latitude on the overexposure side than on the underexposure side. These statements apply to conventional black-and-white and color negative films.

With reversal-type films the exposure latitude is small in both directions, especially when the slide, transparency, or motion-picture film exposed in the camera is the final image to be viewed—often juxtaposed spa-

It is safer to overexpose than to underexpose—with negative type films.

Reversal films have much less exposure latitude than negative films.

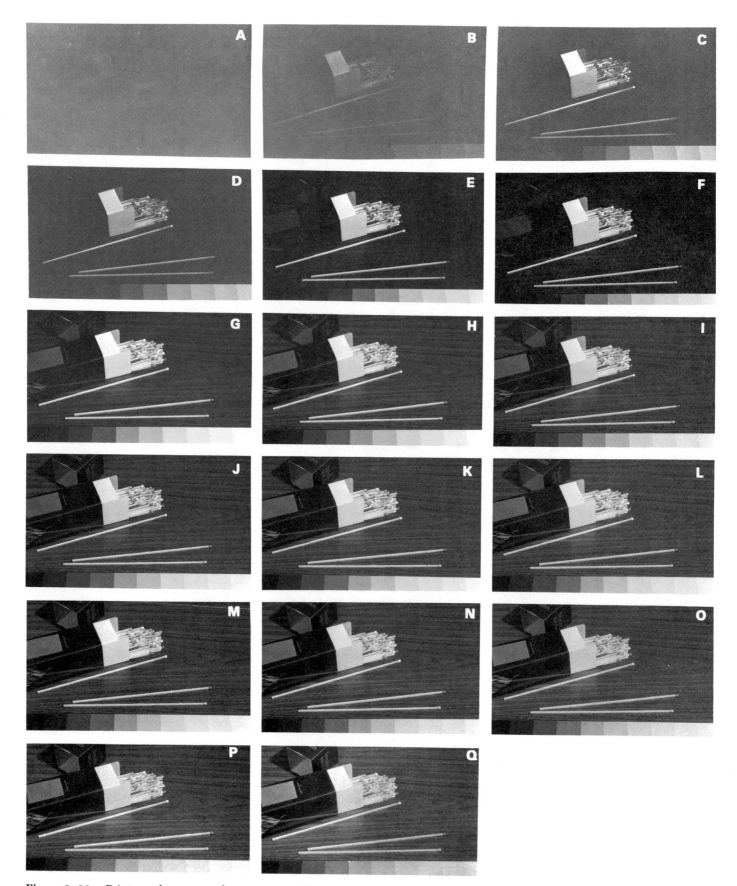

Figure 2–22 Prints made on normal-contrast paper from negatives that received different exposures, ranging from 1/128 normal to 512 times normal in one-stop increments. Print H was made from the normally exposed negative. There is less latitude on the underexposed side due to the loss of shadow detail and the rapid loss of contrast than on the overexposed side.

tially or temporally with other images, making relatively small density differences apparent. Changes of ±1/2 stop in exposure are commonly considered to be the maximum tolerable, with the possible exception of images that are viewed in isolation (as in an otherwise darkened room), viewed in comparison to other images equally underexposed or overexposed, or are to be reproduced photographically or photomechanically. Compensation in reproducing slides and transparencies is generally more successful when they are slightly dark (underexposed) than when they are too light, but this depends to some extent on the relative importance of highlight and shadow detail in each photograph. Because reversal color films have so little exposure latitude, they are especially useful for testing exposure systems.

ALTERNATIVES TO EXPOSURE METERS

Although most photographers, professional and amateur alike, consider exposure meters to be as indispensable to photography as cameras, there are other methods for determining the correct camera-exposure settings. Even trial and error can be used appropriately when a lighting condition remains constant and will be encountered again.

One of the most constant of lighting conditions is direct sunlight between the mid-morning to mid-afternoon hours. A widely used rule for this lighting condition is that the correct exposure is obtained by using a shutter speed equal to the reciprocal of the ISO arithmetic speed of the film at a relative aperture of f/16. With an ISO 125/22° film, for example, the camera exposure would be 1/125 second at f/16. Data sheets supplied with some films include exposure information for other outdoor lighting conditions including hazy sun, light overcast, and heavy overcast (see Figure 2–24). Exposure tables also have been used successfully for tungsten lamps, flashlamps, and electronic flash units where the lamp output and lamp-to-subject distance are constant or are measurable.

One of the most ambitious projects to provide exposure information for a variety of natural lighting conditions is the American National Standard *Photographic Exposure*

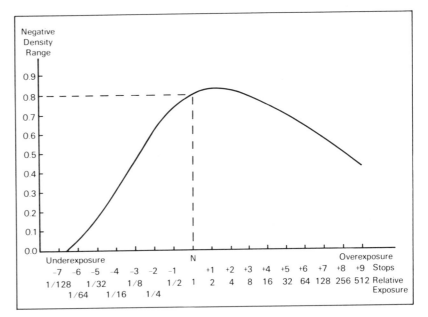

Figure 2–23   The effects of underexposure and overexposure on negative contrast are shown in this curve. Contrast decreases rapidly with underexposure, and overexposure first produces an increase in contrast before it declines at a more gradual rate. The data were obtained from the negatives used for Figure 2–22.

Rule of thumb for exposing film outdoors in sunlight: Shutter speed at f/16 = 1/film speed.

KODACOLOR II Film

Load and unload your camera in subdued light.
DAYLIGHT EXPOSURE: Cameras with automatic exposure controls—Set film speed at ISO (ASA) 100. **Cameras with manual adjustments**—Determine exposure setting with an exposure meter set for ISO (ASA) 100 or use the table below.

Bright or Hazy Sun on Light Sand or Snow	1/125 sec. f/16
Bright or Hazy Sun (Distinct Shadows)	1/125 sec. f/11*

Weak Hazy Sun (Soft Shadows)	1/125 sec. f/8
Cloudy Bright (No Shadows)	1/125 sec. f/5.6
Open Shade or Heavy Overcast	1/125 sec. f/4

*f/5.6 for backlighted close-ups

Light	Film Speed	Filter
DAYLIGHT	**ISO (ASA) 100**	**None**
3400 K photolamps	ISO (ASA) 32†	No. 80B
3200 K tungsten	ISO (ASA) 25†	No. 80A

†For through-the-lens exposure meters, see camera manual.

Courtesy of Eastman Kodak Company

Figure 2–24   Daylight exposure table. (Courtesy of Eastman Kodak Company.)

Guide, ANSI PH2.7-1985. The exposure data for daylight are based on the latitude, time of day, sky conditions, subject classification, and direction of the lighting. Also included are data for aerial photography, underwater photography, photography by moonlight; and photographs of sunsets, rainbows, lightning, lunar and solar eclipses, and aurora borealis. Certain artificial lighting conditions are also covered, including fireworks and television screens.

EXPOSURE METERS

In the past many different types of light-measuring devices or exposure meters have been used, but nearly all contemporary exposure meters are of the photoelectric type (See Figure 2–25).

Ideally, the spectral sensitivity of exposure-meter cells would closely resemble the spectral sensitivity of typical panchromatic films, but unfortunately none of them do. The curves in Figure 2–26 show the spectral sensitivity of selenium, cadmium sulfide, silicon, and silicon blue cells. Variations of spectral sensitivity will be found among cells of the same type due to manufacturing controls, including the use of filters over the cells; and minor variations are even found among

Taking exposure meter readings through filters can lead to exposure errors due to differences in the spectral sensitivity of the meter and the film.

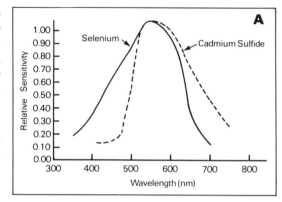

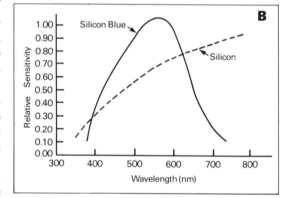

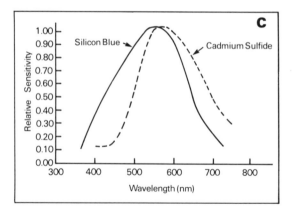

Figure 2–26   Spectral sensitivity of selenium, cadmium, sulfide, and silicon exposure-meter cells. (A) Spectral sensitivity curves for selenium cells and cadmium sulfide cells. Cadmium sulfide cells have much higher overall sensitivity, which is not revealed because the curve heights have been adjusted, but the spectral sensitivity of selenium cells more closely matches that of panchromatic films. (B) Silicon cells have high red sensitivity and low blue sensitivity, but the spectral sensitivity balance is improved in silicon blue cells through the use of filters. (C) A comparison of the spectral sensitivity of silicon blue cells and cadmium sulfide cells.

Figure 2–25   A calculator dial translates the meter readings into combinations of shutter and f-number settings, or exposure values, taking the film speed into account.

meters of the same brand and model due to the variability inherent in all manufacturing processes.

Blue filters, as used in silicon blue meters, can compensate for the cells' inherently low blue sensitivity. However, full correction is not an attractive option, since the filter simply absorbs the colors of light to which the cells are more sensitive, thereby reducing overall sensitivity. Silicon cells, which were introduced in the 1970s, share the high sensitivity of cadmium sulfide cells without the disadvantage of memory. Meter manufacturers are continuing to develop new types of photoelectric cells, such as silicon photodiode and gallium arsenide phosphide photodiode, in an effort to achieve further improvements in performance.

High red sensitivity and low blue sensitivity of a meter cell, compared to the sensitivity of panchromatic film, can cause a difference in density of negatives exposed outdoors and indoors because of the difference in the color temperature of the light sources. A further complication is that the film may not have exactly the same speed with the two types of illumination even though only a single film speed is published for both. The greatest danger of incorrect exposure due to poor spectral response, however, occurs when reflected-light meter readings are taken from strongly colored subject areas and when color filters are used over the camera lens with behind-lens meters. With a meter having high red sensitivity, metering through a red filter would lead to underexposure of the film (see Figure 2–27).

REFLECTED-LIGHT/ INCIDENT-LIGHT EXPOSURE METERS

Figure 2–27   Copy photographs of a black-and-white print and gray scale, exposed according to a through-the-lens exposure meter. The top photograph, made with no filter, is correctly exposed. The bottom photograph, made through a red filter, is underexposed due to the high red sensitivity of the cadmium sulfide cell, which produced a false high reading.

Strictly speaking, all exposure meters can measure only one thing: the light that actually falls on the photoelectric cell. By adjusting the angle over which the cell receives light, and by using neutral-density filters that transmit only a desired proportion of the light (or by incorporating a corresponding adjustment in the calibration system), it is possible for the photographer to obtain correct exposure information (a) when the cell is aimed at the subject as a reflected-light meter; (b) when the cell is placed near the subject and aimed at the camera as an incident-light meter, and (c) when the cell is placed behind the lens in a camera where it receives a sample of the image-forming light that falls on or is reflected by the film.

Reflected-light exposure meters are calibrated to produce the correct exposure when the reading is taken from a medium-tone area.

Figure 2–28   The white circle on the post on the right represents the approximate area measured with a 1° spot meter on a photograph made with a normal focal length lens where the angle of view is approximately 53°.

An 18%-reflectance gray card is widely used as an artificial or substitute midtone. Some hand-held reflected-light meters have acceptance angles that are approximately the same as the angle of view of a camera equipped with a normal focal length lens, which is approximately 53°, although 30° is more typical. The acceptance angle of reflected-light exposure meters is defined as the angle where the response is one-half the on-axis response.[4]

When a reflected-light meter is aimed at a scene from the camera position, it integrates the light from the scene's various parts and gives an average reading. Such a reading will produce the correct exposure only with scenes that have a balance of light and dark tones around an appropriate midtone. The meter can be moved closer to the subject to make a reading from a single subject area or from

---

[4]ANSI, General Purpose Photographic Meters (photo-electric type) (PH2.12-1961).

a substitute midtone area. An alternative is to limit the angle of acceptance so that a smaller subject area can be measured from the camera position. Spot attachments are available for some reflected-light meters, and some meters are designed as spot meters with angles of acceptance as small as 1° (see Figure 2–28). Fiber-optics probes are also offered as accessories for some meters, permitting readings to be made from very small areas.

A more appropriate description of reflected-light exposure meters is *luminance* meters, since such meters are used to measure other than reflected light, including light transmitted by transparent and translucent objects (such as stained-glass windows) and light emitted by objects (such as night advertising signs and molten metal). Also, some persons prefer to use the term *light* meter rather than exposure meter, but the terms *reflected-light meter* and *exposure meter* have become entrenched in the photographic vo-

A typical angle of acceptance for general purpose exposure meters is 30 degrees. Some spot meters have an angle of acceptance as small as 1 degree.

Another name for reflected-light exposure meter is luminance meter.

cabulary through common usage. Actually, since the spectral response curves of exposure meters do not (and should not) match the spectral response curve of the standard observer, the meters do not truly measure light.

Behind-lens meters also can be designed either to integrate the light over a large angle to provide an average reading, or to measure the light from smaller areas. Some cameras have "center-weighted" meter systems, others measure off-center areas, and some average separate readings from different areas, such as sky and foreground with a typical landscape composition (see Figure 2–29).

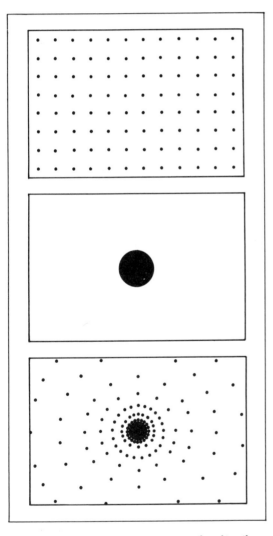

Figure 2–29 The three most popular distributions of exposure-meter sensitivity within the picture area in 35 mm single-lens reflex cameras are: averaging (top), spot (middle), and center-weighted (bottom).

If a hand exposure meter is to be used for both reflected-light and incident-light readings, an adjustment must be made when the meter is turned around so that the cell is receiving the higher level of light falling on the subject rather than just the 18% that is reflected from a typical scene. Dividing 100% by 18% produces a ratio of approximately 5.6:1. Therefore, a neutral-density filter that transmits 1/5.6 of the light, placed over the cell, would convert the reflected-light meter to an incident-light meter that would produce the same exposure with a subject that reflects 18% of the incident light (see Figure 2–30).

It is appropriate to use a flat receptor on incident-light meters that are to be used for photographic copying, so that the illuminance will vary on the meter cell and the flat original exactly the same when the direction of the lighting changes. Such a meter is identified as a *cosine-corrected* meter because the cosine law of illumination states that the illuminance changes in proportion to the trigonometric cosine of the angle between the light path and a perpendicular to the surface. Copying lights are typically placed at an angle of 45° rather than near the camera, to avoid glare reflections. The cosine of 45° is approximately 0.7, which means the illuminance on the original and the cosine-corrected meter cell will be only 0.7 what it would be if the lights were placed directly in front of the original on the lens axis at the same distance (see Figure 2–31).

Some general-purpose exposure meters provide a flat diffusor to place over the cell that transmits the proportion of light needed to convert the meter to an incident-light meter, as well as to provide cosine-corrected readings for copying purposes. Since the cosine of 90° is zero, flat receptors could not be used with side-lighted and back-lighted three-dimensional subjects and scenes with the meter aimed at the camera. Such a meter could be aimed at the key light provided that a suitable adjustment is made, with a calculator dial for example, to compensate for the change in effectiveness of the illumination, from 100% for front lighting to 50% for 90° side lighting to 0% for 180° back lighting.

In 1940, however, Don Norwood introduced a more satisfactory solution: an incident-light exposure meter with a hemi-

Flat diffusors are recommended for incident-light exposure meters for copying work, and hemispherical diffusors are recommended for photographing three-dimensional subjects.

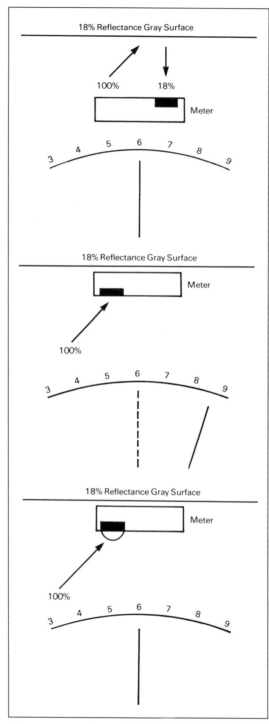

Figure 2–30 Conversion of a reflected-light exposure meter to an incident-light exposure meter. (Top) A reflected-light exposure meter measures 18% of the incident light reflected from a subject of average reflectance. (Middle) A reflected-light exposure meter used as an incident-light exposure meter, without modification, measures 100% of the incident light and produces a false high reading. (Bottom) Adding a diffuser that transmits 18% of the incident light produces the same correct reading as that obtained with the reflected-light exposure meter.

spherical diffusor over the cell that automatically compensated for changes in the direction of the key light with the meter aimed at the camera.[5] The Norwood exposure meter operated on the principle that as the key light was moved toward the side from a front-lighted position, it illuminated less and less of the hemispherical diffusor, producing proportionally lower readings. With 90° side lighting, for example, the key light illuminated half the diffusor, and the illuminance reading was 50% of the front-lighted reading—the equivalent of a one-stop change in the camera exposure settings (see Figure 2–32). Today almost all meter manufacturers have adopted the hemispherical diffusor for incident-light exposure meters intended for use with three-dimensional subjects and scenes. With some, however, care must be taken in positioning the meter so that the meter body does not shield the diffusor from the key light in back-lighted situations.

METER SCALES

Early photoelectric exposure meters were commonly calibrated in photometric units such as footcandles for illuminance and candelas per square foot for luminance readings, or in numbers that represented relative photometric units. The numbers on these meters formed ratio scales such as 1-2-4-8-16, etc., so that each higher number represented a doubling of the light, or the equivalent of a one-stop change (see Figure 2–33). Some other types of markings have been used, including f-numbers that correspond to whole-stop changes, but most contemporary meters with calibrated scales use a simple interval series of numbers such as 1-2-3-4-5-6, etc., where each higher number again represents a doubling of the light, or a one-stop change (see Figure 2–34).

If the numbers on a meter interval scale are *light values* in the additive system of photographic exposure (APEX), then the relationship of the four factors involved in camera exposure—f-number, exposure time, scene illuminance, and film speed—is expressed as:

[5]Norwood, "Light Measurement for Exposure Control." *Journal of the SMPTE*, Vol. 54 (1950), pp. 585–602.

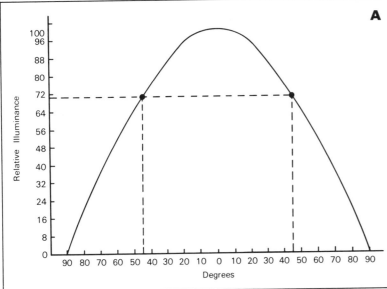

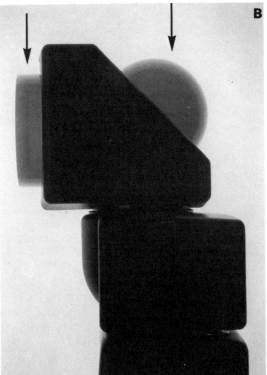

Figure 2–31 The cosine law of illumination. (A) According to the cosine law of illumination, the illuminance for light falling on a flat surface varies with the cosine of the angle of incidence—a relationship represented by the solid curve. The broken lines indicate the relative illuminance for 45° copy lighting. (B) Both flat (cosine-corrected) and hemispherical diffusers are provided with this exposure meter.

$$\begin{bmatrix} \text{Aperture value } + \\ \text{Time value} \end{bmatrix} = \begin{bmatrix} \text{Light value } + \\ \text{Speed value} \end{bmatrix}$$
$$= \text{Exposure value, or}$$
$$A_v + T_v = L_v + S_v = E_v$$

Thus with a meter reading light value of 5 (30 foot-lamberts) and a film speed value of 5 (ISO 100) the exposure value is 5 + 5 or 10. Therefore, any combination of aperture and time values that adds up to 10 can be used, such as an aperture value of 3 (f/2.8) and a time value of 7 (1/125 second). The values for other f-numbers, exposure times, film speeds, and scene illuminance readings

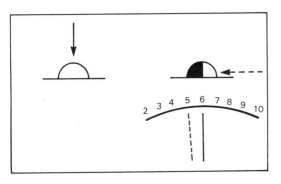

Figure 2–32 Front lighting and 90° side lighting of a hemispherical diffuser, resulting in a one-stop decrease in the meter reading.

Figure 2–33 The numbers on this meter dial are proportional to the amount of light falling on the cell. Since each number is double the next lower number, the numbers form a ratio scale.

Figure 2–34 Even though the numbers on this meter dial form an interval scale with equal increments of 1, each number represents double the amount of light of the next lower number.

are given in Table 2–3. Most interval-scale meters use an arbitrary set of numbers beginning with 1 for the lowest mark on the scale rather than using APEX light-value numbers, but some do include exposure value (E_v) numbers—a combination of light values (L_v) and film-speed values (S_v)—on the calculator dial or elsewhere.

An average outdoor scene has a luminance ratio of approximately 128 to 1, or a luminance range of approximately seven stops.

The f/16 rule can be used as one check on the accuracy of an exposure meter. A meter also can be checked against other meters with a standard light source.

LIGHT RATIOS

Since exposure-meter readings are commonly used to determine lighting ratios and scene luminance ranges or ratios, it is important to note the difference between ratio (1-2-4-8) scales and interval (1-2-3-4) scales on meters and to be able to use either. With a ratio scale meter the scene luminance ratio can be determined by taking reflected-light readings from the highlight and shadow areas and dividing the larger number by the smaller. For example, 256 (highlight) divided by 2

(shadow) is 128, or a luminance ratio of 128:1. Similarly, the lighting ratio for a studio portrait can be determined by taking an incident-light or a reflected-light reading for the key light plus the fill light and a second reading for the fill light alone, so that readings of 60 and 15, for example, represent a 4:1 lighting ratio. A comparison of shadow and highlight reflected-light readings is shown below for the two types of scales:

1 - 2 - 4 - 8 - 16 - 32 - 64 - 128 - 256 - 512

1 - 2 - 3 - 4 - 5 - 6 - 7 - 8   - 9 - 10

  S   H

The 128:1 luminance ratio scene on the ratio scale can most conveniently be expressed as a 7-stop (9 − 2) scene on the interval scale. Seven stops can be converted to a 128:1 ratio by raising 2 to the seventh power, 2^7 (i.e., $2 \times 2 \times 2 \times 2 \times 2 \times 2 \times 2$) = 128. (Using logarithms, $7 \times \log 2 = 7 \times 0.3 = 2.1$, and the antilog of 2.1 is approximately 128.)

METER ACCURACY AND TESTING

It is unrealistic to assume that all exposure meters are accurate, even when they are new. Variability affects all aspects of the photographic process, including different exposure meters produced by the same manufacturer that are supposed to be identical. The important question is whether the variability is small enough so that it can be ignored or whether some compensation should be made. How much error should be tolerated in an exposure meter is not an easy question to

Table 2–3 Conversion formulas and data for the aperture value, time value, light value, and speed value in the additive system of photographic exposure (APEX)

$A_v = \text{Log}_2\ (F\text{-}N)^2$	A_v	0	1	2	3	4	5	6	7	8	9	10	
	$F\text{-}N$	1	1.4	2	2.8	4	5.6	8	11	16	22	32	
$T_v = \text{Log}_2\ 1/T$	T_v	0	1	2	3	4	5	6	7	8	9	10	
	T	1	1/2	1/4	1/8	1/15	1/30	1/60	1/125	1/250	1/500	1/1000	Second
$B_v = \text{Log}_2\ B/6$	B_v	0	1	2	3	4	5	6	7	8	9	10	
	B	6	12	25	50	100	200	400	800	1600	3200	6400	Footcandles
$S_v = \text{Log}_2\ S/3$	S_v	0	1	2	3	4	5	6	7	8	9	10	
	S	3	6	12	25	50	100	200	400	800	1600	3200	ASA (Arithmetic)

answer, since the exposure latitude in a specific picture-making situation depends upon a number of factors, including the type of film being used and how critical the viewer of the resulting image will be. In any event, photographers commonly test their new meters against other meters of known performance, and professional photographers usually feel insecure unless they have more than one meter in their possession at all times.

Calibrated light sources are available for testing exposure meters and other light meters, but because of their cost they are more suitable for use by manufacturers and testing laboratories than by individual exposure-meter owners. Sunlight can be used as a standard light source provided the various factors that can affect the illuminance, such as time of day and sky conditions, are taken into account. If one only wants to make a comparison between a certain exposure meter and other meters, preferably of known performance, a light source merely needs to be constant and does not need to be calibrated. A transparency illuminator is especially well suited to being used for comparing exposure meters, since placing the meters against the center of the illuminator eliminates variability due to differences in angle of acceptance, distance from the light source, and shadows of the meters (see Figure 2–35).

The frequency histogram in Figure 2–36 shows the result of a comparison of approximately 200 new and used exposure meters of various brands. All meters were set for the same film speed, and the indicated exposure times for a relative aperture of f/16 were recorded. Even though the total range of exposure times corresponds to a ratio of approximately 16:1, the equivalent of four stops, about 90% of the readings fell within a 2:1 ratio, or a one-stop range. Since all of the meters that had zero settings were properly zeroed, the only method of compensating automatically for consistently high or consistently low readings is to make an adjustment in the film-speed setting. Thus if a meter reads one stop too high, it should be set for one-half the published arithmetic film speed.

Another method of testing an exposure meter is to expose film in a camera according to a meter reading and to make additional exposures bracketing the indicated exposure

Figure 2–35   A transparency illuminator being used as a standard light source to check the calibration of an exposure meter.

by one-third- or one-half-stop changes, and then adjust the film-speed setting to compensate for any difference between the metered exposure and the best exposure based on the photographic results. Photographers sometimes refer to this testing process as deriving a personalized exposure index. Since photographers do not always come up with consistent results when they repeat this test, it is important to note the possible influence of the following factors:

1. Development
2. Increased lens-to-film distance
3. Reciprocity law failure
4. Shutter accuracy
5. Shutter efficiency
6. Flare
7. Lens transmittance
8. Film speed
9. Filters
10. Color temperature of the light source

METHODS OF USING REFLECTED-LIGHT EXPOSURE METERS

Not only do photographers have a choice between reflected-light and incident-light exposure meter readings, but they also have a

Compensation can be made for an exposure meter that reads consistently high or low by adjusting the film speed setting.

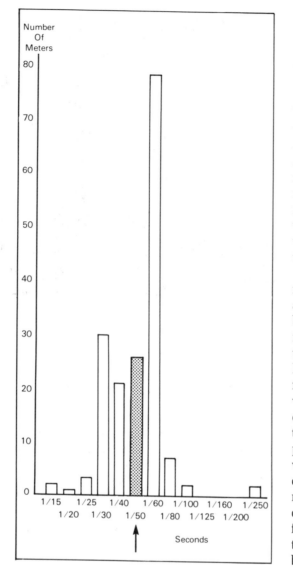

Figure 2—36   Frequency histogram showing the results of the testing of approximately 200 exposure meters on a standard light source. Although the variability represents a range of four stops, 90% of the meters fell within a one-stop range.

choice of various types of reflected-light readings. The types of reflected-light readings that will be considered here are midtone, keytone and Zone System, calculated midtone, camera position, and behind-lens camera meter readings.

Midtone Readings

It may strike some readers as strange that, whereas film speeds for conventional black-and-white films are based on a point on the toe of the characteristic curve where the density is 0.1 above base plus fog density, which corresponds to the darkest area in a scene where detail is desired, exposure-meter manufacturers calibrate the meters to produce the best results when reflected-light readings are taken from midtone areas. Meter users who are not happy with this arrangement can make an exposure adjustment (or, in effect, recalibrate the meter) so that the correct exposure is obtained when the reflected-light reading is taken from the darkest area of the scene, the lightest area, or any area between these two extremes. This type of control is part of the keytone method and the Zone System.

Since reflected-light meters are calibrated for midtone readings, however, it follows that any area used for a reflected-light reading will be reproduced as a midtone on the film, even if the area is the lightest or darkest area in the scene, if no adjustment is made. To illustrate with negative-type film, taking a reflected-light reading from a white subject area will cause this area to be recorded as a medium-tone density, which is considerably thinner than it would be in a correctly exposed negative. In other words, the negative will be underexposed. In terms of the film characteristic curve, all of the tones will be moved to the left on the log exposure axis, causing the midtone of the scene to move from the straight line down onto the toe, and the shadows to move from the toe onto the base-plus-fog part of the curve where detail will be lost.

A typical white surface reflects about 90% of the incident light, which is five times as much as is reflected from an 18%-reflectance gray card, so that taking the reading from a white area causes the film to be underexposed by a factor of five, or a little more than two stops. With the negative-positive process this underexposed negative still can be printed so that the white area from which the reading was taken appears white; but the loss of contrast, and especially the loss of detail in the darker areas, cannot be fully corrected. With reversal color films where the slide or transparency is the end product, and with instant or one-step picture materials that produce prints directly, taking a reflected-light reading from a white area again reproduces that area as a medium tone—which now is

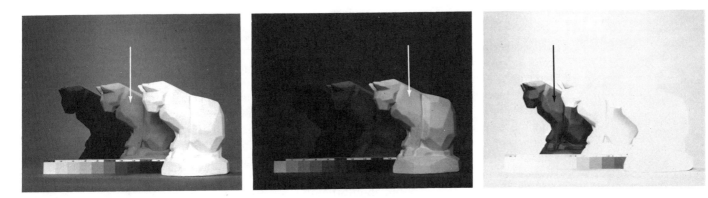

Figure 2–37   Reflected-light exposure meter readings. (Left) The black-and-white plaster cats correspond in reflectance to the two ends of a gray scale, the gray cat represents a midtone. A reflected-light meter reading from the gray cat reproduces it as a realistic medium gray and produces the best exposure for the entire scene. In the original Polaroid print there is tonal separation of all 10 steps in the gray scale and detail in the front planes of all three cats—although not in the shadow underside of the black cat or the rim-lighted top edge of the white cat. (Middle) A reflected-light reading from the white cat results in it being reproduced as a medium gray rather than white, and the photograph as a whole is underexposed. (Right) A reflected-light reading from the black cat results in it being reproduced as a medium gray rather than black, and the photograph as a whole is overexposed.

denser than it would be with normal exposure. Since such reversal materials have little exposure latitude, they are especially useful for demonstrating the differences in results of the various methods of using exposure meters.

The photographs in Figure 2–37 illustrate the results when reflected-light readings are taken from a midtone area, a light area, and a dark area. It can be observed that the areas from which the readings were taken are all reproduced as the same medium density but that the overall effect is satisfactory only when the reading was taken from the midtone area.

One of the problems with the midtone reflected-light method is the difficulty of selecting an appropriate medium-tone area from which to take the reading. If we think in terms of a copying situation in which the subject tones range from white to black, as in a gray scale, a gray that appears to be midway between white and black will have a reflectance of about 18%, or a density of about 0.74 (see Figure 2–38).

Since the photographic midtone agrees with the visual midtone, it is appropriate to use an 18%-reflectance gray card as an artificial midtone; such neutral test cards are made for this purpose. In copying situations the neutral test card should be positioned parallel to the original so that the change in illuminance with the angle of the light sources, as specified by the cosine law of illumination, is the same on the original and the test card.

Taking a reflected-light meter reading from a light area, without adjustment, results in underexposure.

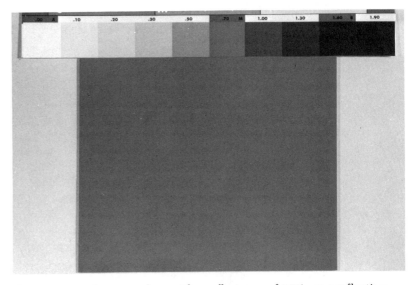

Figure 2–38   A gray surface with a reflectance of 18%, or a reflection density of about 0.74, is perceived as being midway between white and black and serves as an appropriate photographic midtone.

Detail can be obtained in a shadow area by taking a reflected-light reading in that area and decreasing the indicated exposure by four stops.

A neutral test card can be used as an artificial midtone with three-dimensional subjects as well, provided the card is now aimed midway between the camera and the dominant light source rather than directly at the camera (see Figure 2–39). Even though neutral test cards have a more or less dull surface, care must be used, when taking reflected-light meter readings, to angle the meter so as to avoid mirrorlike reflections of the light source. It is also advisable to position the meter close enough to avoid including any lighter or darker area behind the card but not so close as to produce a shadow on the card. Spot attachments or spot meters make it possible to obtain accurate readings at greater distances than with conventional reflected-light meters.

Keytone and Zone System Readings

White-card film speeds were commonly provided on film data sheets in addition to the conventional film speeds in past years, and are still provided for some films designed especially for copying. Since a typical white surface reflects about 90% of the light, or five times as much as the 18%-reflectance neutral test card, the white-card film speed would be only one-fifth the gray-card film speed with arithmetic film speeds.

Reasons for using a white-card exposure meter reading in preference to a gray-card reading are that white surfaces are more read-

ily available than 18%-reflectance gray surfaces, and in low light levels it might be possible to obtain a white-card reading when an exposure meter is not sufficiently sensitive to obtain a reading from a gray card. An alternative to changing the film-speed setting on the exposure meter for white-card readings is to place a marker in the appropriate position on the meter calculator dial and then set that marker rather than the normal marker opposite the meter reading.

When it is especially important to retain shadow detail in a photograph, the keytone method can be used: A reflected-light meter reading is taken from the shadow area and an appropriate mark is made on the calculator dial opposite that reading. (If no adjustment were made, the shadow area would be reproduced as a medium tone, and the negative overall would be much too dense.) The original Weston exposure meter had a U-Position or shadow marker four stops below the normal arrow. The same effect can be obtained with any reflected-light exposure meter by reading the shadow area and then giving the film one sixteenth the exposure or four stops less exposure than that indicated by the normal marker.

A logical extension of the keytone method, beyond calibrating the calculator dial for white-card or highlight readings and for shadow readings, is to calibrate it for each stop between these limits. Such a calibration is an important part of the Zone System, whereby seven gray patches of varying densities are associated with the seven-stop range on the meter dial that represents a normal scene having a luminance ratio of about 1:128. The lightest and darkest patches represent the useful density range of the printing paper, and each patch indicates the predicted density on the print for the corresponding luminance value in the scene. The photographer, however, is not limited to striving for a faithful tone reproduction. If, for example, it is desired to depict snow somewhat darker than normal to emphasize the texture, the patch having the desired density is placed opposite the meter reading from the snow. This procedure is especially useful when using color transparency film where the transparency is the end product, and therefore printing controls are not available (see Figure 2–40). Adjusting film development to com-

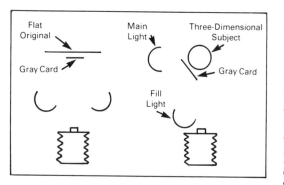

Figure 2–39 In copying setups the gray card should be positioned parallel to the original. With three-dimensional subjects the gray card should be angled midway between the main light and the camera.

Figure 2–40 Silhouettes and semisilhouettes are dramatic examples of situations where facsimile tone reproduction is not the objective, as with this photograph of white golf balls. With the Zone System, the photographer can determine the reproduction tone for a given subject area at the time the film is exposed.

pensate for high-contrast and low-contrast scenes is also an important feature of the Zone System.

Calculated Midtone Readings

In practice it is often difficult to scan a scene and subjectively select an appropriate midtone area from which to take a reflected-light exposure meter reading. The scene may contain light and dark tones but no midtones, and even if there is an appropriate midtone area, it may be difficult to identify due to the influence of the colors of objects, an imbalance in the size of light and dark areas, local brightness adaptation of the visual system, and other factors. An objective method of arriving at a midtone reading is to take separate

readings from the lightest and darkest areas where detail is desired and to select a middle value.

Some care is necessary because the middle value is not an arithmetic average of the highlight and shadow luminances. The scales on most hand meters are set up so that each higher major marking and number represent a doubling of the luminance, or the equivalent of a one-stop change. If the meter has an interval scale of numbers (1-2-3-4-5-6-7), and if the shadow reading is 1 and the highlight reading is 7, the calculated midtone value is the middle number, 4. With such interval scales the same midtone value is obtained by adding the highlight and shadow values and dividing by 2: $\frac{7 + 1}{2} = 4$. If the meter has a ratio scale of numbers (1-2-4-8-16-32-64), and the shadow reading is 1 and the highlight reading 64, the calculated midtone value is again the middle number, in this example 8. Attempting to calculate the midtone value by averaging the shadow and highlight readings with the ratio scale would result in a large error since the arithmetic average of 1 and 64 is 32.5, not 8.

No change in procedure is required for hand meters or behind-lens meters that read directly in f-numbers (for a given shutter speed) or in shutter speeds (for a given f-number), since both scales are ratio. Therefore a correct midtone value is obtained by taking a shadow reading and a highlight reading and selecting the middle number on the f-number or shutter-speed scale (see Figure 2–41).

A bonus of the calculated-midtone method is that by taking readings in the darkest and

f 2.8	SHADOW	1 8 sec.
f 4		1/15
f 5.6		1 30
f/8	MIDTONE	1/60
f 11		1/125
f 16		1 250
f 22	HIGHLIGHT	1 500

Calculated midtone reflected-light exposure meter readings can be obtained by averaging highlight and shadow readings.

Figure 2–41 The calculated midtone is found with a shutter-priority system (left) by noting the f-number that is midway between the f-numbers indicated for a shadow reading and a highlight reading. With an aperture-priority system (right) the shutter speed midway between the indicated speeds for shadow and highlight readings is selected.

The success of camera-position reflected light readings depends upon a reasonable balance of light and dark tones in the scene being photographed.

lightest areas where detail is desired, it is easy to determine whether or not the scene is normal in contrast. The original Weston exposure meter had shadow and highlight markings on the calculator dial to indicate that the luminance ratio for an average scene is 1:128 (or a luminance range of seven stops or seven zones). Later studies suggested a slightly higher ratio of 1:160, which still rounds off to seven stops in terms of whole stops.

Various options are open to the photographer if, for example, a scene is discovered to be considerably more contrasty than the average scene:

1. By adjusting the exposure, detail can be retained in either the highlights or the shadows by sacrificing detail in the other areas.

2. It is possible in some situations to reduce the luminance ratio with a fill light or a reflector, in which case the midtone value should be recalculated.

3. Negative contrast can be lowered by reducing film development so that the negative will print on normal-contrast paper. Since the effective film speed varies with the degree of development, it is necessary to make an adjustment in the film speed or the exposure.

4. Since conventional negative-type films have more exposure latitude than required for an average scene, the contrast adjustment can be postponed to the printing stage. The excess latitude is mostly on the overexposed side, however, so an exposure adjustment may be required to retain shadow detail.

5. With reversal-type color films, contrast can be reduced with controlled flashing. Some slide copiers are designed to provide this option. The same effect can be achieved with any camera by giving the film a second exposure, using an out-of-focus image of a gray card and an appropriate reduction of the exposure.

Other terms applied to exposure methods that are similar or identical in concept to the calculated-midtone method are *brightness-range*, *luminance-range*, *luminance-ratio*, and *log luminance-range*.

Camera-Position Readings

The camera-position or integrated-light method consists of aiming a reflected-light exposure meter at the scene from the camera position. Assuming that the meter's angle of acceptance is about the same as the angle of view of the camera-lens combination, the meter provides a single reading that represents an average of the luminances of the different subject areas being included in the photograph. Correct exposures should result with scenes that are well balanced with respect to light and dark areas. Scenes in which light areas predominate produce false high readings, which lead to underexposure. Scenes in which dark tones prevail produce false low readings and overexposure (see Figure 2–42).

A large proportion of photographs made on negative-type films using this method are satisfactory because most of the scenes photographed are reasonably well balanced in light and dark areas, because of the generous exposure latitude of the film, and perhaps even because of the generous acceptance latitude of the photographers. It is not difficult to recognize scenes that deviate dramatically from a balance of light and dark areas, where corrective action should be taken. Also, exposure-meter instruction manuals usually caution the user to tilt the meter downward to avoid including bright sky areas in outdoor meter readings.

Camera Meters

Although selenium-cell photovoltaic-type exposure meters have been built into small-format cameras, the introduction of the more sensitive cadmium sulfide cell and other meters of the photoconductive type made it possible to reduce the cells to a more appropriate size. Camera meters can be divided into two basic types: those with the cell mounted on the surface of the front of the camera and those with the cell located behind the camera lens (see Figure 2–43). Meters with surface-mounted cells perform in a manner similar to that of a reflected-light hand meter that is aimed at the subject from the camera position to make an integrated-light reading. The an-

Figure 2–43 Surface-mounted exposure-meter cell on a between-the-lens shutter 35 mm camera.

gle of acceptance of the surface-mounted cell can be controlled with a baffle or lens, and the meter can be coupled with the camera-lens diaphragm or shutter. It is possible to place a diffusor over the cell and take incident-light readings by turning the camera around and holding it in front of the subject, but this has not been a widely used feature.

Behind-lens meters are better suited to focal-plane shutter cameras than to cameras having between-the-lens shutters that—at least with small format cameras—are open only when the film is being exposed. There has been considerable experimentation with the location, size, and shape of the behind-lens cell or cells. Cells have been placed in various positions on the prism of single-lens reflex cameras, on the mirror, behind the mirror, and off to one side where the cells receive and measure light reflected from a mirror or beam splitter, the shutter curtain, or even the film itself (see Figure 2–44).

Many technical problems had to be solved with behind-lens meters, such as preventing light that enters the viewfinder from behind the camera from reaching the cell; preventing polarization of the image light being mea-

No compensation is required for the increased lens-to-film distance for closeups with behind-lens exposure meters.

Figure 2–42 Camera-position reflected-light readings. (Top) Camera-position reflected-light exposure meter readings produce the best results with scenes that are balanced in light and dark tones. (Photograph by John Johnson.) (Middle) Predominately light scenes produce inflated meter readings, which result in underexposure. (Photograph by John Johnson.) (Bottom) Dark scenes produce false low readings which result in overexposure. (Photograph by Yu Yin Tsang.)

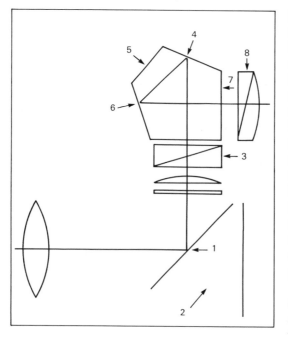

Figure 2–44   A variety of positions have been used for the exposure-meter cell in single-lens reflex cameras by different manufacturers.
1. Behind the mirror. Measures direct light from the lens through clear lines or partially transmitting mirror.
2. Behind the mirror. Measures light reflected from the film during the exposure.
3. Between the focusing screen and the pentaprism. Measures light diverted by a beam-splitter.
4–7. On the pentaprism.
8. On the eyepiece. Measures light diverted by a beam-splitter. Difficulties of behind-the-lens metering can include inflated readings due to light entering the eyepiece from the rear, false readings with polarizing filters due to polarization of light by the metering system, and inconsistencies in diverting a fixed proportion of light to the meter cell.

sured (which would give inaccurate readings when a polarizing filter is placed on the camera lens), and compensating for the change in the size of the diaphragm opening between the time the light is measured and when the film is exposed. There are also many variations concerning the size, shape, and location of the picture area being measured, including integrated, center-weighted, off-center weighted, spot, and multiple cells.

Although behind-lens meters generally offer less precision with respect to measuring the light from selected parts of a scene, and although they are restricted to reflected-light

type readings, they offer certain advantages over hand meters in addition to convenience. Since behind-lens meters essentially measure the light falling on the film, variations in the light due to increasing the lens-to-film distance for closeup photography, flare light, and variations in transmittance of different lenses set at the same f-number are automatically compensated for. Some camera manufacturers also claim that their meters compensate for the absorption of light by color filters placed over the lens, but until the spectral response of the meter cells is brought into closer agreement with the spectral sensitivity of typical panchromatic films than they are at the present time, they should not be depended upon for this function.

In addition to this use with small-format cameras, behind-lens meters are also available for use with view cameras. By attaching the light-sensitive cell to the end of a movable probe in a film-holder type frame, the cell can be positioned to take readings in any selected area while observing the image on the ground glass, or multiple readings can be taken for the calculated midtone method (see Figure 2–45).

INCIDENT-LIGHT READINGS

Incident-light meter readings have long been popular with professional motion-picture and still photographers, because such readings can be made quickly and give dependable results. Typical instructions for using an incident-light exposure meter are simply to hold the meter in front of the subject and aim it at the camera. There is no need to ponder whether the scene has an equal balance of light and dark areas because the meter measures the light falling on the subject rather than the light reflected by the subject, and the meter is calibrated for a subject having a range of tones from black to white with a medium tone of 18% reflectance. In addition, there is no need to worry about the position of the dominant light source with incident-light meters having hemispherical diffusors because such meters automatically compensate for changes in the direction of the light and the corresponding change in the relative sizes of highlight and shadow areas.

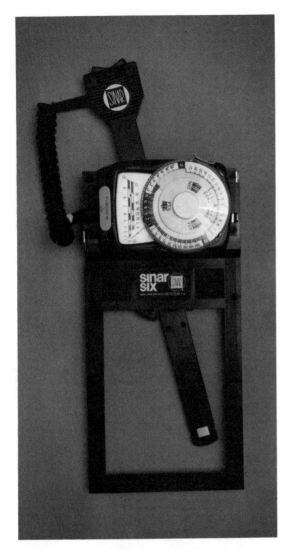

Figure 2–45 A view-camera exposure meter that uses a movable probe to obtain readings in selected areas at the film plane.

Instruction manuals for exposure meters sometimes warn the user that the meter can only measure light, it can't think for the photographer; this applies to incident-light as well as reflected-light exposure meters. One potential source of error is associated with the time-saving observation that when the light falling on the subject is the same as that falling on the camera, as when both are in direct sunlight, the reading can be taken at the camera position. It is obvious that when the camera is in the shade of a tree or building, whereas the subject is in direct sunlight, the incident-light reading must be taken in the sunlight; but it is not so obvious, when

a light source is close to the subject, that readings can be dramatically different at the subject and one foot in front of the subject. Thus, the closer the light source is to the subject, the more important it is to take the incident-light meter reading as close to the subject as possible. The inverse-square law of illumination can be ignored with sunlight because of the great distance of the sun, but it should not be ignored indoors when a light source is within a few yards of the subject.

Although the point has already been made that it is not necessary to consider the balance of light and dark areas in a scene for incident-light readings, it is advisable to modify the exposure indicated by an incident-light meter when the main subject is either very light or very dark in tone. It is advisable to increase the exposure by one-half to one stop with black or very dark objects and to decrease the exposure by one-half to one stop with white or very light objects. Since it is possible to hold detail in an entire gray scale or a scene containing both white and black areas, even with reversal color films, which have limited exposure latitude, it is necessary to explain why the exposure should be adjusted when the subject is predominantly light or dark.

Incident-light exposure meters with hemispherical diffusors should be positioned close to the subject and aimed at the camera.

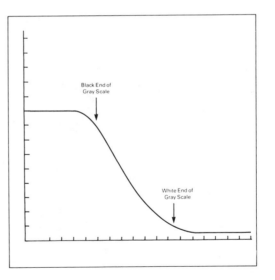

Black End of Gray Scale

White End of Gray Scale

Figure 2–46 Since the white and black ends of a gray scale photographed on reversal-type film fall on the shoulder and toe of the characteristic curve where the slope is lower, some contrast is lost in both of these areas compared to the intermediate tones, which fall on the steeper central part of the curve.

Referring to the characteristic curve for a reversal-type film shown in Figure 2–46, it will be noted that the white end of a correctly exposed gray scale falls on the toe and the black end falls on the shoulder. Some contrast is therefore lost in both of these areas, compared to the intermediate tones, which fall on the central part of the curve. This compression of contrast in the lightest and darkest areas is acceptable when they are relatively small in area; but when either becomes the main area of interest, it is better to adjust the exposure to move that area onto or closer to the straight line in order to increase the detail and contrast (see Figure 2–47).

No single method of using an exposure me-

Reflected-light type exposure meter readings are appropriate for subjects that emit or transmit light rather than reflect light.

ter is appropriate for all picture-making situations. Incident-light readings cannot be made when the image-forming light is either emitted or transmitted by the subject rather than being reflected. Examples of emitted light are molten metal, flames, and fluorescent and phosphorescent objects. Transmitted light is encountered when transparent or translucent materials are illuminated from behind, as with stained-glass windows, some night advertising signs, a photographic transparency placed on an illuminator for copying, and sunsets.

There are, on the other hand, situations where incident-light meter readings are more appropriate than reflected-light readings. Probably the most commonly encountered

Properly made reflected-light meter readings and incident-light meter readings should indicate the same camera exposure settings with normal scenes.

Figure 2–47   Although incident-light meter readings (or reflected-light meter readings from a neutral test card) produce the optimum exposure for subjects having a normal range of tones as represented by a gray scale, exposure adjustments are recommended for high-key scenes and for low-key scenes.
(Top left) White object on light background with normal exposure. (Top right) One stop less exposure. (Bottom left) Black object on dark background with normal exposure. (Bottom right) One and one-half stops more exposure.

situation of this type is where the subject or the subject areas are too small to be read accurately even with a narrow-angle spot-type reflected-light meter. This problem occurs not only in the fields of closeup photography and photomacrography but also when photographing a detailed object such as a black line drawing on a white background, where the light and dark tones are not balanced in size and the individual areas are too small to be measured (see Figure 2–48).

In most picture-making situations the photographer has the choice of making either reflected-light or incident-light exposure meter readings. With meters that are designed to be used in both modes, it is a valuable exercise for the photographer to make comparison readings. When the readings do not agree, comparison photographs will reveal which method is superior in that type of situation. Reflected-light readings from an 18%-reflectance neutral test card are essentially the same as incident-light readings and should consistently indicate the same camera exposure.

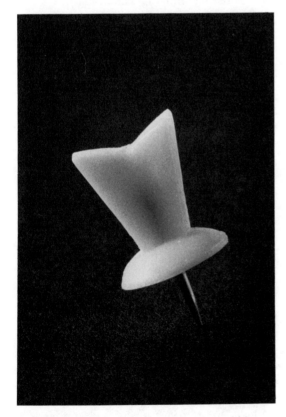

Figure 2–48   A situation where it would be difficult to obtain a dependable reflected-light exposure meter reading due to the small size and the light tone of the push pin.

DEGREE OF DEVELOPMENT AND EFFECTIVE FILM SPEED

Published film speeds for black-and-white films are based on a specified degree of development. Developing the film to a higher contrast index tends to increase the film's effective speed, up to a limit, and developing to a lower contrast index tends to decrease the effective speed. In practice it is not uncommon for photographers to develop film to higher and lower contrast indexes than those indicated in the film speed standard.

Film development is sometimes altered to compensate for the difference in printing contrast between condenser and diffusion enlargers, to compensate for variations in scene contrast, and to compensate for variations in the amount of camera flare light in different situations. Adjustments in the effective film speed can be made either by altering the film-speed setting on the exposure meter, or by using the published film speed and then incorporating the correction in the choice of camera shutter speed and f-number settings.

The nomograph shown in Figure 2–49 makes it possible for the photographer to determine the recommended development contrast index and the exposure adjustment needed to compensate for changes in the type of enlarger, scene contrast, and flare light. Lines have been drawn on the nomograph for an example in which a condenser enlarger is used, the subject has a luminance range of eight stops, and the camera flare is moderate. The scale on the right side indicates that the film should be developed to a contrast index of 0.39, and the exposure should be increased by two-thirds of a stop. Testing procedures for deriving personalized development and exposure data for these variables are provided in some of the Zone System references.

FLASH GUIDE NUMBERS

Guide numbers provide a means of determining the f-number setting required for correct exposure with flash and electronic-flash light sources. The guide number can be defined as the product of the f-number and the

A downward adjustment in the film-speed setting is recommended when developing film to a lower than normal contrast index.

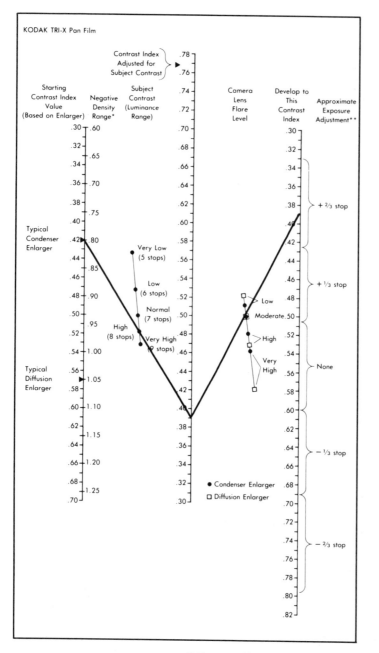

© Eastman Kodak Company 1976

* These are the typical negative density ranges that result when normal luminance range subjects (7 stops range) are exposed with moderate flare level lenses and developed to the contrast index shown in the left scale.

** Some film and developer combinations may require more or less exposure adjustment than these average values shown here, especially when the adjusted contrast index is less than .45 or greater than .65. Some of the finest-grain developers cause a loss in film speed that must be considered in addition to the losses caused by developing to a lower contrast index.

Figure 2–49   Contrast-control nomograph. (© Eastman Kodak Company 1976.)

distance that produce the correct exposure for a given set of conditions, including the light source, reflector, film speed, shutter speed, and reflectance of the surroundings. In practice, the guide number is divided by the flash-to-subject distance to determine the correct f-number; or if the photographer wants to use a certain f-number, the guide number is divided by the f-number to determine the correct flash-to-subject distance. Since the guide-number system is based on the assumption that the illumination from the flash follows the inverse-square law, which in turn is based on a point source of light without a reflector, some variation in exposure may be encountered at small and large distances.

Additional complications are encountered in attempting to determine the correct exposure with guide numbers when using bounce flash (where the reflectance of the reflecting surface must be taken into account) and multiple flash (where the cumulative effect of two or more flash sources at different distances and angles, and sometimes different intensities, must be calculated).

FLASH METERS

There has been little demand for exposure meters to be used with expendable flashbulbs, as it would be necessary to fire a flashbulb for the measurement before firing a second flashbulb to expose the film. Conventional exposure meters cannot be used with flash and electronic flash since the meters are designed to measure the light from a source of constant intensity rather than a source of rapidly changing intensity and short duration. Electronic-flash meters work on the principle of measuring the total amount of light that falls on the photocell over an appropriately short period of time by charging a capacitor and then measuring the charge (see Figure 2–50). Some electronic-flash meters have required a sync cord connection with the flash unit to be able to measure only the light from the flash plus ambient light for the approximate period that the camera shutter would be open. Other meters are able to achieve this without a physical connection with the flash unit, and meters are now available that can be used both as a conventional exposure meter and as an electronic-flash

Figure 2–50   An electronic-flash exposure meter.

meter. Whereas certain electronic-flash meters can be adjusted to integrate the flash plus ambient light over a time interval equal to the selected shutter speed, others operate only over a fixed instantaneous interval, such as 1/60 second.

An alternative to measuring the light from an electronic-flash unit with a preliminary trial flash and a meter is to measure the light reflected from the subject by a sensor built into the flash unit and then terminate the flash when the sensor system indicates the film has received the correct amount of light. The process by which the duration of the flash is shortened to provide the correct exposure is called *quenching*. Early quenching systems achieved the effect by dumping the remaining capacitor charge when sufficient light had been emitted, thus wasting the unused charge and shortening battery life. More recent units have incorporated a means of breaking the circuit between the capacitor and the tube when a signal is received from the metering system, without wasting the remaining charge in the capacitor. When quench-controlled electronic-flash units are used for closeup photography, the flash duration tends to be extremely short. The short duration provides excellent action-stopping capabilities for the photographing of rapidly moving objects; but due to reciprocity effects

the density, contrast, and color balance of the photographic image can be affected (see Figure 2–51).

COLOR-TEMPERATURE METERS

Color-temperature meters are closely related to photoelectric exposure meters even though they are designed to measure the spectral quality of light rather than the quantity. An ingenious photographer can use a conventional exposure meter as a color-temperature meter by taking readings in sequence through a red filter and a blue filter and setting up a conversion chart or dial based on the relative values of the two readings. One type of color-temperature meter contains two photocells and circuits, with a red filter over one cell and a blue filter over the other, and a calibrated dial to indicate the relative currents in the two circuits in terms of degrees Kelvin or mireds (see Figure 2–52). Color-temperature meters of this type work satisfactorily only with light sources that approximate a blackbody such as tungsten lamps. Three-filter (red, green, and blue) color-temperature meters provide more useful information with other types of light sources.

Some electronic flash units have built-in sensors that terminate the flash when the system indicates that the film has received the correct exposure.

Color-temperature meters operate by comparing the relative amounts of two or three different colors of light in the incident light.

Figure 2–51   A bullet photographed in the process of cutting a playing card in half demonstrates the action-stopping capability of short-duration electronic-flash light sources. (Photograph by Andrew Davidhazy.)

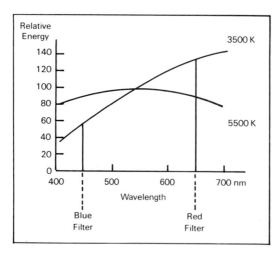

Figure 2–52   In a two-filter color-temperature meter the relative amounts of blue light and red light are measured. The ratio of blue to red is approximately 1:1 for the 5500 K illumination, but approximately 1:2.4 for the 3500 K illumination.

REVIEW QUESTIONS

1. Identify the correct formula: *(p. 32)*
   A.
   $$\text{f-number} = \frac{\text{effective aperture}}{\text{focal length}}$$
   B.
   $$\text{f-number} = \frac{\text{image distance}}{\text{effective aperture}}$$
   C.
   $$\text{f-number} = \frac{\text{focal length}}{\text{image distance}}$$
   D.
   $$\text{f-number} = \frac{\text{focal length}}{\text{effective aperture}}$$

2. As the opening in the diaphragm is made smaller, the f-number . . . *(p. 33)*
   A. increases
   B. decreases

3. Select the number that does not represent a whole stop in the conventional f-number series. *(p. 33)*
   A. 64
   B. 4
   C. 1.8
   D. 5.6
   E. 1.4

4. Most lenses for 35 mm cameras are made so that they cannot be stopped down beyond f/22 because . . . *(p. 34)*
   A. the exposure times would be too long to hand-hold the camera
   B. it is difficult to make diaphragms with such small openings
   C. diffraction would degrade the images
   D. of reciprocity effects

5. The minimum resolving power that is considered adequate in a print to be viewed at a distance of 10 inches is approximately . . . *(p. 34)*
   A. 10 lines/mm
   B. 28 lines/mm
   C. 35 lines/mm
   D. 64 lines/mm
   E. 80 lines/mm

6. The ratio of the amount of light transmitted by f/1.4 and f/1.2 lenses at the maximum diaphragm openings is approximately 1 to . . . *(p. 36)*
   A. 1.20
   B. 1.26
   C. 1.30
   D. 1.36
   E. 1.40

7. A 50 mm focal length lens is set at f/8. If the image distance is 100 mm, the exposure factor is . . . *(p. 36)*
   A. 1
   B. 2
   C. 4
   D. 8
   E. 16

8. If an f/1.4 lens that transmits only half of the light that it would if no light were lost due to reflection and absorption is recalibrated in T-numbers, f/1.4 would become . . . *(p. 37)*
   A. T/1
   B. T/1.6
   C. T/1.8
   D. T/2
   E. T/2.8

9. According to the rule of thumb, the slowest shutter speed that is safe to use with a hand-held camera equipped with a 55 mm focal length lens is approximately . . . *(p. 37)*
   A. 1/15 sec.
   B. 1/30 sec.
   C. 1/60 sec.
   D. 1/125 sec.
   E. 1/250 sec.

10. When accurate between-the-lens shutters are used at high speeds with the diaphragm stopped down, the effective exposure time is . . . *(p. 39)*

A. longer than the marked exposure time
B. the same as the marked exposure time
C. shorter than the marked exposure time

11. Overexposure of black-and-white film can cause . . . (p. 44)
 A. an increase in contrast (only)
 B. a decrease in contrast (only)
 C. either an increase or a decrease in contrast
 D. None of the above. Exposure does not affect contrast.

12. If one were to test an exposure meter by exposing film in a camera, it would be best to use . . . (p. 45)
 A. a negative type black-and-white film
 B. a negative type color film
 C. a reversal type color film

13. If a black-and-white photograph is made on panchromatic film with a red filter on the camera lens according to a behind-lens meter that has high red sensitivity, the resulting negative will probably be . . . (p. 49)
 A. underexposed
 B. correctly exposed
 C. overexposed

14. The term "luminance meter" is a synonym for . . . (p. 50)
 A. reflected-light meter
 B. incident-light meter
 C. ultraviolet radiation meter
 D. infrared radiation meter

15. When using an exposure meter with a hemispherical diffuser with a three-dimensional subject, it is recommended that the meter be aimed . . . (p. 52)
 A. at the main light source
 B. at the camera

C. midway between the main light source and the camera

16. The following series of numbers (1, 2, 4, 8, 16) is identified as . . . (p. 52)
 A. a ratio scale
 B. an interval scale
 C. a hybrid scale

17. If a reflected-light exposure meter reading is taken from a white area in a typical scene, the resulting negative or transparency will be . . . (p. 57)
 A. underexposed
 B. correctly exposed
 C. overexposed

18. Using a neutral test card as an artificial midtone for an exposure meter reading with a three-dimensional subject, the card should be aimed . . . (p. 58)
 A. at the camera
 B. at the dominant light source
 C. midway between the camera and the dominant light source

19. With an exposure meter that has an interval scale of numbers (1-2-3-4-etc.), the calculated midtone reading for a scene in which the shadow reading is 2 and the highlight reading is 8 is. . . (p. 59)
 A. 3
 B. 4
 C. 5
 D. 6
 E. 7

20. An advantage behind-lens meters have over hand-held meters is. . . (p. 62)
 A. greater sensitivity
 B. better spectral response
 C. automatic compensation for changes in lens-film distance
 D. that they do not require batteries

3 Variability and Quality Control in Photography

David Spindell. *Rebus.*

INTRODUCTION TO THE CONCEPT OF VARIABILITY

No two things or events are ever exactly alike.

Photographers are by necessity experimenters. Although they are not thought of as scientists operating in a research atmosphere, they most certainly are experimenters: Every time an exposure is made, and the negative is processed, a genuine experiment has been made. There is no way the photographer can precisely predict the outcome of such an experiment in advance of doing the work. The objectives of the experiment derive from the purpose of the assignment, and the procedures flow from the photographer's technical abilities. The success of the experiment will be determined by evaluating the results. Finally, conclusions are reached after interpreting the results, which in turn shape the performance for the next assignment/experiment. It is the last step in the sequence, namely the interpretation of test results and the conclusions reached, that we will be concerned with in this section.

A great deal of what is written about the practice of photography consists of opinions. When someone states that "black-and-white photographs are more artistic than color photographs" or that "condenser enlargers are better than diffusion enlargers," it represents an expression of personal judgment. Such statements are properly referred to as *subjective* because they arise from personal attitudes. Because opinions are statements of personal feelings, they are neither right nor wrong and, consequently, are always valid. The problem with subjective opinions is that they lead to conclusions that are not easily tested and analyzed. Therefore, when one relies solely on personal opinion, the possibilities for obtaining new insights into the photographic process become very limited.

Because of the potential ambiguity of subjective opinions, it is often preferable to use numerical expressions that have been derived from measurements. Such numbers can be considered *objective* in that they are generated outside of personal attitudes and opinions. For example, the blackness of a photographic image can be measured with a densitometer and the result stated as a density value. Such a numerical expression of its appearance can be obtained by independent workers and therefore verified. Although there may be many subjective opinions about the blackness of the image, there can be only one objective statement based on the measurements. Consequently, a goal of anyone who is experimenting is to test opinions with facts that usually are derived from measurements.

Whenever an experiment is performed and a result obtained, there is a great temptation to form a conclusion. For example, a sample roll of a new brand of film is tested and found to give unacceptable results; most photographers would not bother to purchase another roll. On the other hand, if the temperature of a processing bath was found to be 68° in the bottom of the tank, many people would act as if it were the same everywhere. Such conclusions should be resisted because of a fundamental fact of nature: No two things or events are ever exactly alike. No two persons (not even so-called identical twins) are exactly alike; no two snowflakes are ever identical. Photographically speaking, no two rolls of the same brand of film are exactly alike. No two frames on the same roll, no two areas within the same frame and indeed no two silver halide crystals are ever truly identical. If the inspection is close enough, differences will always be found. Stated more directly: *variability always exists.*

In addition to these differences, time creates variations in the properties of an object. The photographic speed of a roll of film is not now what it was yesterday, nor will it be the same in a few months. Using this point of view, an "object," say a roll of film, is really a set of events that are unique and will never be duplicated. As the Greek philosopher Heraclitus long ago said, "You cannot step in the same river twice."

With this view of the real world, it should be obvious that an essential task of the photographer/experimenter is to determine the amount of variability affecting the materials and processes being used. This will require that *at least two* separate measurements be made before reaching a conclusion about the characteristics of an object or process. A photographer who ignores variability will form erroneous conclusions about photographic materials and processes, and will likely be plagued by inconsistent results.

SOURCES OF VARIATION

In an imaginary gambling situation where a perfect roulette wheel is operated with perfect fairness, the outcome of any particular spin of the wheel is completely unpredictable. It can be said that, under these ideal circumstances, the behavior of the ball is determined by *random* or *chance* causes. These terms mean that there are countless factors influencing the behavior of the wheel, each of which is so small that it could not be identified as a significant factor. Consequently, these numerous small factors are grouped together and identified as chance effects; and their result is termed *chance-caused variability*. In addition to being present in games of chance, chance-caused variation occurs in all forms of natural and human undertakings.

Consider, on the other hand, a rigged roulette wheel, arranged so that the operator has complete control over the path of the ball. Under these circumstances there would be a reason (the decision of the operator) for the fall of the ball into any particular slot. With this condition it is said that there is an *assignable cause* for the behavior of the ball.

The distinction between these types of effects—chance and assignable cause—is important, because it will influence the actions taken. For example, your desire to play the roulette wheel would quite likely disappear if you believed the operator was influencing the results. A similar dilemma arises when, for example, an electronic flash fails to fire. The film is advanced and a second photograph is taken and again the flash fails. If this procedure is continued, it would be based upon the assumption that it was only a chance-caused problem. On the other hand, if the equipment were examined for defects, the photographer would be taking action with the belief that there was an assignable cause for the failure of the flash.

In other words, when it is believed that chance alone is determining the outcome of an event, the resulting variation is accepted as being natural to the system and likely no action will be taken. However, if it is believed that variability is due to an assignable cause, definite steps are taken to identify the problem so that it can be removed and the system returned to a natural (chance-governed) condition. Consequently, it is very important to be able to distinguish chance-caused from assignable-caused results.

Although there is no possibility of predicting the outcome of a single turn of a fair roulette wheel, there is an underlying *pattern* of variation for its long-term performance. If the ball has the same likelihood of dropping into each of the slots, then over a large series of plays each of the numbers should occur as frequently as any other. Thus, the *expected* pattern of variability for the long-term performance of the roulette wheel could be described by a graph as shown in Figure 3–1. This is called a *frequency distribution* graph because it displays the expected frequency of occurrence for each of the possible outcomes. In this case, it is termed a *rectangular distribution*, which arises any time there is an equal probability of occurrence for every outcome. Every one of the 20 slots has an equal chance of receiving the ball.

Suppose now that the roulette wheel is spun many times so that the *actual* pattern of variation can be compared to the *expected* distribution. If the wheel is actually operated 200 times, it is highly unlikely that each number will occur exactly 10 times. Instead, a pattern similar to that shown in Figure 3–2 will probably arise, indicating that the actual distribution only approximates that of the expected (or theoretical) results. It would likely be concluded from these results that the wheel is honest. However, if the actual distribution of 200 trials appeared as shown in Figure 3–3, there would be good cause to suspect something abnormal was happening.

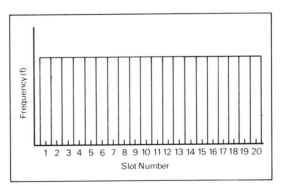

Figure 3–1 Rectangular pattern of variability from the expected long-term performance of a perfect roulette wheel. Each number has an equal chance of occurring.

Variations in a repetitive process can be attributed to chance, an assignable cause, or both.

The expected pattern of variation with repeated spinning of an honest roulette wheel is unlikely to ever occur in practice.

Assignable-cause variability can be corrected, random-cause variability cannot.

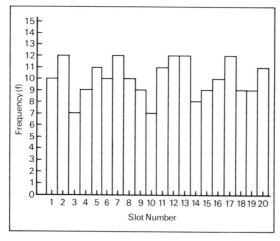

Figure 3-2   Pattern of variability from the actual performance of a roulette wheel indicating that only chance differences are occurring. Although theoretically each number has an equal chance of occurrence, some numbers occur more frequently over the short run.

Thus, a basic method for judging the nature of variation in a system is to collect enough data to construct a frequency distribution (sometimes referred to as a frequency histogram) and compare its shape to that which is expected for the process. If the shapes are similar, the process is behaving properly. A significant difference indicates an abnormal

The normal-distribution curve represents the expected pattern of random-cause variability with a large amount of data.

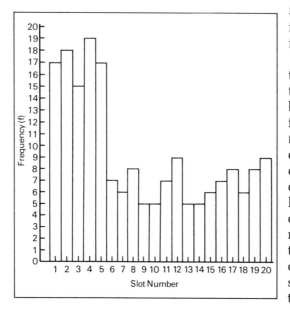

Figure 3-3   Pattern of variability from the actual performance of a roulette wheel indicating assignable-cause influence. The lower numbers are occurring with much greater frequency.

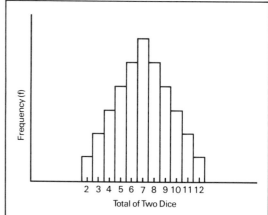

Figure 3-4   Triangular pattern of variability from the expected long-term performance of two randomly thrown dice.

condition that would usually be identified and corrected.

In the game of dice, if a single die is thrown repeatedly in a random manner, a rectangular distribution is also to be expected. In other words, each of the six numbers is equally likely to appear on every roll. For the total of two dice, a quite different pattern is expected since the probability of rolling a seven is greater than for any other number. In other words, there are more combinations that total seven than any other number ($1+6$, $2+5$, $3+4$, $4+3$, $5+2$, $6+1$). The expected pattern is referred to as triangular distribution and is shown in Figure 3-4.

If this concept is extended to the totals of three dice, five dice, etc., the expected distributions begin to look like the symmetrical bell-shaped curve shown in Figure 3-5. This frequency pattern is called the *normal distribution*. It is of exceptional importance because whenever there is a multiplicity of outcomes with many factors affecting each occurrence (as there almost always is in real-life events), this is the pattern suggested by chance. Consequently, when evaluating the results of tests and experiments using photographic materials, the normal distribution of data is the expected condition. If the results of the tests give something other than this bell-shaped pattern, it is an indication of assignable-cause problems and appropriate action should be taken.

Notice that to obtain a useful picture of the pattern of variation, more than a few samples

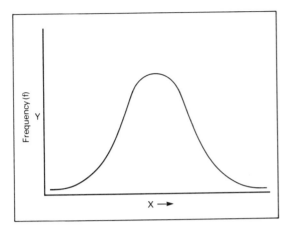

Figure 3–5 Expected pattern of variation when there is a multiplicity of outcomes with chance governing the process. This bell-shaped curve is called the normal distribution.

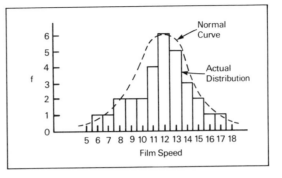

Figure 3–7 Frequency histogram of film speeds (30 samples).

must be collected. Statistical theory suggests that *at least 30 samples* should be used to construct the distribution before making any inferences, because the pattern that develops as additional samples are taken tends to stabilize at 30. For example, a slow film was tested and five samples gave speeds of 10, 12, 10, 14, and 14, producing the histogram shown in Figure 3–6. Although the resulting pattern is not a bell-shaped curve, there are too few samples to draw a valid conclusion. An additional 25 samples added to the first five could give the distribution illustrated in Figure 3–7. Here the pattern approximates the normal distribution, indicating that the differences in film speeds are only the result of chance-caused effects.

Suppose the results of testing 40 different shutters at the same marked shutter speed gave a set of data distributed as shown in

Figure 3–8. The pattern exhibited differs markedly from the normal distribution; thus it is inferred that something unusual is occurring. The two peaks in the histogram suggest the presence of two sources (or populations) of data, such as two different types of shutters.

Consider the histogram shown in Figure 3–9, which is the result of testing the same meter 35 times against a standard light source. Notice that the distribution is asymmetrical—the left-hand "tail" is longer than that on the right. Again, it should be concluded that there is an assignable cause in the system. Perhaps because of a mechanical problem the needle is unable to go above a given position, hence the lack of data on the high side. In both of these instances, because the bell-shaped distribution did not occur, a search for the assignable cause is necessary.

Two numbers, one representing an average value and the other the variability, can summarize the information obtained from a large number of measurements.

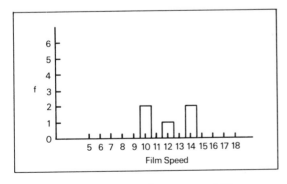

Figure 3–6 Frequency histogram of film speeds (five samples).

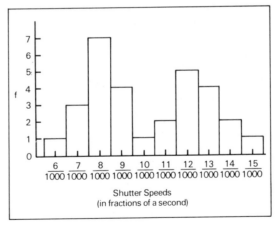

Figure 3–8 Frequency histogram of 40 different shutters tested at the same marked shutter speed of 1/125 (8/1000) second.

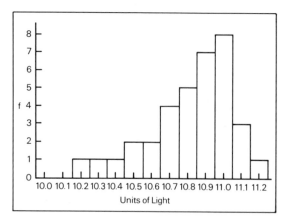

Figure 3–9 Frequency histogram from testing the same meter against a standard source 35 times.

THE NORMAL DISTRIBUTION

After a histogram has been inspected and found to approximate the normal distribution, it is possible to compute two numbers that will completely describe the important characteristics of the process.

The first of these is the arithmetical *average* or *mean*. This value is obtained by adding all of the sample data together and dividing by the number of samples taken. The resulting number is the value around which the rest of the data tend to cluster. In other words, it is a measure of the *central tendency* of the distribution. There are two other measures of central tendency that are sometimes used to describe a set of data. The *mode*, which is defined as the most frequently occurring value in a data set, is located at the peak of the histogram. The *median* is that value which divides the data set into two equal halves. Of all three measures of central tendency, the average is by far the most often used. It is sometimes useful to represent variability with the simpler concept of the relationship between the smallest value and the largest value in a set of data. For data that are measured with interval scales, such as temperature, length, and weight, the difference between the largest and smallest values is identified as a *range*. For data that are measured with ratio scales, such as the contrast of a scene, the largest value divided by the smallest value is identified as a *ratio*. Thus the contrast of a certain scene might be ex-

The values of the mean, the median, and the mode will be the same for a set of data only when the data conform to a normal distribution.

pressed as having a luminance ratio of 128 to 1.

The second number of interest is called the *standard deviation* and represents a measure of the variability of the system. To obtain this number, each piece of data is compared to the average (or mean) and the average difference or deviation for all of the data is determined. This value is a measure of the width of the normal distribution. If the normal distribution is very narrow, the variation is small; if it is very wide, the variation is large. Thus, the standard deviation is a direct measure of the *amount of variability* in a normal distribution.

Figure 3–10 illustrates the relationship between the average and the standard deviation for a hypothetical normal distribution. Notice that if the distribution should shift left or right, this would be reflected in a change of the average since it means that the central position of the system has changed. If the width of the distribution were to become narrower or wider, the standard deviation would change because the amount of variability in the system would be changing. Taken together, these two values can provide a useful set of numbers for describing the characteristics of any system tested.

POPULATIONS VS. SAMPLES

At this point, it is necessary to distinguish between populations and samples. If, for example, a new brand of developer were being marketed, it would be intelligent to test it

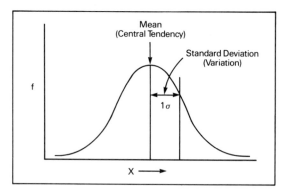

Figure 3–10 Theoretical normal distribution illustrating the relationship between the mean and the standard deviation.

prior to use. To obtain a crude estimate of the variability affecting this product, at least two samples should be tested. If it is desired to determine the pattern of variation, at least 30 should be tested. Even if 30 tests were performed (which would be quite a task) there would be countless more bottles that were not tested. Thus the term *population* refers to all possible members of a set of objects or events. Additional examples of populations are: the ages of all United States citizens; the weights at birth of all newborns; the ISO/ASA speeds of all sheets of Brand X film; the temperatures at all possible points in a tank of developer; the accuracy of a shutter over its entire lifetime.

In each case, an extremely large number of measurements would have to be made to discover the population characteristics. Seldom, if ever, are all members of a population examined, primarily because it is usually impossible to do so. Further, by examining a properly selected group of *samples* from the population, almost as much insight can be obtained as if all members had been evaluated. Herein lies the value of understanding some basic statistical methods. By obtaining a relatively small number of representative samples, inferences can be made about the characteristics of the population. Such is the basis of much scientific investigation, as well as political polls.

If the sample is to be truly representative, it must be selected from the population with great care. The principle to be followed is this: Every member of the population must have an equal chance of being selected. When taking sample density readings from a roll of film after processing, for example, it is dangerous to obtain the data always from both ends and the middle. It is also dangerous never to obtain data from the ends and the middle. The samples must be selected so that in the long term data are obtained from all parts of the roll with no area being consistently skipped.

Although this principle appears simple, the method for carrying it out is not. The most common approach for ensuring that all members of the population have been given an equal opportunity is to use a *random sampling* method. Random means "without systematic pattern" or "without bias." Human beings do not behave in random ways but are strongly biased by time patterns and past experiences. Consequently, special techniques must be employed to avoid the non-random selection of samples. Techniques such as drawing numbered tags from a box, using playing cards, tossing coins, etc., can be used to minimize bias in obtaining samples. Tables of random numbers are also very useful in more complex situations.

It should be obvious, then, that there is a distinction between the average of the population and the average of the sample data. The symbol μ (Greek letter *mu*) is used to refer to the population average, while the symbol $\overline{X}$ (X bar) is used for the sample average. Likewise there is a difference between the standard deviation of the population and that of the sample data. The symbol σ (Greek letter *sigma*) is employed for the standard deviation of the population, and the symbol *s* identifies the sample standard deviation. If the sample data are representative of that in the population, $\overline{X}$ and *s* will approximate μ and σ, respectively. Also, as the number of samples taken increases, the conclusions reached become more reliable. Table 3–1 displays the symbols and formulas employed.

DESCRIPTION OF THE NORMAL DISTRIBUTION

The graph of the normal distribution model is a bell-shaped curve that extends indefinitely in both directions. Although it may not be apparent from looking at Figure 3–11, the curve comes closer and closer to the horizontal axis without ever reaching it, no mat-

Samples used to calculate the mean and the standard deviation must be selected on a random basis.

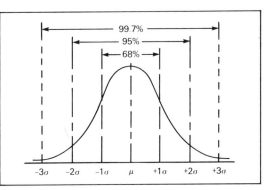

Figure 3–11   The normal distribution illustrating areas contained under the curve for ± 1, 2, and 3 standard deviations.

Table 3–1   Statistical symbols

Terms	Population (Parameter)	Sample (Statistic)
Number of observations	N	n
Mean (average)	μ (mu)	$\overline{X}$
Standard deviation	σ (sigma)	s
A single observation	X	X
Formulas		
Mean (average)	ΣX/N	ΣX/n
Standard deviation	$\sqrt{\dfrac{\Sigma(X-\mu)^2}{N}}$	(Formula I) $\sqrt{\dfrac{\Sigma(X-\overline{X})^2}{n-1}}$

Where Σ = sum

(Formula II) $\sqrt{\dfrac{n\Sigma X^2 - (\Sigma X)^2}{n(n-1)}}$

ter how far it is extended in either direction away from the average.

An important feature of the normal distribution is that it is symmetrical around its mean. In other words, if the curve in Figure 3–11 were folded along the line labeled μ, the two halves of the curve would coincide. Thus, to know the mean (μ) is to know the number that divides the population into two equal halves. Additionally, it can be seen that μ identifies the maximum point of the distribution. Consequently, the mean (μ) represents the best measure of the central tendency of the population.

When describing the spread of the normal distribution, the standard deviation (σ) is most useful. The standard deviation (σ) locates the point of inflection on the normal curve. This is the point where the line stops curving downward and begins to curve outward away from the mean. Since the curve is symmetrical, this position will be located at equal distances on either side of the mean (μ). The total area contained under the curve would include all members of the population and is therefore equal to 100%. The standard deviation (σ) can be used to describe various portions of the population in conjunction with μ as follows:

1. The inflection points lie at ±1σ from the mean (μ). Between the inflection points are found approximately 68% of all members of the population. About 32% will be located beyond these points with 16% on the left and 16% on the right.
2. Between ±2σ from the mean occur approximately 95% of the population. About

> **Standard deviation is a single number that represents an average deviation from the mean for all of the individual numbers in a set of data.**

5% will be located beyond these points equally distributed on either side.
3. Between ±3σ from the mean are found nearly 99.7% of the population. Only 0.3% will lie outside of these points.
4. As the distances on either side of the mean become greater, the percentage of the population contained therein likewise increases. However, it will never reach 100% because the curve never touches the horizontal axis.

Thus, a normal curve is completely specified by the two numbers, or parameters, μ and σ. In effect, μ locates the position of the curve along the X axis, and σ specifies the spread of the curve. If the sample data obtained approximate the shape of the normal curve, and the sample size is large enough, the calculated values of $\overline{X}$ and s for the sample can be used the same way as were μ and σ for the population.

For example, suppose the repeated testing of a light meter against a standard source produced a set of data that approximated the normal distribution, and $\overline{X}$ is found to be 15.0 and s is equal to 1.0. With this information, it can be inferred that 68% of the readings will be between 14.0 and 16.0 due to chance, 95% of the readings will be between 13.0 and 17.0 due to chance, 99.7% of the readings will be between 12.0 and 18.0 due to chance, while only 0.3% will be less than 12.0 and greater than 18.0 due to chance.

If the meter were to be checked again and gave a reading of 13.5 (1.5 standard deviations from the mean), the decision most probably would be that the meter was operating normally. In other words, the amount of variation in the reading is not greater than that which is suggested by chance. However, if the meter gave a reading of 18.5 (3.5 standard deviations from the mean), this would be viewed as an unusual occurrence and one that would call for a careful study of the equipment. This is because the difference is now greater than that which is suggested by chance, since only 0.3% of the readings will exceed ±3s due to chance.

This method of describing the characteristics of the materials and processes of photography minimizes guesswork and allows decisions to be made from facts.

MEASUREMENT METHODS AND SUBJECTIVE EVALUATIONS

All the concepts addressed thus far have dealt with data resulting from objective measurements. Variables such as density, temperature, weight, and shutter speed are evaluated by using measuring instruments, and the results are expressed in relation to scales of numbers. These numbers are assigned on the basis of standardized units such as degrees Fahrenheit, centimeters, grams, etc. Such measurements are referred to as objective because they are derived from an external source (the measuring instrument) and are verifiable by other people using similar instruments.

Although objective measurements are commonly used to describe the physical properties of photographic materials and processes, they cannot be used to give information about the human perceptions of such properties. For example, when considering the tonal quality of a black-and-white print, reflection density measurements can be used to describe the print densities and contrast. However, these data will give little insight into the way a person would perceive the tonal quality of the print, as that is a subjective experience. Obviously, the subjective quality of the photographic image is the final test of most photographic endeavors, thus it can be extremely useful to learn the meaningful relationships between objective measurements and subjective perceptions.

When one works with subjective concepts, it is the perception of the physical event that is being considered rather than the physical event itself. Nevertheless, numbers may be assigned to these perceptions to allow for their scaling and evaluation. Measurement, in the broadest sense, can be defined as the assignment of numbers to objects, events, or perceptions according to rules. The fact that numbers can be assigned under different rules leads to the use of different kinds of scales. All measurement scales can be classified as one of the following.

Nominal Scale

Nominal scales are essentially categories that are labeled with names, as when photographs are placed into categories such as portrait, landscape, indoor scene, outdoor scene, etc. A nominal scale would also be obtained through the assigning of numbers to baseball players since the number serves as a name. This is the most primitive of measurement scales, and therefore such data are not very descriptive.

Ordinal Scale

Ordinal scales are those in which categories are ordered along some variable and numbers are used to represent the relative position of each category. When the results of a horse race are given as first, second, and third, an ordinal scale is being used. If photographs were rated according to their acceptability, the following ordinal scale could be used: (1) excellent, (2) acceptable, and (3) unacceptable. Notice that there is no attempt to indicate how much better (or worse) the images are, just as there is no distance differentiation made between the winner of a horse race and the runner-up. All that is expressed is the order of finish. This approach is often referred to as rank-order or, more simply, "ranking" and is frequently used for scaling subjective responses. The graininess categories—extremely fine, very fine, fine, medium, moderately coarse, and coarse—constitute an ordinal scale.

Interval Scale

An interval scale is a refinement of the ordinal scale whereby the numbers given to categories represent both the order of the categories and the magnitude of the differences

It can be extremely useful to learn if meaningful relationships exist between objective measurements and subjective perceptions.

Rulers and thermometers use interval scales.

between categories. Arithmetically equal differences on an interval scale represent equal differences in the property being measured. Therefore, such scales can be thought of as linear since they possess a simple, direct relationship to the object being measured. The Fahrenheit and Celsius temperature scales are excellent examples of interval scales; thus, a temperature of 80 F is midway between temperatures of 70 F and 90 F. The interval scale is truly a quantitative scale.

Ratio Scale

F-numbers and shutter-speed numbers on cameras represent ratio scales.

Numbers on ratio scales increase by a constant multiple rather than by a constant difference as with interval scales. Thus, with the ratio scale of numbers 1—2—4—8—16, etc., the multiplying factor is 2. Incident-light exposure meters that measure directly in lux or footcandles, and reflected-light exposure meters that measure directly in candelas per square meter or candelas per square foot, typically have ratio scales where each higher number represents a doubling of the light being measured. Some exposure meters use arbitrary numbers rather than actual light units on the measuring scale. If the arbitrary numbers are 1—2—3—4—5, etc., the scale is an interval scale.

Logarithms make it possible to avoid large numbers by converting ratio scales to interval scales.

2 plus 2 can equal 8 if the numbers represent exponents to the base 2.

Thus the type of scale is determined entirely by the progression of the numbers, and not by what the numbers represent. A meter could even have two sets of numbers on the same calibrated scale, one a ratio scale of light units and the other an interval scale of arbitrary numbers. Other examples of ratio scales are shutter speeds (1—1/2—1/4—1/8—1/16, etc.) where the factor is 2; f-numbers (f/2—2.8—4—5.6—8, etc.) where the factor is the square root of 2 or approximately 1.4, and ISO film speeds (100—125—160—200—250, etc.) where the factor is the cube root of 2 or approximately 1.26.

The human visual system is most precise when making side-by-side comparisons.

Interval scales of numbers are somewhat easier to work with than ratio scales, especially when the ratio-scale sequence is extended and the numbers become very large (e.g., 64—128—256—512—1,024, etc.) or very small (but never reaching zero). Converting the numbers in a ratio scale to logarithms changes the ratio scale to an interval scale. For example, the ratio scale 10—100—1000—10,000 converted to logarithms be-

comes the interval scale 1—2—3—4. This simplification is a major reason why logarithms are used so commonly in photography, where density (log opacity) is used in preference to opacity, and log exposure is used in preference to exposure in constructing characteristic (D-log H) curves. DIN film speeds are logarithmic and form an interval scale compared to ratio-scale ASA film speeds. The ISO film speed system uses dual scales (e.g., ISO 100/21°) where the first number is based on an arithmetic scale and the second is based on a logarithmic scale. The APEX exposure system is also based on logarithms to produce simple interval scales (1—2—3—4—5, etc.) for shutter speed, lens aperture, film speed, light, and exposure values.

Exponential Scale

In addition to the four types of scales discussed above—nominal, ordinal, interval, and ratio—which cover all of the normal subjective measurement requirements related to visual perception and most of the objective measurements used in photography, there is another type of scale. The sequence of numbers 2—4—16—256, etc., is extended by squaring each number to obtain the next. This is called an exponential scale since each number is raised to the second power or has the exponent of 2.

Table 3–2 relates many types of photographic data to the appropriate measurement scale.

PAIRED COMPARISONS

One of the most basic techniques for evaluating subjective judgment is to present the observer with two objects (for example, two photographic images) and require a choice to be made on the basis of a previously defined characteristic. The two alternatives may be presented simultaneously or successively, and no ties are permitted.

In its simplest application, this method is used to compare only two samples, but it can also be applied in experiments designed to make many comparisons within a multiple of samples. The results are expressed as the number of times each sample was preferred for all observers. If one of the two samples is selected a significantly greater percentage of the time, the assumption may be made that it is superior relative to the characteristic being evaluated.

For example, consider the problem of evaluating two black-and-white films for their graininess. Since graininess is a subjective concept, a test must be performed using visual judgment. The two films are exposed to the same test object and processed to give the same contrast. Enlargements are made from each negative to the same magnification on the same grade of paper. The two images are presented side by side and each observer is asked to choose the image with the finer grain. A minimum of 20 observers is usually needed to provide sufficient reliability for this type of experiment.

If the results of such an experiment are 10 to 10, then it is concluded that there is no difference between the two samples. But if the results are 20 to 0, it can be concluded that one of the images actually had finer grain. A problem arises with a score such as 12 to 8 because such a score could be the result of chance differences only. Thus the purpose of this test is to distinguish between differences due only to chance and those resulting from a true difference.

The conclusions from these tests must be based upon the probabilities of chance differences. Table 3–3 may serve as a guide for making such decisions. Column n refers to the number of observers. The columns titled Confidence Level identify the degree of confidence indicated when it is concluded that a true difference exists. The numbers in the body of the table give the maximum allowable number of times that the less frequent score may occur. Therefore, in this example where 20 observers were used, the less frequently selected photograph may be chosen only five times if 95% confidence is desired in concluding that a difference exists. If a higher level of confidence is desired, say 99%, then the less frequently selected photograph can be chosen only three times. Notice that a score of 12 to 8 would lead to the conclusion that there was no significant difference between the two images even at the lower 90% confidence level. The price that is paid to obtain greater confidence in the decision is that a greater difference must be shown between the two samples. If the number of observers were increased to 50, then only a

Table 3–2 Types of measurement scales

Type of Scale	Examples of Photographic Data
1. Nominal	A. Emulsion identification numbers. B. Classification of photographs into categories. C. Serial numbers on cameras. D. Social security numbers of famous photographers. E. Names of colors, such as red, green, and blue.
2. Ordinal	A. The scaling or grading of photographs along a rank order such as first, second, and third place or A, B, C, etc. B. The rating of the graininess of photographs using the categories microfine, extremely fine, very fine, fine, medium, moderately coarse, and coarse. C. The ordering of photographs along a continuum such as lightest to darkest, largest to smallest, flat to contrasty.
3. Interval	A. Temperature readings; degrees on a thermometer. B. Hue, value, and chroma numbers in the Munsell Color Notation System. C. Wavelength (nanometers) of electromagnetic radiation. D. Contrast index values.
4. Ratio	A. Shutter speeds. B. F-numbers. C. Exposures in a sensitometer. D. ISO/ASA film speeds.
5. Exponential[a]	A. Depth of field vs. object distance. B. Depth of field vs. focal length. C. Illuminance vs. distance from a source (inverse-square law). D. Cosine law of light falloff in a camera.

[a]Although these represent exponential relationships, numbers are not normally presented in exponential scales due to the difficulty of interpolating.

Table 3–3 Maximum allowable number of wins for the less-frequently chosen object in a paired comparison test

n	Confidence Level		
	90%	95%	99%
8	1	0	0
9	1	1	0
10	1	1	0
12	2	2	1
14	3	2	1
16	4	3	2
18	5	4	3
20	5	5	3
25	7	7	5
30	10	9	7
40	14	13	11
50	18	17	15
75	29	28	25
100	41	39	36

32–18 vote (or approximately a 2:1 ratio) would be needed. The increase in sample size provides greater sensitivity, which allows for better discrimination.

The paired comparison approach may be extended to include more than two samples or objects. If three or more objects are to be compared, the observer must be presented with all possible pairings. Again, a choice must be made each time. Such a method produces a rank order for the objects judged. This technique leads to the use of an ordinal scale of measurement.

Consider, for example, the same negative printed on four different brands of black-and-white paper; the desire is to determine which print looks best. The prints are identified by the letters A, B, C, and D. The observer is presented with all possible pairings and each time is asked to state a preference. With four prints there is a total of six possible pairings. The results of such a test are shown in Table 3–4.

Before the rankings can be determined it is necessary to find out if the observer's judgments were consistent. Inconsistency is shown by the presence of one or more triads in the results. These triads (or inconsistencies) can be located by arranging the letters in a triangle and connecting them with short straight lines as shown in Figure 3–12. In example A, when prints A and B were compared, B was preferred so an arrow is drawn toward B. When prints A and C were compared, C was preferred so an arrow is drawn toward C. Likewise, for prints B and C, B was chosen and the arrow points toward it. The resulting pattern illustrates a consistent set of judgments. If, however, the three arrows move in a clockwise or counterclockwise direction as seen in Figures 3–12B and C, an inconsistency has been located. In Figure 3–12C for example, when prints A and C were

When determining possible differences between two similar photographs, the effects of guessing can be eliminated by using a paired-comparison test.

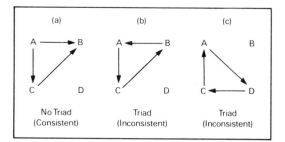

Figure 3–12   Location of triads (inconsistencies).

compared, A was preferred, thus A is better than C. When prints A and D were compared, D was preferred; thus D is better than A. However, when prints C and D were compared, C was selected as being better, which is inconsistent with the earlier decisions.

Once a judge has been shown to be consistent, the rank order of preference can be determined. This is achieved simply by determining the number of times each print was selected. The data for this example are summarized in Table 3–5. The prints also may be located along a scale based upon their number of wins. In this example, print B was judged to be the best as it has the highest rank. Prints C, A, and D were the runners-up in that order. Thus simple statements of preference have been transformed into a rank order along an ordinal scale. In this example, only one observer was used; however, it would be possible to have many observers and average the results. Practically any subjective characteristic can be evaluated in this fashion.

Table 3–4   Results from preference test

	Pair	Preference
1.	A vs. B	B
2.	A vs. C	C
3.	A vs. D	A
4.	B vs. C	B
5.	B vs. D	B
6.	C vs. D	C

Table 3–5   Number of wins and rank order of preference test results

Number of Wins	
A	1
B	3
C	2
D	0

```
        D   A      C      B
        |___|_____|_____|
        0   1      2      3
```

Rank Order
(Ordinal Scale)

TEST CONDITIONS

When performing subjective evaluations, perhaps the single greatest problem to overcome is the influence of the experimental conditions. It is critical that an observer's judgments be based only upon the properties of the objects being evaluated and not some outside factor such as a natural bias. For example, it is known that observers tend to prefer images placed on the left-hand side over images on the right. Also, images located above tend to be preferred over images placed below. In addition, there is a learning effect that occurs as a test continues. Consequently, images presented later in a test will be judged differently from those seen earlier, since the observer will be more experienced. The most common method for avoiding these difficulties is the randomizing of items; i.e., the item to be judged must *not* be viewed consistently in the same position in space or time. If true randomization is achieved, each item will have an equal chance of being chosen, and the choice will be primarily the result of the observer's opinion of the item's inherent qualities.

REVIEW QUESTIONS

1. The difference in contrast between grade 2 and grade 3 of the same printing paper would be identified as . . . *(p. 73)*
   A. chance-cause variability
   B. assignable-cause variability
   C. number-cause variability
2. Throwing two dice repeatedly and adding the numbers for each throw would be expected to produce a distribution pattern that is . . . *(p. 74)*
   A. rectangular
   B. basically rectangular, but with random variations
   C. triangular
   D. bell-shaped
3. We expect to find a normal distribution of data . . . *(p. 74)*
   A. always, when we have enough data
   B. when only human error is involved
   C. when only machine error is involved
   D. when chance is causing the variation
   E. when we made very precise measurements

4. A term that is *not* a measure of central tendency for a set of data is . . . *(p. 76)*
   A. average
   B. median
   C. range
   D. mean
   E. mode
5. If two normal distribution curves differ only in the widths of the curves, they would have different . . . *(p. 76)*
   A. means
   B. frequencies
   C. standard deviations
   D. inclinations
6. The minimum sample size that is recommended for a serious study of variability is . . . *(p. 76)*
   A. 1
   B. 2
   C. 30
   D. 100
   E. 1000
7. The selection of a sample from a population of data should be on the basis of . . . *(p. 77)*
   A. taking half from the beginning and half from the end
   B. taking all from the center
   C. a biased selection process
   D. a random selection process
8. The proportion of the total data included between one standard deviation above the mean and one standard deviation below the mean is . . . *(p. 78)*
   A. 50%
   B. 68%
   C. 86%
   D. 95%
   E. 99.7%
9. The series of f-numbers for whole stops on camera lenses represents . . . *(p. 80)*
   A. a nominal scale
   B. an ordinal scale
   C. an interval scale
   D. a ratio scale
   E. an exponential scale
10. In a paired comparison in which observers are asked to compare the graininess of three prints, viewed two at a time in all three combinations, one person selects A over B, B over C, and C over A. That person's choices were . . . *(p. 82)*
    A. consistent
    B. inconsistent

4 Photographic Sensitometry

Neil Beckerman. *A Pair of Pears*. Copyright © 1989 by Neil Beckerman.

INTRODUCTION TO TONE REPRODUCTION

Photographers generally consider the images they produce as blending visual creativity with technical expertise. That these two elements are given equal attention is not surprising in view of the history of art and photography. The most successful and creative artists have generally not been limited by a lack of technical capability. Indeed, the ability to control the medium is an important trait among successful artists.

The study of sensitometry provides the necessary understanding of the technical characteristics of photographic films and papers. It deals with all aspects of the photographic process from the original subject to the finished image. Within that process are many steps and, consequently, many opportunities for control. In order to learn some of the fundamental conditions of black-and-white photography, the steps have been simplified to those shown in Figure 4–1. In this case it is assumed that the photographic process performs in a "black box" fashion and is always capable of technical excellence. Thus, attention may be focused on the properties of the subject and the resulting print. In this fashion we can obtain an answer to a question often asked by photographers: "How well were the tones in the scene reproduced in the print?"

Perhaps the most obvious way of answering this question is to return to the original scene with print in hand and make a subjective judgment. There are many problems with this approach, among them the transitory nature of most subjects and the lack of proper viewing conditions for the print. In addition, it is most difficult to obtain specific information with this method.

To obtain objective data about the subject, we need some way of measuring the tones it contains.

The alternative approach is to obtain objective data that are representative of the tones in the scene and the tones in the print. It is convenient to think of the subject tones as *input data* and the print tones as *output data*. Once these data are in hand, we can objectively describe the subject and the print, and then compare them. A graphical representation as shown in Figure 4–2 is appropriate. The shape of the resulting plot will indicate the tone-reproduction characteristics for this process.

To obtain objective data about the subject, we need some way of measuring the tones it contains. Areas of low lightness or reflectance are shadow tones, while those of high lightness are the highlight tones. The technical name for these reflected-light values is *luminance*, which is a measure of the intensity of the light per unit area. The basic unit of measurement is the candela per square foot. Therefore, the shadow areas have low luminance values while the highlights have high luminance values.

Most photoelectric meters are equipped to measure the subject luminances by using the meter in the reflected-light mode and pointing it at the area to be measured. The readings are usually in the form of arbitrary units that are proportional to the luminance in candelas per square foot.

If only two tones in the subject are measured, the darkest and the lightest, it is possible to determine the scene ratio by dividing the smaller into the larger. For example, if the shadows have a luminance of 2 candelas per square foot and the highlights have 200 candelas per square foot, the ratio would be 100:1.

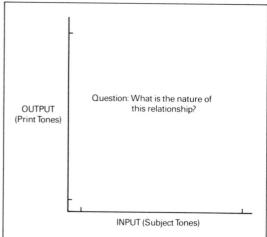

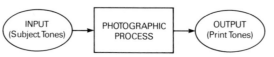

Figure 4–1   Black box model of approach to the photographic process.

Figure 4–2   Graphic model for a tone-reproduction study.

The luminance ratio provides valuable information about the overall *contrast* of the scene. By measuring numerous outdoor scenes, the average luminance ratio was found to be approximately 160:1. Consequently, scenes with ratios close to this are referred to as *average* or normal in contrast. If the subject luminance ratio significantly exceeds 160:1, the scene has more than normal contrast and is referred to as a *contrasty* subject. Those subjects with a lower luminance ratio have less than normal contrast and are referred to as being *flat*.

Since one of the purposes for studying the tone-reproduction characteristics of a photographic system is to judge its ability to make visually acceptable images, luminance data are usually transformed into logarithms. The reason for using logarithms is associated with the nature of the human visual system. Experimental evidence indicates that the visual process responds in a nearly logarithmic fashion. Therefore, by taking the logarithms of the subject luminances we have subject (input) data that correspond to the appearance of the scene. Table 4–1 illustrates the relationship between the subject luminances and subject log luminances for a variety of scene contrasts. The methods for finding the logarithm of a number and the antilog (arithmetic value) of a logarithm are addressed in Appendix C.

An alternative method for obtaining input data about the subject is to use a *reflection gray scale* as a standardized reference. A gray scale is nothing more than an organized set of subject tones conveniently displayed with known tonal values. These tonal values have been premeasured on a reflection densitometer, and the values shown are reflection densities. Most gray scales have a total luminance ratio of approximately 100:1 and, therefore, are close to being representative of average outdoor scenes. As input data, the

only significant difference between gray-scale data and subject log luminance values is that the numbers run opposite to each other. This is because reflection density is a measure of the amount of darkening occurring in an image, while luminance is a measure of the amount of light emanating from a surface. Therefore the two concepts are reciprocally related. Although the concept of density will be addressed in detail later, it is important to note that the reflection densities of the gray scale are themselves logarithms, since, by definition, reflection density is the logarithm of the reciprocal of the reflectance.

When seeking how to obtain objective output data about a print, the obvious answer is again the use of a reflection densitometer. With this instrument, we measure the same tonal areas of the print that were measured in the scene. If a gray scale were included in the scene, the measurements would be taken from the reproduction of it in the print. Since density is a logarithmic concept, the resulting data will automatically be in logarithmic form. Thus both the input and the output data are logarithmic values.

We are about ready now to undertake a tone-reproduction study. The sequence of events is as follows:

1. Obtain input data by measuring subject luminances OR include a gray scale.
2. Expose and process the film normally and make a print.
3. Obtain output data by measuring the reflection densities on the print.
4. Compare the print with the subject by plotting the input-output relationship (i.e., a tone-reproduction graph).

The graphic framework of Figure 4–2 can be labeled and numbered in line with the previous discussion as shown in Figure 4–3. The shadows will be plotted in the upper left

The average outdoor scene has a luminance ratio of about 160 to 1, which corresponds to an exposure range of approximately 7 stops.

One method for obtaining input data about the subject is to use a reflection gray scale as a standardized reference.

A tone-reproduction graph for a faithful reproduction of a subject would be a 45-degree straight line.

Table 4–1   Relationship between subject luminance ratios and log subject luminance ranges

Shadow Luminance	Highlight Luminance	Subject Luminance Ratio	Log Subject Luminance Range
2 c/ft²	200 c/ft²	100:1	2.0
0.5 "	500 "	1000:1	3.0
3 "	600 "	200:1	2.3
2 "	100 "	50:1	1.7
2.5 "	350 "	160:1	2.2
5 "	100 "	20:1	1.3

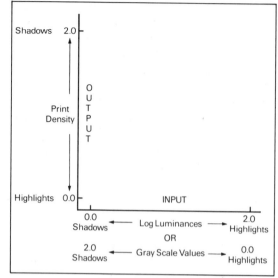

Figure 4–3 Numerical values for input and output characteristics.

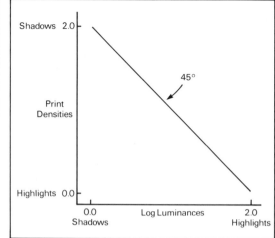

Figure 4–4 Theoretically ideal (facsimile) tone reproduction.

portion of the graph while the highlights will plot in the lower right. Regardless of whether gray-scale data or subject log luminances are used as input, the shadows are plotted on the left side and the highlights on the right.

Before we examine what the process is actually capable of producing, let's hypothesize about the ideal possibilities. One way to think of an ideal reproduction is to consider a print where the tones exactly match those in the scene. Such a result might be more accurately labeled a facsimile reproduction. This is represented in a tone-reproduction graph in Figure 4–4. The 45° line indicates that for every tone in the subject there is a matching tone in the print, and that the tonal differences of the subject have been exactly maintained in the print. In pictorial photographic systems, this condition is impossible to achieve because of the tonal distortions introduced by the camera optics, film, enlarger optics, and paper. Therefore a literal translation of subject to print is unobtainable in pictorial photography. In fact, even if facsimile reproduction were possible, experiments indicate that viewers would not prefer it. It is important to note, then, the difference between an ideal reproduction and a preferred reproduction. Obviously, we are more interested in making prints that are preferred.

Figure 4–5 shows the typical tone-reproduction curve resulting when the photo-

graphic process is run so that it produces a pictorial print that is judged excellent by a large number of people. Notice that the shape is decidedly different from the 45° line that has been drawn in for reference. In general, where the slope is less than the 45° line there is a compression of tones, and where the slope is greater than 45° there is an expansion of tones. Figure 4–5 indicates that the shadow tones have been somewhat compressed. This is not to say that the reproduction is unusual or abnormal but merely that the photographic process contains some inescapable nonlinear characteristics that do not eliminate the possibility of excellent results.

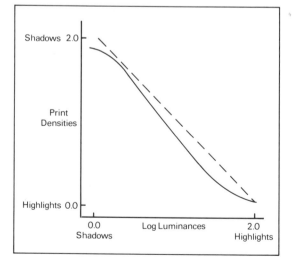

Figure 4–5 Preferred (aim) tone-reproduction curve (pictorial subject).

Notice that the tone-reproduction curve in Figure 4–5 does not reach higher than a print (output) density of 2.0. This illustrates an unavoidable limitation in the photographic process when the end product is a reflection print. Black-and-white photographic papers cannot exceed an output density range of much more than 2.1, or a little greater than a 100:1 ratio. Therefore, the contrast of subjects with luminance ratios greater than 100:1 will necessarily suffer tonal compression in the reproduction. This limitation is primarily the result of first surface reflections of the light illuminating the print and so is not greatly different between papers of similar surface characteristics.

The curve shown in Figure 4–5 is associated with a print that a large number of observers would categorize as excellent. Consequently, it is referred to as the *preferred* (or *aim*) curve because we strive to match it through the system we employ. The preferred curve has been included in Figure 4–6 for comparison. The second print shows less slope in the shadows and midtones than the aim curve, which indicates that there is too much compression of the tones in these regions. In other words, the second print is lacking in contrast in the shadows and midtones and, in the photographer's vocabulary, would be called a flat print. Notice that the curve was compared to the aim curve and *not* the 45° line.

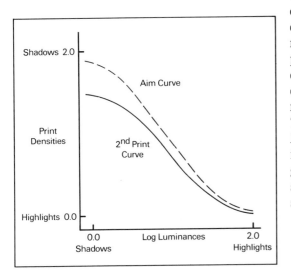

Figure 4–6 Tone reproduction curves for preferred and inferior prints.

PHOTOGRAPHIC EXPOSURE

When a camera is focused on a subject and the shutter is opened, a variety of light levels falls on the image plane at the back of the camera. In the shadow areas the level of light will be quite low; for the highlights the level will be great. In fact, for every different subject tone (luminance) there will be a different light level in the image at the film plane. These light levels are referred to as *illuminances* since we are dealing with the light incident on the film plane, as opposed to luminances, which are related to the light reflected from subject surfaces. In effect, the subject luminances in front of the camera become image illuminances at the back of the camera. The camera's shutter speed determines the length of time the image light will be allowed to fall on the film.

In photographic applications, the term *exposure (H)* is defined as a measure of the quantity of light received by the photographic material. Two variables define this quantity. The first is the *illuminance (E)*, the level of light falling on the surface. This concept is expressed in metercandles (an equivalent term is *lux*), with 1 metercandle being equal to the amount of light falling on a point 1 meter away from a 1-candlepower source. The second variable is the length of *time* (symbol T) the illuminance is allowed to fall on the film, and is expressed in seconds. Consequently, exposure is expressed in metercandle-seconds (or lux-seconds) and is calculated by the formula $H = E \times T$. For example, a gray card registering an illuminance on the film plane of 2 metercandles photographed at a shutter speed of 1/100 second would give an exposure of 0.02 metercandle-second ($1/100 \times 2 = 0.02$). In many references the formula is given as: $E = I \times T$, where E is the symbol used for exposure, I for illuminance, and T for time. However, in 1972 the symbols were changed to those given first, for the purpose of international standardization. The relationship remains the same.

In most ordinary picture-making situations, using this definition, the film receives a multitude of exposures each time the shutter is tripped, with each different exposure related to a different subject tone. It is also important to note that expressing exposure

The preferred or aim tone-reproduction curve represents a photograph that most viewers would classify as having excellent quality.

When film is exposed in a camera it receives as many different exposures as there are tones in the subject.

as a combination of camera settings, such as 1/60 at f/11, is not consistent with the accepted definition. In fact, if only the shutter speed and aperture are known, it is impossible to determine the exposures the film received because the image illuminance values are missing. In practice, a photoelectric meter is typically used to measure the subject luminance, and the meter converts the input to an f-number–shutter speed combination appropriate for the film being used.

When testing a photographic emulsion to determine its sensitometric characteristics, it is necessary to make an accurately known set of exposures. Although cameras are the principal instruments used to expose film in practice, they are not the best choice for film-testing purposes. If the film is to receive a set of known exposures, the illuminances and times must in turn be known, since the exposures will be calculated from them. Because of light falloff and flare problems at the back of the camera, it is most difficult (if not impossible) to calculate accurate illuminance values from subject luminances and the f-number. Although some large-format cameras can be equipped with a meter having a probe to make readings of small areas, the illumination is never uniform at the film plane. Additionally, the shutter speeds marked on most cameras may not correlate closely with the actual exposure times. Unless the shutter has been pretested and found to be consistent and accurate, reliable information about the exposure time will not be obtained. Devices specifically designed to expose film for testing purposes are called *sensitometers*.

Sensitometers provide a systematic series of exposures for the purpose of studying the response of light-sensitive materials.

A sensitometer consists of three major parts as shown in Figure 4–7: a suitable light source, a step tablet, and a shutter. Sensitometers and cameras are similar in that they are both used to expose film. A sensitometer, however, has its own light source and subject (step tablet). Because a sensitometer is not a camera and is not a direct part of the picture-making process, great care is taken so that it simulates actual picture-making conditions. For example, the light source is chosen so that the color of light it produces matches that which is used in reality. Since the most commonly encountered light is daylight, sensitometric light sources that approximate its color quality are used. Usually, a tungsten lamp with appropriate filtration is used because it can be precisely calibrated and is very stable over its lifetime. Some sensitometers make use of a xenon-filled flashtube (electronic-flash) source, which does not require filtration since its spectral output already is close to the color of daylight. Whatever source is used, the illumination reaching the film plane is nearly uniform and can be measured easily.

The step tablet serves as the subject for the sensitometer. Its purpose is to convert the single level of light produced by the lamp into many different levels or illuminances. This simulates the condition in the camera where the film receives varying illuminances due to the tonal differences in the scene. A step tablet is a series of calibrated filters with uniform density increments, which may be made by incorporating particles of carbon in gelatin and copying them onto photographic film. Such transmission density step tablets can be purchased in a variety of sizes and formats. The most commonly used step tablet varies in density from approximately 0.10 to 3.10. When such a tablet is placed in the light path, the light will be attenuated over a log range of 3.0 (3.10 − 0.10 = 3.00). Since densities are logarithmic values, the illuminances reaching the film plane will have a log range of 3.0. By taking the antilog of 3.0, we find that the ratio of illuminances will be 1000:1, which means that the highest illuminance is 1,000 times that of the lowest illuminance. Therefore, with such a step tablet in the sensitometer, the film will receive a ratio of illuminances far greater than that typically encountered in the camera, which is desirable for testing purposes.

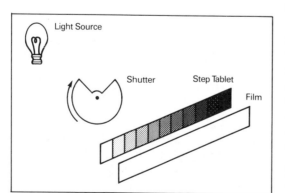

Figure 4–7 The major parts of a sensitometer.

If the tablet contains 11 different equally spaced steps, there will be 10 increments or intervals. Since the total density range is 3.0, the step-to-step difference can be determined by dividing 3.0 by 10, giving 0.30. This means that the steps get progressively denser in 0.30 increments. The antilog of 0.30 is 2, indicating a 2X factor change between successive steps (recalling that addition of logs corresponds to multiplication of antilogs). The step-to-step exposure change is equivalent to one stop, and the entire range covered is equal to 10 stops.

The shutter in a sensitometer performs the same task as the one in the camera: to control the length of time for which the light will strike the film. The shutter in a sensitometer usually is powered by a synchronous motor that operates in a repeatable fashion. A variety of exposure times can be achieved by using differing gear ratios and slit sizes. The actual exposure time should be representative of the times used in practice. Theoretically, any combination of illuminance and time giving equal numerical exposures could be used if the emulsion response depended only on the total amount of light received. Because the film responds differently to long and very short exposure times, an effect identified as *reciprocity law failure* (or *reciprocity failure* or *reciprocity effect*), an appropriate shutter speed must be selected to avoid the effect or an exposure adjustment will be needed. Where the exposing source is a xenon flashtube, the flash duration governs the length of exposure time.

For the most part, photographers do not own or have access to sensitometers. Therefore, the use of a camera for sensitometrically testing film is often unavoidable. In this case a reflection gray scale may be photographed and used as a test object. As seen earlier, reflection gray scales contain tonal values conveniently displayed from light to dark, and they are probably the most common test objects used in photography. Commercially available gray scales have a density range of approximately 2.0, which means a luminance ratio of 100:1 or about seven stops from the lightest to the darkest patch. The actual log values (reflection densities) are usually marked for each patch. If the gray scale has 10 different patches, then the film will receive 10 different exposures when the shutter is tripped.

This approach to sensitometry invariably leads to less precision in the resulting data for many reasons. First, when the gray scale is photographed, a nonlinear error is introduced as a result of optical flare in the camera, and there is a lack of image illuminance uniformity due to light falloff toward the corners. Further, the lighting of the original gray scale must be uniform and arranged to prevent reflections, including the subtle uniform reflections of light-colored walls or ceilings.

PROCESSING

After the film has been properly exposed, the next major step is to process the image. The most important phase of the processing operation is the development step. When exposed film is developed, the latent image is amplified by as much as 10 million to 1 billion times. It is this tremendous amplification ability that makes the silver halide system of photography so attractive, where even weak exposures can yield usable images. The result is a system with great sensitivity or, in the photographer's vocabulary, a *fast film*.

Of the four major factors in development—time, temperature, developer composition, and agitation—the last is the most difficult to standardize. The temperature of processing baths can be measured to an accuracy of $\pm 1/10° F$. The use of a reliable timer allows the length of development time to be controlled to within a few seconds. Prepackaged chemicals have reduced the problems involved in consistent developer composition. Proper agitation is difficult to achieve because, as we will see, there are at least two functions it must serve, and it must be different for different systems.

One of the critical problems in the development of the latent image to metallic silver is obtaining uniform density in uniformly exposed large and small areas. Variations of density in uniformly exposed areas are usually damaging to image quality and are almost always due to nonuniform development. Proper agitation of the developing solution is the best safeguard of uniformity. Let us examine the effects of developing film with a total lack of agitation. Figure 4–8 illustrates such a "stagnant tank" condition. The sequence of events is as follows:

An increase of 0.30 in density in a step tablet produces a decrease in exposure by the equivalent of one stop.

The four major development factors are time, temperature, agitation, and developer composition.

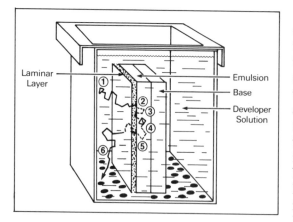

Laminar Layer

Emulsion

Base

Developer Solution

Figure 4–8   Stagnant tank (zero-agitation) development. The numbers correspond to the sequence of events described in the text.

1. As the emulsion is immersed, the molecules of developer move around the tank in random fashion.
2. Since the concentration of developing agent is high in the developer and low in the emulsion, a transfer of the developing agent into the emulsion occurs. This transfer is called diffusion.
3. As the developing agent continues to diffuse into the emulsion it becomes attached to silver halide crystals.
4. If the crystals have received an exposure to light, a chemical reaction occurs that produces metallic silver (the image) and developer reaction byproducts.
5. The developer reaction byproducts diffuse through and eventually out of the emulsion and into the developer solution.
6. Since some of these byproducts are heavier than the developer solution, they tend to drift downward and collect on the bottom of the tank, creating an exhausted layer of developer.

The laminar layer of used developer on the surface of film during development acts as a barrier to fresh developer.

Correct agitation reduces uneven development.

Notice that the physical activity is completely based on the process of diffusion; diffusion of developing agents into the emulsion, and diffusion of development byproducts out of the emulsion. Clearly, agitation can have nothing directly to do with this process of diffusion, a random molecular activity determined by the temperature and alkalinity of the developer, among other things.

We attempt to control the concentration of chemicals at the emulsion surface by agitation. There is always a relatively undisturbed thin layer of developer lying on the surface of the emulsion that is essentially stuck to the gelatin. This layer acts as a barrier, and as it becomes thicker, the process of diffusion is slowed. This explains why development times must be extended when little or no agitation is used during processing. Because this barrier (sometimes referred to as laminar) layer is not the same thickness everywhere, it explains why improper agitation typically leads to uneven densities in uniformly exposed areas. In order for the barrier layer to be minimized, the fresh developer must make contact with the emulsion with a fair degree of force. Consequently, excellent agitation is characterized by vigorousness.

An additional function of agitation is to maintain the uniformity of the processing solution's chemical composition and temperature. Since the development byproducts tend to settle to the bottom of the tank, the agitational motions used should produce a random mix of both fresh and exhausted developer. With this achieved, every part of the emulsion will have access to developer of the same composition. This is why it is generally not desirable to agitate film in a tray or tank in the same direction every time. The result of such nonrandom, directional movement is a lack of uniformity in the image.

None of the agitation methods commonly used completely meet the requirements of vigor and randomness due to factors such as cost, productivity, and ease of operation. The test of proper agitation ultimately rests with the type of image quality desired. In most cases, the typically used methods of tray, tank, and machine processing give acceptable results. However, when problems occur, most likely they are the result of deficiencies in one or both of these two characteristics: vigor and randomness.

There are various methods for processing photographic films and papers, each with strengths and weaknesses. Some of the more commonly encountered techniques are illustrated in Figure 4–9.

Uniformity of development, either within a given piece of film or between sheets or rolls of film, and for any of the methods, depends to a considerable extent on the degree of development. As the degree is increased and gamma infinity (maximum gamma) is approached, variations in time, temperature, and agitation have a decreased effect on image density and constrast. Thus it is more difficult to obtain uniform and consistent development of films that are developed to a low contrast index (as when photographing high-contrast scenes) than when developing to a higher contrast index. Since photographic papers and lithographic films are processed in high-activity developers and reach gamma infinity quickly, uniformity of development is less of a problem with these materials.

DENSITY MEASUREMENT

The most common method for determining the effect of exposure and processing on a sensitometric strip is to measure its light-stopping ability. Such information, properly collected, will give excellent insight into the visual and printing characteristics of the image. When light strikes a photographic film image, some of it is reflected backwards, some is absorbed by the black grains of silver, and some is scattered (i.e., has its angle of travel changed) as a result of bouncing off grains. This is illustrated in Figure 4–10. The light-stopping ability of a silver photographic image is determined by a combination of these three optical occurrences. Notice that the light transmitted through the sample is distributed over a wide angle primarily as a result of the bouncing or scattering effect of the silver grains. For the present, we will consider only the amount of transmitted light in relation to the amount of incident light on the sample.

a) Tray Shuffle

e) Rotary Machine

b) Tray Rock

f) Lift & Drain

c) Spiral Reel & Small Tank

g) Nitrogen Burst

d) Tube

h) Roller Transport

Figure 4–9 Methods of agitation.

Basically, the transmittance (T) of a sample is the ratio of the transmitted light to that which is incident. Therefore:

$$\text{Transmittance } (T) = \frac{\text{Transmitted light}}{\text{Incident light}}$$

Consider, for example, the situation in Figure 4–11 where there are 100 units of light in-

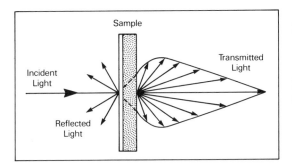

Figure 4–10 Distribution of transmitted light rays caused by light scattering.

Developing film to gamma infinity reduces the unevenness of development associated with poor agitation.

The formulas for Transmittance and Reflectance are similar: T = transmitted light/incident light. R = reflected light/incident light.

cident on the sample and 50 units are transmitted. Thus the transmittance $(T) = 50/100$, which is 0.50 or ½ or 50%. While this approach seems logical and direct, the disadvantage is that the number (transmittance) describing light-stopping ability becomes smaller as the light-stopping ability increases. For samples where the light-stopping ability is great, the transmittance is small.

To overcome this shortcoming, we can calculate the opacity (O) of a sample, which is the ratio of incident light to transmitted light, and thus a reciprocal of the transmittance. The formula is:

$$\text{Opacity } (O) = \frac{\text{Incident light}}{\text{Transmitted light}} \quad \text{OR}$$
$$\text{Opacity} = \frac{1}{T}$$

In the example shown in Figure 4–11, the opacity $(O) = 100/50$, or 2. Because opacity is the reciprocal of transmittance, the numbers denoting an image's light-stopping ability will increase as that ability increases, providing a more logical relationship. However, the potential awkwardness of opacity is revealed when considering the effects of equal increases in light-stopping ability:

Light-Stopping Ability (Expressed in Thickness of Sample)

Thickness of Sample	Opacity/Increments	
1	2	
		> 2
2	4	
		> 4
3	8	
		> 8
4	16	
		>16
5	32	

Notice that as the thickness of the sample increases (and therefore, the light-stopping ability) in equal amounts by adding one sample atop another, the differences between the opacities are unequal as they become progressively greater. The opacities form a geometric (ratio) progression that becomes inconveniently large as the light-stopping ability increases. For example, ten thicknesses would give an opacity of 1,024.

Density equals the logarithm of the opacity or the logarithm of the reciprocal of the transmittance.

Exposure meters and densitometers both measure light, but they express the measurements in different units.

To compensate for this problem there is yet a third expression of the photographic effect called *density*. Density (D) is defined as the logarithm of the opacity. Thus:

$$\text{Density } (D) = \log \text{ of opacity} \quad \text{OR}$$
$$\text{Density} = \log \text{ of } \frac{1}{T}$$

The density of the sample in Figure 4–11 is found as the log of 2, or 0.30. Extending the work from the previous table, the densities of the samples are:

Opacity/Increments		Density/Increments	
2		0.30	
	> 2		>0.30
4		0.60	
	> 4		>0.30
8		0.90	
	> 8		>0.30
16		1.20	
	> 16		>0.30
32		1.50	

The concept of density provides us with a numerical description of the image that is a more useful measure of light-stopping ability. An added benefit is that, as stated earlier, the visual system has a nearly logarithmic response, so our data will bear a relationship to the image's appearance.

Now that we have considered expressions of the image's light-stopping ability, let us turn our attention to the practical problems of actually measuring the effect. Of the *conditions* of measurement, two are most important.

The first difficulty involves the measurement of the transmitted light. As shown in

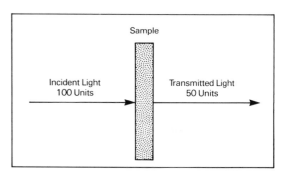

Figure 4–11   The determination of light-stopping ability in a simple system.

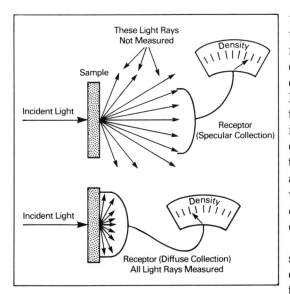

Figure 4–12 Two different conditions for measuring transmitted light.

Figure 4–10, transmitted light rays form a distribution as a result of bouncing off the silver grains. This distribution of transmitted light will be wider for coarse-grained images than for fine-grained images because the larger grain size provides a greater surface area over which the bouncing can occur. Consequently, coarse-grained images scatter more light than fine-grained images.

Regardless of the grain size, the basic question is: Where should the measurement of transmitted light be made? If, as shown in Figure 4–12, the receptor is placed far from the sample, only light transmitted over a very narrow angle will be recorded; this is called *specular* measurement. Alternatively, when the receptor is placed in contact with the sample, all the transmitted light will be collected, with the angle of collection being very large; this is referred to as a *diffuse* measurement. The specular density will be different from the diffuse density taken from the same sample.*

For the companies that design and manufacture densitometers this difference between specular and diffuse density is a major concern, as it is for those who hope to apply their test results to a real picture-making situation. The answer lies in the principle that the testing conditions should simulate those

used in practice. Therefore, if the negative is to be printed on a contact printer where the receptor (photographic paper) is in direct contact with the image, then the densitometer to be used should be similarly designed. For those negatives to be projection printed, the receptor will be far from the image, which indicates the opposite condition. These conditions represent extremes from very diffuse to very specular. Almost all commercially available densitometers are designed to provide an intermediate result termed a *diffuse density* (as opposed to doubly diffuse density).

It is important to realize that density measurements may come from an instrument that does not exactly simulate the particular photographic system being used. Trial-and-error is usually necessary to determine the appropriateness of the data. The relationship between the diffuse density and the specular density for a given sample may be determined and is called the Callier Coefficient or Q factor. Therefore:

$$Q = \frac{\text{Specular density}}{\text{Diffuse density}}$$

The second problem associated with the actual conditions of density measurement is the color response of the densitometer. Photocells are typically used in densitometers to sense the light being transmitted. The color-response characteristics of all photocells are not alike. Further, the color response of most photocells is unlike that of photographic printing papers. This would not be a significant problem if all the images being measured were perfectly neutral. However, we often measure negatives that are decidedly non-neutral as a result of using a fine-grain developer, a staining developer, or some after-treatment of the negative. Indeed, when measuring the densities of color negatives, we never encounter neutral images.

At this point it is appropriate to ask: How will the densities be used? If the purpose is to predict the printing characteristics of a negative, the spectral response characteristics of the printing paper should be simulated. To determine the visual appearance of the image, the spectral response of the human eye should be simulated. In the first case the result is called a *printing density*, in the

Most commercial densitometers measure diffuse density, which is between specular density and doubly-diffuse density.

The color response of densitometers more closely matches the color response of the eye than that of black-and-white printing papers.

*ANSI and ISO now use the term *projection density* instead of *specular density*.

second a *visual density*. These may be achieved by using certain filters in conjunction with the photocells. All commercially available densitometers are equipped to read visual density (sometimes referred to as black-and-white density). Only those specially equipped with the proper filters will read printing densities.

Modern solid-state photoelectric densitometers are excellent measuring devices capable of reading densities over a wide range with a high degree of accuracy. No matter how sophisticated the instrument may be, however, if the conditions of measurement do not simulate the photographic system being used, the resulting data will lack validity.

CHARACTERISTIC CURVES

In order to evaluate and understand the results of a sensitometric test, it is necessary to plot the densities occurring on the test strip in relation to the exposures that were received. The data contained in Table 4–2 illustrate the relationship between the input (exposure) data and output (density) data for a sensitometric test of a typical black-and-white negative film. It can be seen that each of the 11 densities produced is the result of a known exposure. The input data are given as both exposure and log exposure. When graphing the relationship, log exposure is used in preference to actual exposure to describe the input values because it tends to conveniently compress the input scale. The use of a logarithmic scale also makes it easier

Characteristic curves are photo-graphs that show tonal relationships between input (subject) and output (image).

Although film characteristic curves are divided into four parts, only two parts—the toe and the straight line—are used for most photographs.

to determine exposure ratios, which are an important part of the evaluation process. Table 4–2 shows that exposures less than 1.0 lux-second are described by negative logarithms. Although the use of logarithms is addressed in Appendix C, it should be noted that negative logs are frequently encountered in sensitometry because the higher sensitivity of modern-day emulsions requires only small amounts of exposure to yield image density. Consequently, sensitometric exposures of less than 1.0 lux-second are common.

Notice in Table 4–2 that the differences in log exposure are 0.30 each, which is the result of using an 11-step tablet containing step-to-step density differences of 0.30. When plotting the data, actual log exposures or relative log exposures may be used as input data, depending upon what is known about the exposure conditions. The resulting graph will show log (or relative log) exposure on the horizontal axis as input and density on the vertical axis as output. Such a plot is referred to as a *characteristic curve* or D-log H curve. Older references call it the H&D curve after Hurter and Driffield, who first described the technique in the late 1800s, and later the D-log E curve. Figure 4–13 shows the curve resulting from a plot of the data in Table 4–2.

The curve in Figure 4–13 is typical of negative-working camera films. It can be conveniently divided into four major sections as follows:

1. *Base plus fog:* This is the area to the left of point A and is the combination of the density of the emulsion support (base) and

Table 4–2 Relationship between exposure, log exposure, and the resulting density in a set of sensitometric exposures

Density of the Original Step Tablet	Actual Exposure (H) (lux-sec.)	Actual Log H	Relative Log H	Step Number	Densities of the Resulting Strip
3.10	0.0064	$\overline{3}.80$	0.00	1	0.18
2.80	0.0128	$\overline{2}.10$	0.30	2	0.25
2.50	0.0256	$\overline{2}.40$	0.60	3	0.39
2.20	0.0512	$\overline{2}.70$	0.90	4	0.54
1.90	0.1024	$\overline{1}.00$	1.20	5	0.78
1.60	0.2048	$\overline{1}.30$	1.50	6	1.03
1.30	0.4096	$\overline{1}.60$	1.80	7	1.28
1.00	0.8192	$\overline{1}.90$	2.10	8	1.55
0.70	1.6384	0.20	2.40	9	1.73
0.40	3.2768	0.50	2.70	10	1.83
0.10	6.5536	0.80	3.00	11	1.85

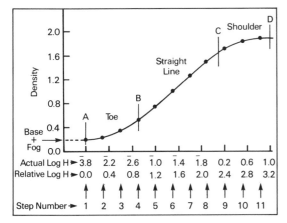

Figure 4–13 The characteristic curve resulting from the data in Table 4–2.

the density arising from the development of some unexposed silver halide crystals (fog). Here the curve is horizontal and incapable of recording subject detail or tonal differences.

2. *Toe.* This is the section between points *A* and *B* and is characterized by low density and constantly increasing slope as exposure increases. Different subject tones will be reproduced as small density differences. It is in this area that shadow detail in the subject is normally placed.

3. *Straight line.* This portion, extending from point *B* to point *C*, is a middle-density region where the slope is constant everywhere. It is also here that the slope of the curve is steepest; thus the subject tones are reproduced with the greatest separation (contrast). For many emulsions, however, the middle section of the curve is quite nonlinear, as will be seen later.

4. *Shoulder.* Located between points *C* and *D*, this is the portion where the density is high but the slope is decreasing with increases in exposure. Ultimately the slope approaches zero, where it is again impossible to record subject detail. Consequently, this section is avoided when exposing pictorial films.

The characteristic curve displays a "picture" of the film's ability to respond to increasing amounts of exposure as a result of a specific set of development and image-measurement conditions. It represents the single most com-

monly used method for illustrating the sensitometric properties of photographic materials and should be clearly understood.

For now we will assume that the subject tones are directly related to the log exposure axis of the graph. If a uniformly illuminated gray card is photographed, it will have only a single luminance value and give a single level of light (illuminance) at the camera back. When the shutter is tripped, the illuminance is multiplied by the exposure time, producing a single exposure value. Therefore, the exposure of that gray card can be represented by a single position along the log exposure axis as seen in Figure 4–14A. If the same gray card is used to make a second photograph but the exposure is increased by a factor of 100 (either by using a wider aperture or a

A change in exposure of one stop, or a ratio of exposures of 1 to 2, corresponds to a change in log exposure of 0.3 on a characteristic curve.

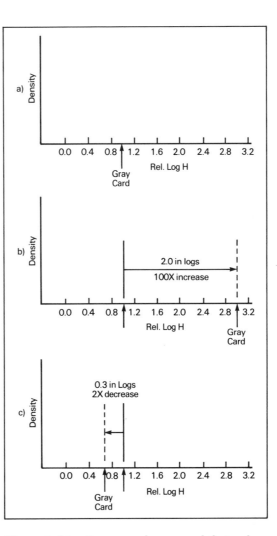

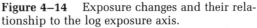

Figure 4–14 Exposure changes and their relationship to the log exposure axis.

slower shutter speed or both) the second exposure will be located to the right of the first at a distance equivalent to 2.0 in logs (i.e., the logarithm of 100) and is shown in Figure 4–14B. If a third photograph is made of the gray card under the same conditions as the first except that the aperture is closed down one stop (exposure reduced by a factor of 2), the resulting log exposure will fall to the left of the first as seen in Figure 4–14C. The separation will be 0.30, which is the log of 2, the equivalent of one stop.

Assume now that we are photographing the face of a model with highlight and shadow areas, illuminated so that the lighter side has eight times the luminance of the darker side, giving a subject-luminance ratio of 8:1. Since there are now two tones in the scene, there will be two image illuminances and, ultimately, two different exposures with one trip of the shutter. Figure 4–15A shows that these two input values will be separated by the log of 8, which is 0.90. If a second photograph

is made of this subject, giving one stop more exposure, both log exposures will shift to the right a distance of 0.30 in logs but still will be separated by 0.90 because the subject luminances have not changed, as shown in Figure 4–15B.

Therefore, the tones of the subject can be related to the log exposure axis of the characteristic curve. As shown above, the ratio of the subject luminances (highlights to shadows) plays a major part in determining the ratio of exposures (or range of log exposures) the film will receive. Also, by changing the camera settings or the level of light on the subject, the level of the exposures may be increased or decreased (shifted right or left) on the log exposure axis.

It should be noted here that the concepts discussed above are only approximately correct because of the image distortion introduced by the optical flare in the camera. What is important at this point is that the conceptual relationship between subject luminances and log exposures be clearly understood. Methods of compensating for flare-related distortions will be introduced later. It should be apparent that errors made in the placement of the subject tones relative to the log exposure axis when the film is exposed can seldom be corrected in later stages of the process.

In reference to the simple two-tone scene described above, exposure and processing produce two different densities. The values of the two densities can be found by extending lines up from the two log exposure positions to the characteristic curve and then to the left until they intersect the density axis. This is illustrated in Figure 4–16A, which shows both log exposures falling on the straight-line section. The actual values of the densities are less important than the density difference because these relate to tonal separation or contrast. Therefore, density differences describe the amount of shadow, midtone, and highlight detail in the negative. In this example the log exposures are separated by 0.90 (an exposure ratio of 8:1), while the resulting densities show a difference of only 0.60. This compression occurs because the slope of the straight-line section is less than 1.0 (45°). In fact, the actual slope may be computed by comparing the output density difference to the input log exposure dif-

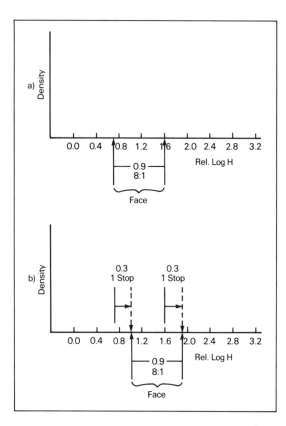

Figure 4–15   Changes in exposure for a simple two-tone scene as related to the log exposure axis.

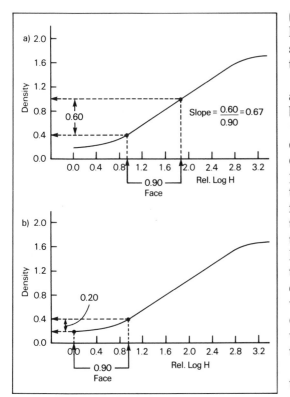

Figure 4–16 Relationship between log exposure and density for two different sections of the characteristic curve.

ference. This leads to the generalized form for finding the slope:

$$\text{Slope} = \frac{\text{Density difference}}{\text{Log exposure difference}} \quad \text{OR}$$

$$\text{Slope} = \frac{\Delta D}{\Delta \text{Log H}}$$

For the situation shown in Figure 4–16A the ΔD is 0.60 and the $\Delta \log H$ is 0.90; 0.60/0.90 is equal to 0.67, which is the slope of the straight-line region. This number (slope) is related to the steepness of the curve and describes the rate at which density will increase as a result of increasing the exposure. A slope of 1.0 indicates that a change in log exposure will yield an equal change in density. With a slope of 0.50, the change in density is only half as great as the change in log exposure. A slope of 2.0 means that the change in density is twice as great as the change in log exposure. The relationship can be restated as follows: Slope relates the negative contrast (density differences) to the subject contrast

(log exposure difference). However, this relationship only applies to the straight-line section of the characteristic curve if indeed there is one.

Where the section of the graph is a curve, as in the toe, there is no simple relationship between log exposure and density. Figure 4–16B shows the same two-toned scene placed entirely on the toe portion of the curve. In other words, the subject was given less exposure, as these log exposures are located to the left of the first. The resulting density difference is 0.20, which is considerably less than the first (0.60) because the slope in the toe is quite low. This means that there is far less tonal separation in this negative than in the first; it has less contrast. If the slope were calculated as before, the resulting number would be related to an imaginary straight line drawn between the two density points and would not describe the actual slopes in the toe, which are everywhere different.

It should be clear from this discussion that the contrast (density difference) of the negative can be changed by changing the camera exposure. This will be the case either when there is no straight-line portion whatever or when the exposures are placed in a nonlinear area such as the toe and shoulder. If the subject tones are placed in the straight-line portion, they will be reproduced in the negative with the greatest possible separation (contrast). As the exposures become less and move into the toe region, the density differences of the shadows will rapidly decrease and the shadow contrast will be reduced. If the exposures are increased and move into the shoulder region, the density differences of the highlights are quickly reduced, causing a loss of highlight contrast.

In practice, the camera is likely to be aimed at scenes that range from about 20:1 to nearly 800:1, with 160:1 being average. Experience shows that excellent negatives are produced when the shadow exposures are placed in the toe of the curve while the midtones and highlights fall in the midsection as shown in Figure 4–17. The shoulder section is almost always avoided because of the longer printing times required and the resulting loss of image quality. Most camera speed pictorial films can accept these ranges and give excellent results. Occasionally, however, scenes with excessive contrast (1,000:1 or greater)

Total negative contrast, or TNC, is another name for negative density range.

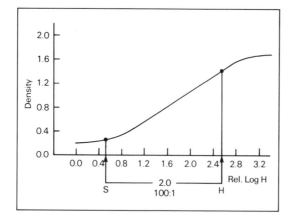

Figure 4–17   Proper placement of log exposures for an average-contrast scene.

are encountered. Special emulsions and developers are generally needed so that proper shadow and highlight detail are maintained in the negative.

Total negative contrast (the density difference between the thinnest and densest areas, also known as the density range) is largely determined by:

1. The subject contrast, which determines to a good first approximation the log H range on the horizontal axis of the D-log H curve.
2. The placement of this range of log exposures, determined by the camera settings.
3. The shape of the specific characteristic curve, which is mainly determined by the choice of emulsion and the development conditions.

The total negative contrast is a number similar in concept to the log of the subject luminance ratio. It describes the log range of output (i.e., the density range) as a result of the previously mentioned conditions. For pictorial photographic negative emulsions, a total negative contrast of approximately 1.1 is considered normal (for diffusion enlargers) and is based on the printing characteristics of the negative. If the total negative contrast reaches 1.6, the negative is considered to have excessive contrast and probably is unprintable, even on a grade one paper. When the total negative contrast is on the order of 0.4, it also will be nearly unprintable but because it is too flat for even a grade five printing paper.

Shadow detail, midtone contrast and highlight detail are important, as noted earlier. A negative can have a large total negative contrast and still be totally lacking in shadow detail, as shown in Figure 4–18A. Likewise, the total negative contrast may be great but the highlights are blocked up, as in Figure 4–18B. Thus we reach again the concept of *slope* as a measure of contrast in different parts of the film D-log H curve, and therefore in different parts of the negative (shadows, midtones, and highlights). A slope of 1.0 means the subject contrast is exactly duplicated in the negative. A lesser slope means the contrast is reduced while a larger slope means it is increased. The correct contrast relationship is determined by the way the negative will be used and, in most cases, the way it will be printed, including the grade

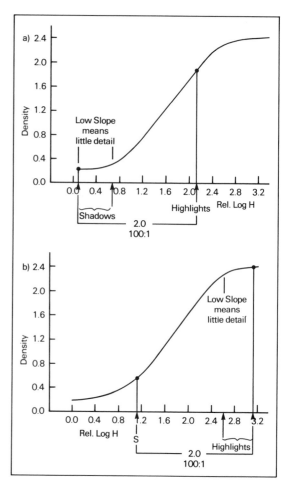

Figure 4–18   The relationship between slopes and density differences for underexposure and overexposure conditions (with more than normal development).

of paper that will be used. Interestingly, a "normal"-contrast negative actually has significantly less contrast than the original subject.

LOG EXPOSURE RANGE

In light of the relationship between subject luminances and camera exposures to the densities of the resulting negative, it should be realized that excellent images can only be made if all the subject tones fall on the characteristic curve where there is sufficient slope. One of the most important measures of the sensitometric properties of the film is the *useful log exposure range*. This is the range of log exposures over which the emulsion can produce adequate separation of densities and is based on the minimum slope required to achieve it. For example, as the camera exposure decreases, the subject tones move farther to the left into the toe and ultimately reach an area where the curve is completely flat (zero slope), resulting in a loss of detail in that portion—typically the shadows—of the negative. The smallest slope necessary to preserve acceptable shadow detail is located at a point on the toe that is the minimum useful point. Experiments using pictorial subjects indicate that the minimum useful point in the toe occurs where the slope is not less than 0.20, as illustrated in Figure 4–19A. Since one cannot find the slope of a point, the 0.20 refers to the slope of a line tangent to (just touching) the toe. For most camera-speed pictorial films this point of tangency usually falls at a density of not less than 0.10 above the base-plus-fog density and is referred to as the *minimum useful density*.

The location of the maximum useful point on the shoulder is known with less accuracy, partly because proper reproduction of the shadows is more important and partly because the highlight area of most scenes falls short of the shoulder portion. However, experience indicates that the minimum useful slope on the shoulder is also approximately 0.20 and is found as shown in Figure 4–19B. Since the shape of the upper portion of the curve is highly dependent upon the degree of development, no generalization can be made about the maximum useful density for the curve.

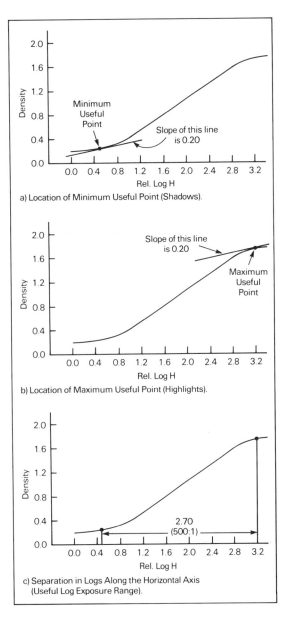

a) Location of Minimum Useful Point (Shadows).

b) Location of Maximum Useful Point (Highlights).

c) Separation in Logs Along the Horizontal Axis (Useful Log Exposure Range).

Figure 4–19 Determination of useful log exposure range. (A) Location of minimum useful point (shadows). (B) Location of maximum useful point (highlights). (C) Separation in logs along the horizontal axis (useful log exposure range).

Figure 4–19C illustrates the concept of useful log exposure range, which is the distance along the horizontal axis between the minimum and maximum useful points. In this example the log range is 2.70, which means the useful exposure ratio is about 500:1 (antilog of 2.70 = 500). As long as the subject tones are placed between these points, the resulting negative will have acceptable detail in all areas. If the ratio of subject tones is so great that it exceeds the useful exposure ratio, either the shadows or the highlights will lack detail in the negative. Consequently, the useful log exposure range is a basic measure of the ability of a given film-developer combination to adequately record the subject tones.

It is safer to overexpose negative film, black-and-white and color, than to underexpose it.

EXPOSURE LATITUDE

Associated with the useful log exposure range is the concept of *exposure latitude,* which relates to the margin for error in the camera exposure. For example, suppose we were using the film-developer combination resulting in the curve shown in Figure 4–19C to photograph a subject with a luminance ratio of 50:1. The subject will provide a log exposure range of 1.70 (log of 50 = 1.70). The useful log exposure range of the film-developer combination is 2.70, leaving a difference in logs of 1.00 (2.70 − 1.70 = 1.00). This means there is a 1.00 log interval "left over" that represents our margin for error. If the darkest shadow in the subject is to be placed at the minimum useful point as shown in Figure 4–20A, the underexposure latitude is zero, while the overexposure latitude is 1.00 in logs, or a factor of 10, or 3 1/3 stops. When the lightest highlight is to be placed at the maximum useful point, the conditions are exactly reversed as illustrated in Figure 4–20B. If the subject exposures are placed directly in the middle of the useful log exposure range as in Figure 4–20C, the under- and overexposure latitude will be equal (1.00/2 = 0.50).

In practice, the shadows of the scene are usually placed at the minimum useful point because this allows for the highest effective film speed. The result is that exposure latitude in the camera is typically overexposure latitude only. Most camera-speed pictorial films have the capability of properly recording scenes with luminance ratios of up to 500:1. As the subject luminance ratio becomes greater, the exposure latitude becomes smaller, which creates a need for more accurate camera exposures. When the subject-luminance ratios become excessive (greater than 500:1), special films and developers are required if excellent images are to be produced. (The effect of camera flare light on image contrast will be considered later.)

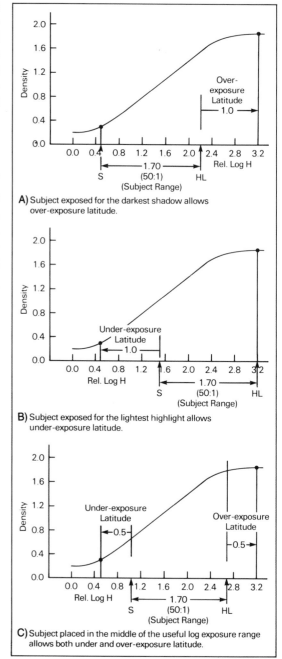

A) Subject exposed for the darkest shadow allows over-exposure latitude.

B) Subject exposed for the lightest highlight allows under-exposure latitude.

C) Subject placed in the middle of the useful log exposure range allows both under and over-exposure latitude.

Figure 4–20   Exposure latitude: The relationship between subject log luminance range and the useful log exposure range. (A) Subject exposed for the darkest shadow allows overexposure latitude. (B) Subject exposed for the lightest highlight allows underexposure latitude. (C) Subject placed in the middle of the useful log exposure range allows both underexposure and overexposure latitude.

Thus far we have considered only those properties that have been related to one curve and one development time. If a variety of development times are used with some less

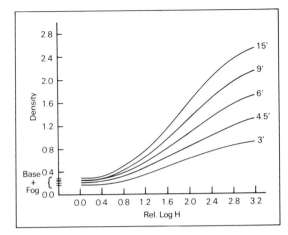

Figure 4–21 A family of curves for five different development times.

and some greater than the "normal," a family of characteristic curves can be plotted as in Figure 4–21. Notice that all of the curves exhibit the same general shape. The following generalizations may be made:

1. Each curve shows a greater density in all areas as development time increases. Notice that base-plus-fog density is similarly affected.
2. The slope in the toe section is small and relatively unaffected by increases in development.
3. The slope of the midsection increases greatly as development time is increased.
4. The slope in the shoulder remains small for all development times, even though the density level changes dramatically.
5. The useful log exposure range gradually becomes smaller as the development time increases, as a result of the higher slope of the straight line and a more prominent shoulder.

Subject tones placed in the toe section of the curve will always be reproduced as small densities but, more significantly, with only slight separation of tone (contrast) regardless of the development time. When the subject tones are placed in the middle part of the curve, changes in development time will have a significant effect on the density differences produced. It is in the straight-line region where the photographer has the greatest control over contrast with development. Those subject tones placed in the shoulder section will result in high density levels, but again

with only little separation of tone for all development times.

Since most pictorial photographic situations involve only the toe and midsection of the characteristic curve, it can be seen in practice that the darker tones (shadows) should govern camera exposure determination, while the lighter tones (highlights) should govern the degree of development. This idea is consistent with a common saying among photographers: "Expose for the shadows and develop for the highlights." It should also be apparent that underexposure will shift the shadow tones farther to the left in the toe, where even lengthy development times will not produce a steep enough slope for the detail to be recorded. Further, the increase in development time will adversely affect the midtone and highlight reproduction by giving them unacceptably high contrast. This is typically the result when photographers attempt to "push" their film speed by underexposing and overdeveloping. This condition is illustrated in Figures 4–17 and 4–18A. Figure 4–17 shows the result of the proper placement of subject tones for a nearly average-contrast scene with normal development. Figure 4–18A shows the same scene but with the subject tones given less exposure, which places the shadow tones far into the toe region. The steeper slope of this curve is caused by an extended development time with the result being an obvious lack of shadow detail and excessive midtone and highlight contrast. The negative could well be unprintable on even the lowest contrast grade of paper.

GAMMA

Of the several available measures of the contrast of photographic materials, *gamma* has the longest record of use. Gamma (γ) is defined as the slope of the straight-line section of the D-log H curve. The formula is as follows:

$$\gamma = \frac{\Delta D}{\Delta \text{Log H}}$$

where ΔD = the density difference between any two points on the straight-line portion of the curve; and

A common saying is to expose for the shadows and develop for the highlights.

Gamma = slope = ΔD/Δlog H.

ΔLog H = the log exposure difference between the two points chosen.

Examples of the determination of gamma are shown in Figure 4–22A and B for two different development times. Gamma is considered to be an excellent measure of development contrast because it is the slope of the straight-line portion that is most sensitive to development changes. It can also be used to predict the density differences that will result in the negative for exposures that fall on the straight-line section. For example, the curve in Figure 4–22A exhibits a gamma of 0.50, which means that the density difference in the negative will be one-half that of the log exposure range. Likewise, in Figure 4–22B where the gamma is 2.0, the density difference will be twice as large as the log exposure range. It is assumed in both cases that all of the exposures are on the straight line part of the D-log H curve.

If the gammas of all of the curves in Figure 4–21 are measured and plotted against the

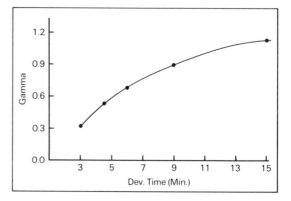

Figure 4–23   Development time vs. gamma chart for family of curves in Figure 4–21.

time of development, the curve shown in Figure 4–23 results. This is referred to as a time-gamma chart and illustrates the relationship between the slope of the straight line and the development conditions. For short development times, the gamma changes very rapidly, and as the time of development lengthens, the gamma increases more gradually. At some point in time, the maximum gamma of which this film-development combination is capable will be reached. This point is labeled gamma infinity and appears to be about 1.10 for these data. Since for this developer gamma increases greatly during short development times, extra care must be taken to minimize processing variability if consistent results are to be achieved. However, if longer times are used, the amount of development-time latitude (margin for error in developing time) increases.

It is important to note here that gamma and negative contrast are different concepts. Gamma refers to the amount of development an emulsion receives, while negative contrast is concerned with density differences. Developing to a large gamma does not necessarily mean that a contrasty negative will result. The reason for this is that many factors in the photographic process affect the contrast of the negative, including the subject's range of tones and the placement of the exposures on the D-log H curve.

Gamma is a measure of development contrast, which is only one of multiple factors that determine negative contrast.

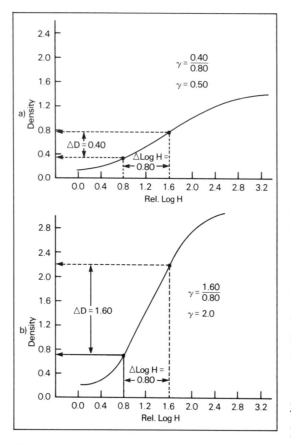

Figure 4–22   The determination of gamma for two different film samples.

AVERAGE GRADIENT

In many cases, photographic materials do not provide a straight midsection, and in others there may be two straight-line sections with different slopes. Clearly, a single measure of gamma has little significance for these materials. Even for films exhibiting a single straight line, the useful portion of the curve usually includes part of the toe in addition to the straight line. In these situations where gamma cannot be used as a measure of slope, the concept of *average gradient* is substituted. An average gradient ($\overline{G}$) is basically the slope of an artificial straight line connecting any two points on the curve, as illustrated in Figure 4–24. The D-log H curve shown here has no straight-line section, but the average gradient can be computed for the artificial line connecting points A and B by using the basic slope formula where slope = ΔD/Δlog H. For this example, then: ΔD = 1.10, Δlog H = 1.60,* therefore:

$$\overline{G}(\text{A to B}) = \frac{1.10}{1.60}$$
$$= 0.69$$

The resulting number is a measure of the average contrast over the specified portion of the curve. Notice that this method can produce a variety of slopes by using different sections of the curve. Therefore, when interpreting average gradient data, one should be careful to note the section of the curve used.

CONTRAST INDEX

Another method for measuring the contrast of photographic emulsions is to determine the contrast index of the characteristic curve. *Contrast index* is an average gradient method with the added feature that the slope is always measured over the most useful portion of the curve. This approach evolved from an evaluation of the printing characteristics of a large number of excellent negatives made on a wide variety of emulsions. The result was a method for locating the most likely used minimum and maximum points on the curve and the determination of the average slope between them.

The actual determination of contrast index requires the use of a transparent meter as an overlay on the D-log H curve. The meter is aligned with the log exposure and density

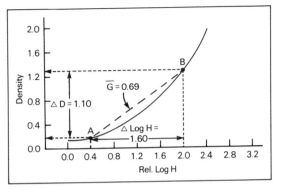

Figure 4–24 Determination of average gradient between two points on the curve.

axes and moved left or right until the curve crosses the small arc on the left side at the same numerical value as it does on the larger arc at the right. The number is the contrast index. The meter and its proper use on a characteristic curve are illustrated in Figure 4–25.

An alternative method is, first, to locate a point on the toe that is 0.10 above base-plus-fog density. Using a compass, strike an arc with the center at this point and a radius of 2.0 in logs so that it intersects the upper portion of the curve. Then draw a straight line between these two points and determine its slope. The resulting number will be a close approximation of the contrast index.

In the manufacturer's published data for a film, contrast index is used almost exclusively as a measure of development contrast. This is because contrast index is a measure of the slope over the area of the curve that will most likely be used when the negative

> Contrast index is the average slope of the most useful part of the film characteristic curve.

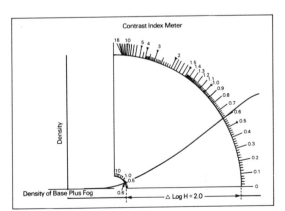

Figure 4–25 The use of a contrast index meter.

---

*Ilford and Fuji use Δlog H = 1.50.

is made. Therefore, unlike gamma, contrast index is an excellent measure of the relationship between the entire log exposure range of the subject and the entire density range of the negative, provided the subject has been exposed on the correct portion of the curve. This means that if the contrast index is 0.50, the total negative contrast (total density difference) will be just one-half the subject log exposure range. For example, an average outdoor scene has a luminance ratio of approximately 160:1 and will provide a log exposure range of nearly 2.20 (log of 160) to the film. A correctly exposed negative, processed to a contrast index of 0.50, will have a total negative contrast of 0.50 × 2.20 or 1.10, and will print easily on a normal grade of paper with a diffusion enlarger.

The important point here is that the required contrast index may be changed to suit the subject log luminance range. Thus if the scene is excessively contrasty, development to a low contrast index will generate a negative that will print on a normal grade of paper. On the other hand, if the scene is unusually flat, the film may be developed to a high contrast index to obtain a normal-contrast negative.

FILM SPEED

Thus far in our discussion of sensitometry and its application to the picture-making effort we have yet to study that concept from which the term *sensitometry* was derived. This is the measurement of the emulsion sensitivity to light or, more commonly, the film speed. The concept of film speed and its relationship to the determination of the proper camera exposure should be clearly understood because, as we have seen, significant errors in the exposure of the film often cannot be rectified by any subsequent steps in the process.

A film speed (sometimes called photographic speed) is simply a number representing the average sensitivity of an emulsion under a given set of test conditions. The number is most commonly intended for use with photoelectric exposure meters to determine the camera settings that yield acceptable negatives and transparencies. Film speed and exposure are reciprocal concepts. For example,

Film speeds are based on the amount of exposure that is required to produce a specified density.

a lower-speed film requires a greater exposure to produce a specified density. Therefore, as the photographic exposure necessary to produce a given quality negative increases, the speed (sensitivity) of the system decreases. The general relationship between speed and exposure is:

$$\text{Speed} = \frac{1}{H}$$

where H = the exposure in lux-seconds to achieve a certain result.

Most pictorial negative films have film speed ratings computed by the manufacturer and published as ISO (ASA) values. This means that the speed has been determined in accordance with the procedures described by the American National Standards Institute and the International Standards Organization. Both organizations are nonprofit operations that exist for the purpose of providing standard procedures and practices to minimize variations in testing. Since there are many sources for variation in the determination of film speed, it is highly desirable to have an agreed-upon method that is accepted by a majority as the "best" technique. This technique prescribes, among other items, the type of exposing light (daylight), the exposure time (1 second to 1/1,000 second), the developer composition (a formula devised for testing purposes that is not commercially available), the type of agitation, and the amount of development (an average gradient of approximately 0.62). Once all these conditions have been met, the speed point is located at a density of 0.10 above base plus fog and the following formula is used:

$$\text{ISO (ASA)} = \frac{1}{H_m} \times 0.8$$

where H_m = the exposure in lux-seconds that gives a density of 0.10 above base plus fog;

0.8 = a constant that introduces a safety factor of 1.2 into the resulting speed value. The film is actually 1.2× faster than the published value, which guards against underexposure.

ISO (ASA) speeds are rounded to the nearest third of a stop using the standard values listed in Table 4–3. Therefore, each ISO (ASA) speed value has a tolerance of ±1/6 stop. Consider the curve shown in Figure 4–26, which is the result of following the ISO (ASA) standard procedures. The determination of speed is as follows:

1. Locate point *m* at 0.10 above the base-plus-fog density, which in this case is at a gross density of 0.20.
2. Beginning at point *m*, move a distance of 1.30 in log units to the right along the log exposure axis.
3. From this position, move up a distance of 0.8 in density units, and if the curve crosses at this point (±0.05), the proper contrast has been achieved and the speed may be computed at point *m*. If not, the test must be rerun using a development time that yields the proper curve shape.

Table 4–3   ISO (ASA) and ISO (DIN) standard speed values

ASA (Arithmetic)	DIN (Logarithmic)
6	9°
8	10°
10	11°
12	12°
16	13°
20	14°
25	15°
32	16°
40	17°
50	18°
64	19°
80	20°
100	21°
125	22°
160	23°
200	24°
250	25°
320	26°
400	27°
500	28°
650	29°
800	30°
1000	31°
1250	32°
1600	33°
2000	34°
2500	35°
3200	36°
4000	37°
5000	38°
6400	39°

In this example, the curve shape is correct, and the log exposure at the speed point is $\bar{3}.63$. By taking the antilog of this value, the exposure is found to be 0.00426 lux-second. Substituting, we have:

$$\text{ISO (ASA)} = \frac{1}{0.00426} \times 0.8$$
$$= 234.74 \times 0.8$$
$$= 187.79$$

Using the values in Table 4–3, the ISO (ASA) speed would be expressed as 200. The manufacturer's published values for a given film type are determined by sampling many emulsion batches and averaging the test results.

The consequence of this approach is a set of film speeds that mean the same thing for all pictorial negative films regardless of the manufacturer. A speed of ISO (ASA) 125 means the same for Brand X as it does for Brand Y because the same test conditions existed. It also means that a given emulsion has only one ISO (ASA) speed, and this value cannot be changed as a result of using a special developer because the conditions of testing have been specified. Any change in these conditions means the standard has not been followed, and, therefore, by definition the ISO (ASA) speed cannot be found. In practice, most photographers will necessarily use conditions different from those outlined in the Standard. For example, if tungsten light is used or a special-purpose developer tried, the circumstances of practice no longer match those of the test. This is an unavoidable pitfall of the ISO (ASA) published speeds; once the "best" method has been agreed upon and

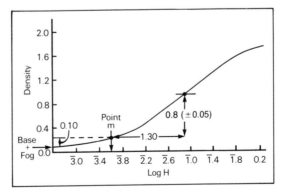

Figure 4–26   The determination of ISO (ASA) speed for a specific sample.

ISO film speeds are rounded to the nearest one-third of a stop.

Separate daylight and tungsten film speeds are published for some films.

standardized, the speeds are only valid for those conditions. It is common for film manufacturers to publish both daylight and tungsten speeds for some nonpanchromatic emulsions.

ANALYSIS OF NEGATIVE CURVES

When surveying the wide variety of black-and-white negative working emulsions available, it becomes an imposing task to describe the properties of each. Fortunately this diverse collection of emulsions can be organized on the basis of contrast (as seen in the curve shape) and film speed. Although there are other considerations for the grouping of films, such as format and image structure, it seems appropriate at this point to consider the sensitometric properties, as they have the greatest influence upon the intended use of the emulsion. Negative working films can be categorized as follows:

Normal-Contrast Films

A contrast index of .42 is considered normal for printing with condenser enlargers while .56 is considered normal for diffusion enlargers.

Normal-contrast films have sensitometric properties similar to those we have already studied. They are intended for making negatives of pictorial subjects with the ultimate goal being a reflection print. They are characterized as medium-contrast films, and they produce contrast indexes of approximately 0.60 with normal development. These films may possess either a short or long toe section as shown in Figure 4–27. In a short-toe film the midsection is reached quickly and shows a long straight line. In practice, this means that a slight amount of underexposure will greatly reduce shadow detail while slight overexposure rapidly increases shadow detail. Since the midsection is long and straight, exposure changes do not affect the midtone and highlight contrast. When considering the long-toe film, small increases or decreases in camera exposure will have only a slight effect on shadow contrast, allowing for more underexposure latitude. However, since the midsection of these materials is seldom a straight line, small changes in the camera exposure can translate into significant changes in midtone and highlight contrast.

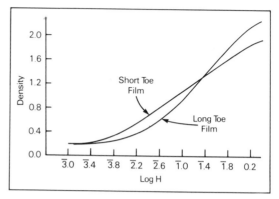

Figure 4–27 Characteristic curves for short-toe and long-toe films developed to approximately the same contrast index.

Normal-contrast films may also be classified on the basis of their speed. The slowest medium-contrast emulsions have an ISO (ASA) speed of 32 while the fastest are on the order of 4000. Associated with the film speed is the ever-present problem of grain size, with the general relationship that as film speed increases, so does the grain size.

When curves are drawn for long-toe emulsions, a "hump" sometimes appears in the midsection. This anomaly is the result of a double-coated emulsion where the shoulder of one leaves off and the toe of the next begins. By combining two different emulsions, both the speed and useful log exposure range can be increased.

Continuous-Tone Copy Films

These materials are used to copy photographic prints and artwork. These films are also useful when a positive transparency is to be made from a negative (or vice versa). It is important to note that the subject luminance characteristics of these types of subjects are considerably different from those of an original scene. For example, the luminance ratio of a black-and-white glossy print seldom exceeds 100:1 and is often much less. Additionally, the luminance differences within the shadows and within the highlights of the print are significantly less than in an original scene. The same conditions exist for paintings and other artwork. Since the subjects are inherently low in contrast, continuous-tone copy films generally have slopes of 1.0 and greater to compensate and are classified as high-contrast emulsions.

Figure 4–28 illustrates two types of continuous-tone copy films. Emulsion A exhibits a curve without a straight-line section, and the slope constantly increases as the exposure increases. This is an extremely useful characteristic for two reasons. First, the highlights in the print will fall on the right-hand portion of the curve where the slope is the greatest, ensuring that the subtle differences in the highlights of the original print will be maintained in the copy negative. Second, because the slope shows large changes as a result of exposure, the contrast may be readily governed by exposure as well as development.

Emulsion B is the type that would be used to make a positive transparency from a negative. Since the resulting image is intended for direct viewing, the slope is greater to provide sufficient visual contrast. The slopes for these films are typically 2.0 or more. The steeper slope means that the useful log exposure range will be short and the tolerance for exposure errors will be quite small. For these materials the midsection is relatively straight and easily controlled by development. Generally, developers with greater-than-normal activity are used.

Since these materials are not used to generate original pictorial negatives, and the conditions of use differ markedly from the standard, ISO (ASA) speed values are meaningless. The manufacturer usually supplies speed data in the form of an *exposure index* that is derived under test conditions (i.e, light sources and developers) similar to those used in practice. The speed points are typically located in the upper midsection near a density of 1.5 because it relates to proper highlight reproduction.

Line-Copy Films

When photographing a line drawing or printed text, an extremely high-contrast emulsion is necessary. These materials are referred to as either line-copy or lithographic (from the graphic-arts vocabulary) films with a typical characteristic curve displayed in Figure 4–29. Slopes of 5.0 and above are common. The extremely steep slope is needed to expand the very short range of exposures received by the film in practice. Although most people think of line drawings and printed text material as high-contrast subject matter, in reality the contrast is quite low. For example, the reflectance ratio of a high-quality printed text seldom exceeds 40:1, and for newsprint it is about 10:1. This is because the inks used do not give a deep black, and the paper is never perfectly white. The lack of intermediate gray tones is the primary reason why these materials are referred to as high contrast. When these ratios are compared to that of an average outdoor scene, it becomes obvious why line-copy films require such a steep slope. Consequently, the useful log exposure range is very short, re-

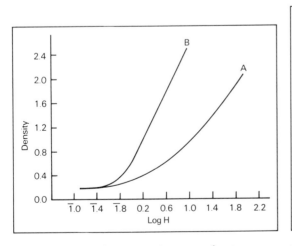

Figure 4–28   Characteristic curves for two types of continuous-tone copy films.

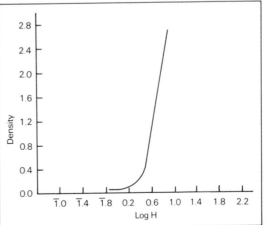

Figure 4–29   Typical characteristic curve for a line-copy film.

sulting in an almost total lack of exposure latitude.

The film speeds for such emulsions are derived from practical tests using the same light sources and developers to be used in practice. The manufacturer gives this information as a point of reference at which the user can begin his or her own tests. Because the developers employed for these films are very powerful, the photographic speed is highly dependent upon the development conditions.

ANALYSIS OF PHOTOGRAPHIC PAPER CURVES

For normal contrast tone reproduction, film curves have an average slope of less than 1.0 and printing paper curves have a slope of more than 1.0.

A characteristic curve for a printing paper can be plotted using the reflection densities of a print of a step tablet.

The procedures involved in the sensitometric testing of photographic papers are essentially the same as those used for negative materials. The details of the testing conditions are changed to more closely simulate practice. For example, tungsten light is used in the sensitometer because that is the source with which most printers are equipped. Also, exposure times are longer (usually between 0.10 and 10 seconds) to approximate those given in practice because reciprocity law failure affects print emulsions as well as films. For cases where a sensitometer is unavailable, a transmission step table may be either contact or projection printed with the actual printer to be used in production. Care must be taken to ensure uniformity of illumination on the easel and accuracy of the timer. If the step tablet is projection printed, optical flare will cause problems similar to those encountered when using a camera and reflection gray scale, as exposure relationships between steps will be distorted. When processing photographic papers, close attention must be given to the details of time, temperature, and agitation because development times are generally short and density is produced quickly.

In the evaluation of photographic papers it is necessary to measure reflection densities. Reflection density is defined as the logarithm of the reciprocal of reflectance (R), where:

Reflectance (R)
$$= \frac{\text{Light reflected from the image}}{\text{Light reflected from the base}}$$

Thus

$$\text{reflection density } (D_r) = \text{Log} \frac{1}{R}$$

The design of a densitometer that measures reflection density necessarily differs from that of a transmission instrument. Since prints are usually illuminated at an angle of approximately 45° and viewed on a plane nearly perpendicular to the image, the American National Standards Institute has standardized this illumination and collection geometry, and all reflection densitometers are based upon it. If the instrument is zeroed on an unexposed, developed, and fixed-out piece of photographic paper, the base plus fog automatically will be subtracted from all future readings. When the instrument is calibrated on a reference white such as magnesium oxide, the readings will include the base-plus-fog density. Both approaches are commonly used.

Figure 4–30 contains a characteristic curve typical of a photographic paper emulsion. The major differences between this curve and that of a pictorial film are as follows:

1. The toe portion is usually longer and extends to a higher density.
2. The slope in the midsection is much steeper, and the straight-line portion is short if present at all.
3. The shoulder quickly reaches a maximum density, which is seldom more than 2.1.

Because of these conditions, the useful log exposure range of photographic paper is sub-

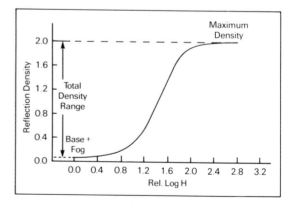

Figure 4–30 Typical characteristic curve for a black-and-white photographic paper.

stantially less than that of pictorial film. However, it should be remembered that the purpose of photographic paper is to receive the transmitted tones of the negative as input, while the purpose of negative film is to receive the reflected tones of the original scene as input. When evaluating the contrast of photographic papers, two types of information are important: the total density (output) range and the useful log exposure (input) range.

PAPER CONTRAST

The *total density range* of a paper is the range from paper base plus fog to the maximum obtainable density (sometimes called maximum black) at the top of the shoulder, as illustrated in Figure 4–30. The minimum density (base plus fog) is determined by the whiteness of the support and the fog picked up during processing. Reflection densities of 0.01 to 0.03 are typical minimum densities. The maximum obtainable density is in part governed by the emulsion type but primarily controlled by the print's surface characteristics. A glossy-surface paper with a good ferrotype finish can give a maximum density of about 2.10, while an unferrotyped finish on the same surface produces a density of about 1.90. Resin-coated papers seldom exceed 2.0. Mat surface papers show a maximum density of about 1.60.

The limit on print density is associated with first-surface reflections, as shown in Figure 4–31. In a typical print-viewing situation, the print is illuminated at 45° and viewed on the perpendicular. For the maximum black patch with the glossy surface, the first-surface reflection is very directional, with the angle of reflectance nearly equal to the angle of incidence. With the eye positioned as shown, the fraction of the light being reflected will not be detected. Since the incident light is either absorbed by the image or reflected at an angle where it is not readily seen, the patch appears black and consequently has a high density. When considering the maximum black patch with a mat surface, it can be seen that the first-surface reflection is quite diffuse and some light will be reflected back to the eye regardless of the viewing angle. The result is a washing-out of the maximum black

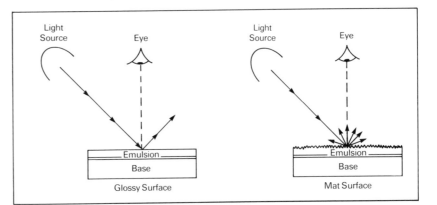

Figure 4–31 First-surface reflection properties of glossy and mat-surface papers.

and a corresponding loss in maximum density. The effect on curve shape of different surfaces for the same grade of paper is shown in Figure 4–32, which demonstrates that the greatest change occurs in the shoulder. The practical consequence of this condition is that *the output range of a photographic paper is controlled primarily by the surface characteristics.*

Unlike the negative film emulsions studied previously, the contrast of photographic papers is not readily controlled by development. Since development occurs rapidly as a result of using a high-activity developer, the basic curve shape is essentially determined after a brief period. The curves in Figure 4–33 result from varying development time, with 1½ minutes considered normal. Although it is apparent that the slope of the curves is changing, development times of less than one minute tend to give a lowered max-

The input range of photographic papers is controlled by the curve slope, which is related to paper contrast grade.

The output range of photographic papers depends primarily on the surface sheen.

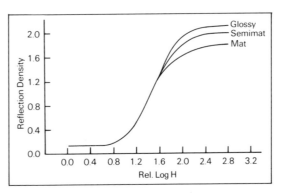

Figure 4–32 The effect of surface characteristics on the curve shape for the same grade of paper.

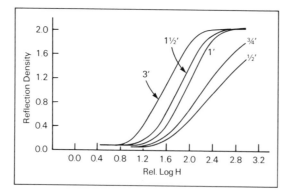

Figure 4–33 The effect of different development times on curve shape for the same grade of paper.

Increasing the developing time of black-and-white papers beyond the minimum recommended time has little effect on the slope of the characteristic curve.

imum density with a mottled appearance and, therefore, unusable images. At a development time of one minute the curve shape is nearly fixed, with increases in development time merely shifting the curve to the left. Thus extended development times don't change the contrast of photographic papers but do increase the speed up to about one stop. If the development time becomes too long, the fog density increases and makes the highlights appear unacceptably gray.

The effects of changing development time on photographic papers may be summarized as follows:

1. Underdevelopment is to be avoided because of the lack of maximum black and the possibility of mottling.
2. Overdevelopment can be used to increase the speed by as much as one stop, but care must be taken to avoid an unacceptable increase in fog density, which ruins the white areas of the print.
3. Since the curve shape is determined early in development, changes in development conditions cannot compensate for an incorrectly chosen grade of paper.

Thus far in our analysis of photographic paper curves we have only considered the output (density) characteristics. To evaluate the input properties requires a determination of the paper's *useful log exposure range*. This concept is similar to that applied to camera films in that minimum and maximum useful points must be located on the curve. The location of these points is based upon the vi-

sual quality of the highlights and shadows produced in the print. Because the tones in the negative are reversed in the print, the highlights are reproduced in the toe area and the shadows in the shoulder of the paper's characteristic curve. Therefore, the minimum useful point in the toe will be based upon proper highlight reproduction. Experiments show that nonspecular highlights in a subject, such as reflections from white surfaces, should be reproduced with a density slightly greater than the base white of the paper. A density of 0.04 above base plus fog is sufficient to give a just-noticeable tonal difference above the base white of the paper, and it identifies the minimum useful point in the toe as shown in Figure 4–34. The base white of the paper is the tone that is reserved for reproducing light sources or the specular (glare) reflections from polished metal or other shiny surfaces. Since these are the brightest parts of the subject, it makes sense that they should be printed at the lightest tone possible.

When considering the reproduction of shadows and thus the maximum useful point, it is necessary to determine the desired amount of shadow detail. Again, many experiments indicate that black-and-white prints judged as excellent consistently exhibit good shadow detail. Those same experiments show that the maximum density of the print was not used to reproduce the shadow detail. Rather, a point farther down on the shoulder is used where there is sufficient slope for maintaining tonal separa-

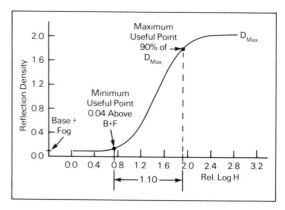

Figure 4–34 The location of minimum and maximum useful points, and the resulting useful log exposure range.

tion. This point is typically located at a density equal to 90% of the maximum density and denotes the maximum useful point on the curve. However, portions of a scene that are very dark and contain no detail of interest are printed at the maximum density so that they may be reproduced as a solid black in the print. To summarize:

1. The minimum useful point in the toe is based upon the reproduction of nonspecular highlights and is located at a density of 0.04 above the paper's base plus fog.
2. The maximum useful point in the shoulder is based upon the reproduction of shadow detail and is located at a density equal to 90% of the maximum density.
3. The log exposure interval between the two points is defined as the useful log exposure range (sometimes just log exposure range) of the paper. For example, in Figure 4–34 the log exposure range was determined to be 1.10.

The characteristic curves for a family of papers of different contrast grades are illustrated in Figure 4–35. The major difference between the curves is in the range of log exposures over which they extend. A grade 0 emulsion (sometimes referred to as a "soft" paper) responds over a long range of log exposures, while a grade 5 (sometimes referred to as a "hard" paper) covers a much shorter log range. All of the grades are capable of generating the same range of densities (output) because they all show the same minimum and maximum densities, which is the case when the paper surfaces are the same. The contrast grade numbers 0, 1, 2, 3, 4, and

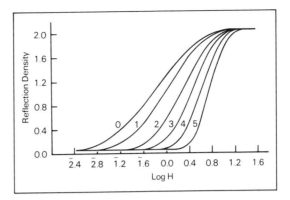

Figure 4–35 Characteristic curves for a family of papers of different contrast grades with the same surface.

5 represent a system for classifying the contrast of a paper based upon the log exposure range. Some manufacturers use the ANSI standard terms for describing emulsion contrasts. Table 4–4 contains the contrast-grade numbers and ANSI standard terms that relate to the log exposure range. Notice that each grade of paper can have a variety of log exposure ranges. For example, a grade 2 paper can have a log exposure range of 0.95 to 1.15, which no doubt explains at least some of the differences between papers of the same grade and surface from different manufacturers.

The current American National Standard governing the sensitometric properties of black-and-white reflection print materials specifies the use of the *Standard Paper Range Number (R)* in lieu of the useful log exposure range. The value of R is determined by multiplying the useful log exposure range (rounded to the nearest 0.10) by 100. For example, if the useful log exposure range for a

Prints made on different brands of contrast grade 2 printing papers will not necessarily have the same print contrast.

Table 4–4 Relationship between paper grade, paper log exposure range, and density range of negatives for diffusion and condenser enlargers

Contrast Grade Number of Paper	ANSI Descriptive Term	Log Exposure Range of Paper	Density Range of Negative Usually Suitable for Each Log Exposure Range or Grade Number (Diffusion Enlarger)	Density Range of Negative Usually Suitable for Each Log Exposure Range or Grade Number (Condenser Enlarger)
0	Very Soft	1.40 to 1.70	1.41 and higher	1.06 and higher
1	Soft	1.15 to 1.40	1.15 to 1.40	0.86 to 1.05
2	Medium	0.95 to 1.14	0.95 to 1.14	0.71 to 0.85
3	Hard	0.80 to 0.94	0.80 to 0.94	0.60 to 0.70
4	Very Hard	0.65 to 0.79	0.65 to 0.79	0.49 to 0.59
5	Extra Hard	0.50 to 0.64	0.64 and lower	0.48 and lower

certain grade of paper were 1.07, it would be rounded to 1.10 and multiplied by 100, giving R = 110. If the digit in the hundredths position is 5 or higher, the number is rounded up; if it is 4 or lower, it is rounded down. The resulting R value has no standard relationship to the grade numbers assigned by the manufacturers to their products and, therefore, does not reduce the differences between the same grade of paper from different manufacturers.

It should be evident by now that the density range of the negative relates to the log exposure range of the paper it is to be printed on. In fact, the negative's density range actually determines the range of log exposures the paper will receive, and thus the paper grade required. A contrasty negative is characterized by a large density range, and so it requires a soft grade of paper that can respond over an equally large range of log exposures. On the other hand, a flat negative has a narrow density range, and it requires a hard grade of paper with a short log exposure range so that a full scale of tones will result with excellent shadow and highlight reproduction.

This situation is illustrated in Figure 4–36, where the density range of the negative is the same as the log exposure range of the paper, and the print has been correctly exposed. The nonspecular highlight in the negative will produce a density of approximately 0.04 above the base white of the paper, and the detailed shadow area of the negative will

The best print quality will generally occur when the density range of the negative equals the log exposure range of the paper.

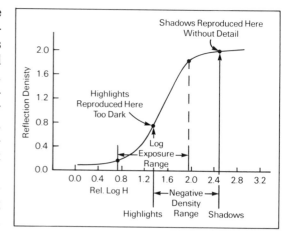

Figure 4–37 Correct match of negative density range and paper log exposure range, with print overexposed by two stops.

generate a density near 90% of the maximum black.

If the same negative and paper grade were used again, with a two-stop increase in print exposure, the condition illustrated in Figure 4–37 occurs. The result is a print without tonal separation in the shadows, as the shadow densities in the negative are located too far into the shoulder of the paper. Also, the highlights will be too dark because the highlight densities in the negative will print too high on the toe of the paper curve. If the same negative is printed on a grade of paper that is too hard, the condition in Figure 4–38 results. Here the print exposure has been adjusted to give adequate shadow detail, as the shadow densities of the negative are printing at the maximum useful point. How-

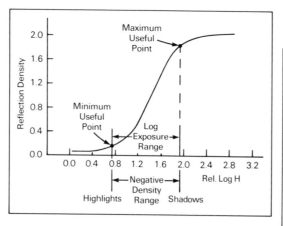

Figure 4–36 The relationship between the density range of the negative and the useful log exposure range of the paper for a correctly exposed print.

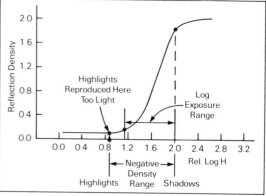

Figure 4–38 Negative printed on a paper grade that is too hard, printed for the shadows.

ever, the highlight densities extend far to the left of the minimum useful point with a complete blocking up of the print highlights. If the print exposure is increased to give better highlight reproduction, the shadows will print too far into the shoulder and begin to block up. Consequently, the odds of achieving a good print with this combination are slim. At this point we can summarize by saying that for pictorial images, the best print quality will generally occur when the density range of the negative equals the log exposure range of the paper, and the print is correctly exposed.

CONDENSER AND DIFFUSION ENLARGERS

There are two basic types of enlargers: diffusion and condenser. The diffusion enlarger employs a large integrating sphere or ground glass to illuminate the negative with scattered light, and is classified as a diffuse system. The condenser enlarger makes use of a set of lenses (condensers) above the negative so that the light from the bulb will be more concentrated when it strikes the negative, and is thus a specular system. When determining a negative's density range for the purpose of selecting the proper grade of paper, the similarity between the optical designs of the densitometer and the enlarger should be considered.

Since nearly all commercially available densitometers measure diffuse density, the resulting values usually can be directly applied to diffusion enlargers. However, as discussed earlier in this chapter, black-and-white negatives have greater density and contrast in a specular system. Consequently, the densities obtained from a diffuse-reading densitometer must be adjusted upward to convert them to specular densities. The exact value of this conversion factor (also referred to as a Q-factor) will depend primarily on the specularity of the enlarger and is best determined by testing the specific equipment. Many tests of condenser enlargers indicate that this factor is approximately 1.3. This means, for example, that a negative with a diffuse density range of 1.10 will actually have a density range of 1.43 (1.10 × 1.3) in a condenser enlarger. To compensate for this dis-

crepancy, Table 4–4 lists the diffuse density ranges for negatives when printing in either diffusion or condenser enlargers. Notice that the diffuse density range of a negative to be printed on a grade 2 paper in a diffusion enlarger can be between 0.95 and 1.14, while the diffuse density range of a negative to be printed on a grade 2 paper in a condenser enlarger is between 0.71 and 0.85.

The curves in Figure 4–39 show the difference in print contrast between a diffusion enlarger and a condenser enlarger equipped first with an opal bulb and then with a clear bulb, using a negative made on a coarse-grained film.

THE EFFECTS OF ENLARGER FLARE AND SCATTERED LIGHT

Flare light has the effect of reducing contrast in the enlarger as well as in the camera. With cameras, shadow contrast and detail are reduced by flare, but with enlargers the highlights are most affected (except with reversal processes).

Flare light in an enlarger originates with the light that bounces around inside the enlarger body and lens before reaching the printing paper as a uniform level of light. The amount of enlarger flare light will vary with the efficiency of the lens coating and the en-

To minimize a loss of contrast due to flare light when using an enlarger, the negative should be masked down to the area being printed.

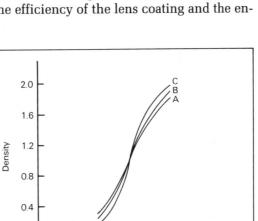

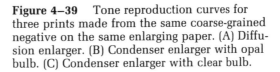

Figure 4–39   Tone reproduction curves for three prints made from the same coarse-grained negative on the same enlarging paper. (A) Diffusion enlarger. (B) Condenser enlarger with opal bulb. (C) Condenser enlarger with clear bulb.

larger light-baffling system, and with the proportion of shadow-to-highlight areas in the negative. It will also depend upon whether the opening in the negative carrier is masked down to the area of the negative that is actually being printed. Because enlarger flare reduces the contrast primarily in the highlight areas, it has a more damaging effect on the image quality when it is excessive than camera flare, which reduces contrast mostly in the shadow areas.

TESTING ENLARGERS FOR FLARE LIGHT

The existence of flare light in an enlarger can be demonstrated easily. Place a small piece of opaque black cardboard in a glass negative carrier without masking off the excess area around the card. Now make a print using a typical combination of exposure time and f-number. Since the black cardboard transmits no light, any light reaching the easel in the corresponding area can be attributed to flare.

Figure 4–40 Although a square black card place in the enlarger's negative carrier transmitted no light, considerable flare light fell on the corresponding area of the easel. The white ring was produced by placing an opaque object on the enlarging paper to protect it from the flare light.

Figure 4–40, made in this way, shows considerable density in the image area of the square card compared to the smaller area, which was protected by an opaque object on the paper. Masking down the area around the black card in the negative carrier will reduce the amount of flare light reaching the paper. Masking off the entire surround will reduce the flare light to zero.

An enlarger can also be tested for flare light under normal printing conditions by placing a small opaque object on a negative in the carrier. The object should be small enough so that it does not significantly reduce the total amount of light transmitted through the negative. It will be easier to detect small amounts of density produced by flare light if a second opaque object is placed on the printing paper. This second object should cover half of the projected spot of the object on the negative.

More can be learned about the nature of enlarger flare by making projection prints from a step tablet with and without masking off the area around the step tablet. A side-by-side comparison of the prints will reveal differences between the two images. Measuring the density of each step with a reflection densitometer enables the construction of characteristic curves demonstrating the differences graphically, as shown in Figure 4–41.

Curve A represents a high-flare situation where the area around the step tablet was not masked off. The excess area was masked down to within a quarter of an inch of the step

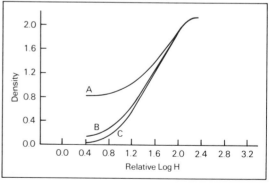

Figure 4–41 Characteristic curves showing the effects of masking off the area around a step tablet in an enlarger negative carrier. (A) No masking. (B) Masked down to 1/4 inch of the step tablet. (C) Masked down to the edge of the step tablet.

tablet for curve B, and completely masked for curve C. It is evident that the enlarger flare most affects the print's lighter tones, and that the flare density increases dramatically with the size of the unmasked area around the step tablet.

In some prints the decreased image contrast produced by enlarger flare light results in improved image quality. In these prints the contrast would have been too high without the flare light. Thus the availability of controlled, variable flare light can be useful.

PAPER SPEEDS

The speed of a photographic paper is inversely related to the minimum exposure necessary to achieve an excellent print. As was shown to be the case in the earlier discussion of pictorial film speeds, experimental evidence indicates that print exposure should be based upon *shadow detail reproduction*. Since the shadows fall on the shoulder of the paper curve, the maximum useful point, located at 90% of the maximum density, is used as the speed point. The speed is referred to as a "shadow" speed, with the following formula being used:

$$\text{Shadow speed} = \frac{1}{H} \times 10,000$$

where H = the exposure in lux-seconds needed to generate a density that is 90% of D max;

10,000 = a large constant that will convert the speed values to a convenient size.

The shadow speeds of the curves for the family of paper grades illustrated in Figure 4–35 are shown in Table 4–5, which indicates that the shadow speed decreases as the

contrast of the paper (paper grade) increases. This means that a grade 4 paper is slower than a grade 2, and appropriate compensation must be made in the exposure time when changing paper grades. The actual speeds of various brands of the same grades of paper will be different, but the general relationship is that the higher the paper grade, the slower the speed.

An alternative method for determining the paper speed is based upon the exposure necessary to produce a *midtone* with a density of 0.60 above base plus fog. This is the current ANSI standard procedure for deriving the paper speed and is predicated on the assumption that proper midtone reproduction is most important. The resulting speed value is called the *ANSI Paper Speed* (sometimes ASAP) and is based on the following formula:

$$\text{ANSI Paper Speed} = \frac{1}{H} \times 1,000$$

where H = the exposure in lux-seconds needed to generate a density of 0.60 above base plus fog;

1,000 = a large constant that will convert the speed value to a convenient size.

Table 4–5 shows a comparison of Shadow Speeds and ANSI Paper Speeds for six grades of paper. If print exposure times are based on the speed numbers when changing paper grades, Shadow Speeds will produce matched shadow densities and ANSI Paper Speeds will produce matched midtone densities. Some brands of photographic paper show curves similar to those illustrated in Figure 4–42. Since all the curves cross at a density of 0.60 above base plus fog, they will all show the same ANSI Paper Speed. The shadow speed obviously will be different since the position of the shoulder is changing. When printing to a midtone with these emulsions, no change in print exposure would be required as a function of changing paper grades. However, when printing for the shadows, significant changes in print exposure would be needed when changing grades. Paper speed numbers can be used to calculate the new printing exposure time when changing from one paper to another. Conversion tables or dials are also available from some paper manufacturers.

Enlarger flare light, or intentional flashing, can improve the quality of prints made from contrasty negatives.

Traditionally, low contrast grades of printing paper have had higher speeds than high contrast grades.

Table 4–5   Speeds for a family of paper grades

Contrast Grade Number of Paper	Shadow Speed	ANSI Paper Speed
0	5000	1600
1	4000	1000
2	2500	500
3	2000	300
4	1600	200
5	1280	160

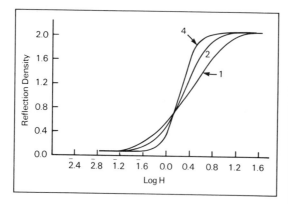

Figure 4–42 Family of paper contrast grades showing the same ANSI paper speed but different shadow speeds.

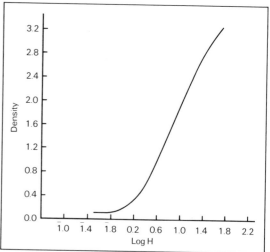

Figure 4–43 The characteristic curve of a film positive emulsion.

changing from one paper to another. Conversion tables or dials are also available from some paper manufacturers.

ANALYSIS OF POSITIVE FILM CURVES

Negatives can be printed onto print film materials to obtain positive images on a transparent base that can be viewed by projection or with transmitted light from an illuminator.

An alternative to printing on photographic paper is to use a relatively high-contrast film, which will yield a positive transparent image that can be viewed by either back illumination or projection. The films used for this purpose are called *film positive emulsions* or release print films. They typically have slow speed, very fine grain and inherently high contrast with gammas varying between 1.8 and 2.8, depending upon the developer formulation and development time. The characteristic curve for such an emulsion is shown in Figure 4–43, where it can be seen that the slope is quite steep.

In principle, the production of film positives from negatives is similar to that of reflection prints except that the contrast of the emulsion being printed upon can be controlled readily by development. This eliminates the need for various contrast grades. Experience indicates that the minimum useful point on the film positive curve for good highlight reproduction is approximately 0.30 above the density of base plus fog. If the highlights are reproduced with a greater density than this, they appear too dark, and if the density is any less, they look "washed out." Although these films are capable of achieving densities over 3.0, the shadows are best re-

produced at densities of 2.7 and less depending upon the contrast of the negative. Since the density range of which the film positive is capable is far greater than that of photographic paper, subjects with high luminance ratios can be faithfully reproduced. The relationship between the density range of the negative and the film positive curve is illustrated in Figure 4–44 for a correctly exposed and processed image. Notice that the

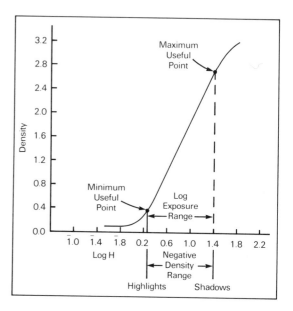

Figure 4–44 The relationship between the density range of the negative and the useful log exposure range of the film positive emulsion for a correctly exposed image.

slopes in both the highlight and shadow regions of the curve are greater than those of a photographic paper. The resulting film positive will have greater tonal separation in those areas than a reflection print made from the same negative.

There is no standard method for determining the speed of positive films, but speed points based upon either highlight or shadow reproduction are commonly used to calculate speed values. The highlight method involves determining the exposure necessary to produce 0.3 above base plus fog density using the following formula:

$$\text{Speed} = \frac{1}{H} \times 1,000$$

The midtone method employs the same formula, except that the exposure necessary to generate a density of 1.0 above base plus fog is used. Shadow speeds are seldom used for these films because the steep slope in the shadow region of the curve assures that good shadow detail will be maintained. Therefore, print exposures are generally based upon either highlight detail or midtone density. Regardless of which speed method is used, the speed of film positive emulsions are highly dependent upon the degree of development. Figure 4–45 shows a family of curves for different development times of a film pos-

itive emulsion. The curve shifts to the left, indicating an increase in speed, as development time lengthens. Thus close attention must be given to the details of development if consistent results are to be obtained.

By printing on film positive emulsions, the photographer can overcome the single greatest limitation of the negative-positive system: the low maximum density produced by photographic papers. The wider density range of the film positive results in images with greater physical and visual contrast. Not only will the tonal range appear greater, but the shadows and highlights will have more detail and brilliance. Ideally, film positives should be viewed on a transparency illuminator in a darkened room with the surround masked off to obtain the greatest visual contrast. If the image is placed in a projector and viewed on a screen under typical conditions, the image contrast is drastically reduced as a result of both optical flare and ambient room light reaching the screen. Even under these conditions, the contrast of projected images still exceeds that of reflection print images.

ANALYSIS OF REVERSAL FILM CURVES

The generation of a positive, transparent image may also be accomplished through a procedure called *reversal processing*. In this method, a film that was exposed in the camera is first developed to a negative, without fixing. In the next step, a bleach bath dissolves the negative silver image leaving unexposed silver halide in amounts opposite to those of the negative. This unexposed silver halide is exposed (fogged) by flashing to light (or is chemically fogged) and then developed to give the desired positive image. The second development is followed by fixation, washing, and drying. The result is a black-and-white transparency produced on the original film exposed in the camera.

The characteristic curves for reversal materials are plotted in the same manner as those of negative working emulsions except that the direction of the curve is reversed. This is necessary because with reversal materials the density decreases as the exposure becomes larger. Consequently, the shadows and darker tones are reproduced as high densities while

The basic steps of reversal processing are to develop a negative image, bleach the negative image, fog the remaining silver halides, and develop the positive image.

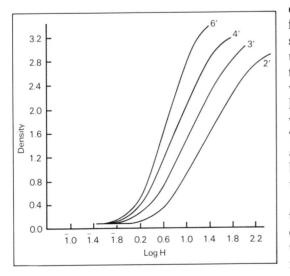

Figure 4–45 Characteristic curves for a variety of development times of a film positive emulsion.

the highlights and lighter tones yield small densities. The D-log H curve for a typical reversally processed film is shown in Figure 4–46. The curve exhibits a steep middle section, which is necessary to give sufficient contrast to the image, since it will be viewed directly. This results in a shortened useful log exposure range, meaning that the exposure latitude is small and scenes with excessive contrast cannot be properly reproduced. In fact, most reversally processed emulsions do not exceed a 1.90 useful log exposure range. Although many negative working films may be reversally processed, emulsion manufacturers usually design or designate specific films to be processed using the reversal method.

Since the reversal process involves more major steps than the negative process, the influence of processing changes on the curve shape is more difficult to ascertain. The effect of varying the time of first development while all other steps are held constant is illustrated in Figure 4–47. The first development time has a significant effect upon all of the sensitometric parameters of the film. As the first development time is lengthened, the maximum density of the positive image becomes lower due to an increase in the fog density at the negative stage. The minimum density also decreases slightly because of the increase in the shoulder density at the negative stage. Increases in the slope of the curve are most noticeable in the toe region, indicating a small gain in contrast.

Photographic effects can produce unexpected responses to exposure of photographic materials.

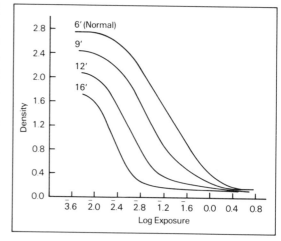

Figure 4–47   The characteristic curves resulting from different first development times for reversally processed film.

The effect on speed of altering the first development time is shown in Figure 4–48. The speed point was located at a density of 1.0 above base plus fog, and the speed was calculated using the following formula:

$$\text{Speed} = \frac{1}{H} \times 10$$

The graph indicates that the speed of reversally processed films can be increased by up to 10 times by lengthening the time in the first developer. However, the speed increase is coupled with a decrease in the maximum obtainable density, which results in poor shadow reproduction. When the maximum density falls below 2.0 for transparencies, shadow-detail reproduction is usually unacceptable. The increased contrast in the midtones may also contribute to an inferior result.

INTRODUCTION TO PHOTOGRAPHIC EFFECTS

Most of the time, the response characteristics of photographic materials and processes are quite predictable based upon the principles discussed thus far. However, there are conditions where the resulting photographic image does not have the characteristics that were anticipated. Although the making of photographs is a relatively simple task, the re-

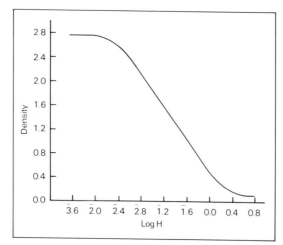

Figure 4–46   Characteristic curve of a reversally processed film.

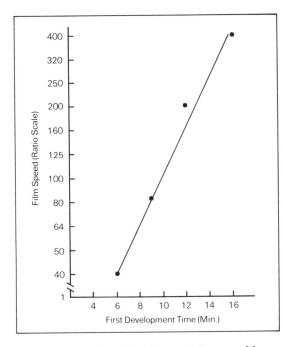

Figure 4–48 First development time vs. film speed for reversally processed film.

sponse properties of photographic materials are quite complicated. An understanding of conditions that can lead to unexpected results will be most beneficial in mastering the photographic process. Some of these photographic effects are considered in the sections which follow.

RECIPROCITY EFFECTS

In 1862, two scientists, Robert Wilhelm Bunsen and Henry Enfield Roscoe, proposed a general law for photochemical reactions stating that the end result of a primary photochemical reaction is simply dependent upon the total energy used, regardless of the rate at which the energy is supplied. Photography involves a photochemical reaction in which light is used to expose film or paper to form a usable latent image, and chemistry is used to develop a visible silver or dye image. The end product (density), however, is not dependent simply upon the total energy (exposure), but also on how that exposure is distributed in terms of illuminance and time. For example, two pieces of film can receive equal exposures that may result in different densities because the exposure times and il-

luminances are quite different. Exposure (H) is a product of illuminance (E) and time (t). An illuminance of 2 lux for 1/100 second produces the same exposure (.02 lux-seconds) as an illuminance of .002 lux for 10 seconds. Because the *response* may be different for these two exposures, the reciprocity law of Bunsen and Roscoe is said to have failed.

Early literature described this situation as "reciprocity law failure" (RLF). Current literature simply refers to it as "reciprocity failure" or "reciprocity effect" to distinguish it from Bunsen and Roscoe's reciprocity law, which holds true for *primary* photochemical reactions. The law is considered valid for x-ray and gamma-ray exposures with silver halide emulsions, and for light exposures with certain non-silver processes such as diazo and selenium electrostatic processes. Reciprocity failure in photography simply means that the *response* (density) of a photographic material cannot be predicted solely from the total exposure a film has received (except over a small range of exposure times). For pictorial films, the range over which reciprocity failure is negligible is from about 1/10 second to about 1/1,000 second. Beyond these time limits, reciprocal combinations of exposure time and illuminance, which give the same total exposure, will not give the same density response. As an example, if a meter indicates camera exposure settings of f/5.6 and 1/30 second, but because greater depth of field is needed the lens is stopped down to f/45 and the film is exposed for two seconds, the resulting densities will not be the same even though the combinations of camera settings were equivalent. The two-second exposure will result in a thinner negative. Density will be less for exposure times longer than 1/10 second or shorter than 1/1000 second. In other words, the film will be underexposed even though the subject was metered correctly, the right camera settings were used, and development was correct. This is an example of reciprocity failure or the reciprocity effect.

In addition to the loss in density already noted, reciprocity failure can change image contrast. The extent of the changes in density and contrast depends upon the combination of illuminance and exposure time and the charactersitics of the emulsion. The reciprocity curve in Figure 4–49 shows the relationship between the total exposure (on the vertical axis) required to obtain a given density of 0.6 and the various combinations of illuminance (horizontal axis) and time (45° lines). If there were no reciprocity failure, the curve would be a straight horizontal line, and the same exposure, regardless of the reciprocal combinations of illuminance and time used, would give the same density. This is true over a limited range of illuminances and times as indicated by the bottom part of the curve, which is almost flat. (Although reciprocity curves will generally differ from one product to the next, they will have the same shape except for certain special emulsions.)

LOW-ILLUMINANCE FILM RECIPROCITY

Low-illuminance reciprocity failure typically results in a decrease in negative density and an increase in gamma.

Reciprocity failure is often encountered in low-light-level photography, and it is commonly the reason for thin, underexposed negatives. The sensitometric effect of low-illuminance reciprocity failure for a pictorial film can be seen in Figure 4–50. All three films were given the same exposure, employing reciprocal values of illuminance and time to allow the use of three different exposure times. The same conditions of development were used for each. If there were no reciprocity failure, the three curves would have the same shape and be superimposed. The fact that the 1-second and 10-second curves are displaced rightward indicates a progressive loss of film speed as the exposure

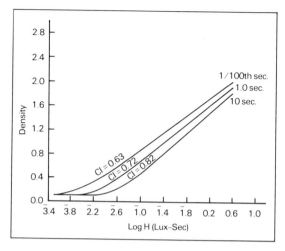

Figure 4–50   Low-illuminance (long exposure time) reciprocity failure.

time increases. In other words, when using exposure times of 1 second and longer for this emulsion, additional exposure is required to obtain the expected image *density*.

Adjustments need to be made in development as well as exposure for long exposure times at low light levels. The reason for this can be seen in Figure 4–50. In addition to the displacement of the curves, the curves differ in slope.

As the exposure times reach 1 second and longer, there is a significant increase in slope and thus image contrast. Whereas an exposure time of 1/100 second with normal development produced a contrast index of 0.63, an exposure time of 10 seconds with the same normal development produced a negative with a contrast index of 0.82—about a 30% increase in contrast. This is caused by the fact that low-intensity reciprocity failure affects the lower density areas more than the higher density areas. Thus, in order to obtain the expected image contrast in the negative, a *reduction* in development is necessary. Since altering the degree of development also affects density, the reciprocity failure exposure factor will be larger than it would be if the degree of development is not adjusted for the anticipated increase in contrast with long exposure times.

Low-illuminance reciprocity failure requires an *increase* in exposure along with a *decrease* in development to adjust for contrast. If only the exposure correction is made, the negative may have to be printed on a lower

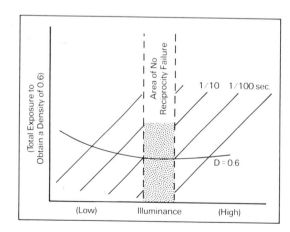

Figure 4–49   Reciprocity curve. A straight horizontal line would represent no reciprocity failure.

grade of paper than usual to adjust for the increase in contrast index of the negative.

HIGH-ILLUMINANCE FILM RECIPROCITY

Exposure of films for times shorter than 1/1,000 second, as with some electronic flash units, results in decreased density. The decrease is largest in the highlight areas, with negative-type films producing a lower-contrast image. An exposure increase and an *increase* in development are both needed to compensate for this.

The characteristic curves in Figure 4–51 illustrate the effect of high-illuminance reciprocity failure. All three films received the same total exposures, using reciprocal values of illuminance and time to obtain shorter exposure times. Development conditions were kept constant. Notice that the 1/10,000- and 1/100,000-second curves are also displaced rightward, indicating a loss of film speed. It is important to notice that under these conditions the curve shape changes in a manner opposite to that with low illuminance. Specifically, the slope of the curve decreases at exposure times of 1/10,000 second and less, indicating a loss of contrast. Observe, however, that contrast increases with increasing exposure time, for both low illuminance and high illuminance reciprocity failure. If only an exposure adjustment is made, the negative would have to be printed on a higher contrast grade of paper to adjust for the loss of contrast in the negative.

BLACK-AND-WHITE PAPER RECIPROCITY

Since reciprocity effects are exhibited by *all* photographic emulsions, they can be seen easily by printing on any photographic paper and using a series of increasing exposure times, as shown in Figure 4–52. A photographic step tablet was first printed at a relatively short exposure time of 10 seconds at f/5.6, then for each subsequent print the exposure time was increased as the illuminance was proportionally decreased by stopping the lens down and adding neutral density filters. The density decreased progressively even though *each test strip received the same series of photographic exposures*—as the illuminance decreased, the exposure time increased, and the development remained the same.

Noticeable reciprocity failure with printing papers is seldom encountered when using normal negatives. It is a problem, however, when printing a severely overexposed and/or overdeveloped negative to a large magnification. With the aperture wide open, the usual way to increase exposure is by lengthening print exposure *time*. Because of reciprocity failure, increasing print exposure times does not give the expected increase in image density.

High-illuminance reciprocity failure typically results in a decrease in negative density and gamma.

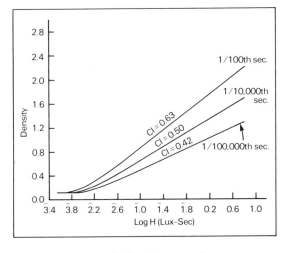

Figure 4–51 High-illuminance (short exposure time) reciprocity failure.

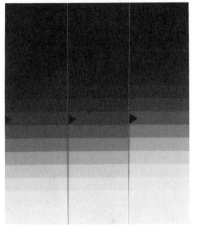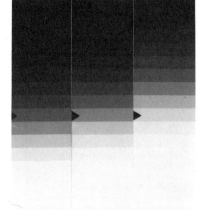

Figure 4–52 Each set of strips received the same equivalent exposure, but for different combinations of time and illuminance. The set to the right exhibits more reciprocity effect.

Reciprocity failure also presents a problem when a high-density area in the negative has to be manipulated by "burning-in" or where high magnifications are required such as in printing photo murals or in printing small areas of a negative to a large print size.

In addition to a loss in speed with printing-paper reciprocity failure one can also expect a loss in contrast. This is illustrated in Figure 4–53. The 240-second (four-minute) curve has been adjusted for the speed loss so that the loss in contrast is more apparent. Whereas one can expect reciprocity failure to cause an increase in contrast for films exposed for relatively long times (low illuminances), a decrease in contrast for print papers can be expected.

COLOR PAPER RECIPROCITY

Plate I shows six color prints made from the same color negative. The prints received identical total exposures but with increasing exposure times (4, 8, 16, 32, 64, 128 seconds) and decreasing illuminances. Color balance and print density were based on the four-second exposure time used for print No. 1. At about 16 seconds the overall density begins to drop and continues to drop as the exposure time increases. Over a range of four seconds to just over one minute, however,

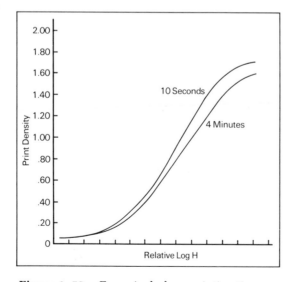

Figure 4–53 Excessively long printing times result in a decrease in paper contrast as well as speed. (The loss in paper speed has been adjusted to emphasize the loss in contrast.)

there is no noticeable shift in color balance. In terms of reciprocity effect this is an exceptionally good color printing paper. No filters are required to adjust for long printing *times*, simply an increase in exposure.

COLOR FILM RECIPROCITY

Figure 4–54 illustrates reciprocity failure curves for a color reversal film. The curves represent constant density lines for the cyan, magenta, and yellow dye layers of the film. Multilayer color films require the balancing of at least three different emulsions so that their reciprocity failure characteristics, over an extended range of exposure times, are similar. For the color film represented in Figure 4–54, the reciprocity failure characteristics are similar over a range of exposure times between approximately 1/10 and 1/1,000 second. (This assumes that processing is in accordance with the manufacturer's recommendations.) When exposure times longer than 1/10 second are used, adjustments to the exposure given each of the three layers must be made. This is usually done by adding the proper color filter and increasing the exposure.

SABATTIER EFFECT

Photographic film or paper that is partially exposed to image-forming light and partially developed (but not fixed) can, if re-exposed with uniform illumination, developed, and fixed, result in a partial reversal of tones. The reversal is most noticeable in the areas re-

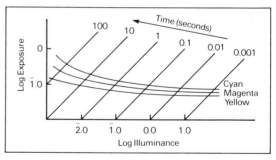

Figure 4–54 Reciprocity failure for a reversal color film. The reciprocity characteristics are made to match over a limited range of exposure times. Beyond those times the three different emulsions exhibit different degrees of reciprocity failure.

ceiving the least exposure: the shadow areas for film and the highlight areas for paper. In addition to a partial reversal, the edge or contour of a shape is often strongly enhanced. This dark line effect, referred to as a Mackie line, and partial reversal of tones can be seen in Figure 4–55. To obtain the Sabattier effect, no special equipment or solutions of any kind are required:

1. Give the film or paper a somewhat less than normal image exposure.
2. Develop for about half the normal development time.
3. Rinse in water (do not use a stop-bath or fixing solution).
4. Reexpose by fogging to light for a brief time.
5. Return the film or paper to the developer and develop for the other half of the normal time.
6. Rinse, fix, wash, and dry as you normally would.

(For single sheets of film or paper the rinse step can be eliminated. The second exposure can be made while the film or paper is in the developer. Allow the solution to settle for about 10 seconds, however, before making the second exposure.)

Some experimentation may be required to get the desired results. The ratios of first to second exposures and first to second developing times are important variables. Figure 4–56 shows that decreasing the first development time and increasing the second development time while keeping the fogging re-

Figure 4–55   The Sabattier effect (top). Note the partial reversal of tones and the black line along some of the contour.

exposure constant increases the amount of tone reversal.

The Sabattier effect can appear if the fogging exposure is accidentally given during first development, as with unsafe "safelights." Applications of this effect have

Figure 4–56   Gray scales showing the Sabattier effect. Gray scale prints made from Kodak Super XX Pan Film processed for various combinations of first and second development times. (Films all received the same first and second exposures.)

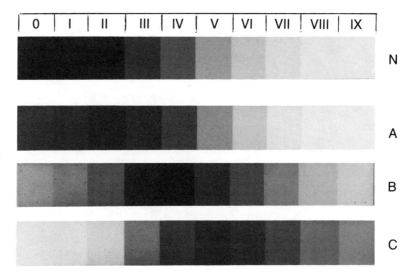

Gray Scale	Development Times		Total (minutes)
	1st Dev.	2nd Dev.	
N	6	0	6
A	4	2	6
B	3	3	6
C	2	4	6

Figure 4–57 An example of solarization due to extreme overexposure of the sun. Photography by Les Stroebel.

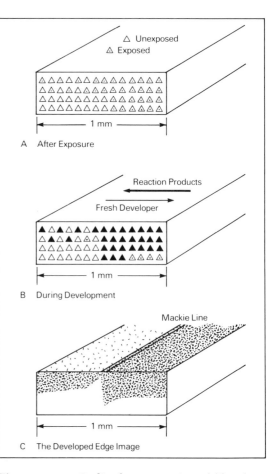

Figure 4–58 Stylized cross section of film that has been exposed and developed to show how the exchange of fresh developer and reaction products of development increases the density difference at the very edge of an image.

mainly been associated with interesting pictorial effects, but the line effect has been used, for example, to enhance the sharpness of spectrographic images. When the Sabattier effect is used with color negative or reversal materials, changes in hues, lightness, and saturation can be produced by varying the spectral energy distributions of the two exposures.[1]

SOLARIZATION

Increasing exposure increases density until a maximum density is reached. Extreme overexposure can result in a loss of density and reversal of the heavily exposed areas. Figure 4–57 shows a severely overexposed negative that resulted in a complete reversal of the sun. The degree of solarization depends on the particular film or paper and the developer used.

MICRO-EDGE EFFECTS

Edge effects can be studied by examining what occurs *within* an emulsion at the edge or boundary between high and low exposures, noting that edge effects apply only for micro-distances of less than 1mm. The sequence of events that takes place is illustrated in Figure

Some modern films are difficult to solarize, especially if processed in a developer containing a silver-halide solvent.

Development edge effects increase the contrast at the border between high-density and low-density areas.

4–58. During development, the high-exposure area produces a high concentration of reaction products, which restricts development in the area of high exposure and, by diffusion, in the adjacent area of low exposure. In the low-exposure area, there is a low concentration of reaction products and a high concentration of fresh developer. The diffusion of fresh developer from the low-exposure area to the adjacent high-exposure area will cause less decrease in density just inside the high-exposure area. Density is actually decreased on both sides of the edge (Figure 4–58C), but the contrast is higher near the edge than elsewhere in the high- and low-exposure areas. This contrast amplification on a microscale at the edge is seen as an increase in contrast on a macroscale that includes the larger areas.

The schematic illustration in Figure 4–58C corresponds to the microdensitometer trace

[1]Eastman Kodak Co., *Darkroom Expressions*, 1984, p. 64.

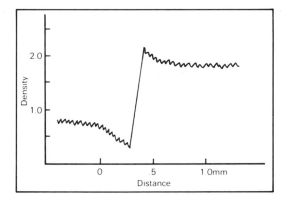

Figure 4–59   Edge effects. A microdensitometer trace of an image shows an increase in the density difference at the very edge. (The film was developed without agitation to maximize the effect.) The denser area is identified as a border effect and the thinner area is known as a fringe effect.

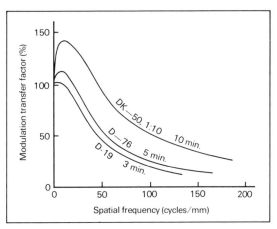

Figure 4–60   Modulation transfer functions for Kodak Plus-X Pan Film developed in Kodak developers D-19 and D-76 with brush agitation and in dilute Kodak developer DK-50 without agitation. (Thomas, W., *SPSE Handbook of Photographic Science and Engineering*, 1973, p. 956.)

of density vs. distance in Figure 4–59. The increase in density on the high-density side of the edge appears as a faint dark line and is called a *border effect*, while the decrease in density on the low-density side is seen as a corresponding faint light line and is called a *fringe effect*. Since these effects are a characteristic of the photographic process, they cannot be eliminated; nonetheless, they can be minimized or maximized. Maximization increases contrast, acutance, and modulation transfer functions (MTF) at low frequencies as shown in Figure 4–60. When photography is used as a photometric instrument, edge effects should be minimized, since they not only affect the change of density at the edges, but, as we shall see, they also can alter the shape of adjacent small images.

ADJACENCY EFFECTS

When two or more edges are adjacent and in micro-proximity, the interaction results enhance the contrast of both edges. The Eberhard effect, illustrated in Figure 4–61, shows microdensitometer traces across four photographic line images of decreasing width from about 2,000 microns (2 mm) to several microns. Note the enhancement of the edges and the increase in density as the line images narrow.

When two or more very small images of high exposure are adjacent to an area of low exposure, the shape of the developed image

will change, and the distance between image centers will increase with increased development. This has been named the Kostinsky effect. Such effects are encountered in photographing close spectrum lines and double stars. They also can be a problem in resolving-power measurements. Figure 4–62 is a representation of how two small areas of similar shape might be altered.

DEVELOPMENT EFFECTS

One of the byproducts of development is bromide (such as potassium bromide), which acts

Development byproducts can cause uneven development unless removed from the emulsion with efficient agitation.

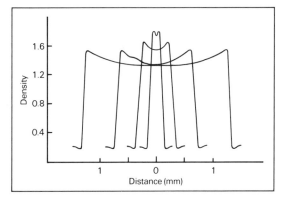

Figure 4–61   The Eberhard effect. A microdensitometer trace showing how areas of different size given the same exposure and development produce different densities. The highest density is reached at an image width of about 0.1 mm.

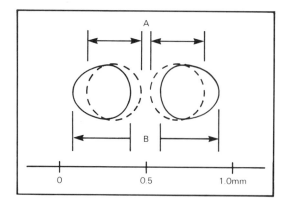

Figure 4–62   The Kostinsky effect. Development is restricted in the area between two small images adjacent to each other. This results in a slight displacement (from A to B) and distortion of the images. (The dotted line represents the optical image, solid lines the developed image.)

to inhibit or restrain development. High concentrations of bromide restrainer are formed in areas of high exposure. Unless the concentration is dispersed by effective agitation, the restrainer will tend to restrict further development in these areas of the film, and, because of diffusion, it will restrict development in the areas immediately adjacent. This results in low-density streaks or streamers (*bromide streaks*), Figure 4–63. When dark streaks are caused by poor agitation it is because areas of low exposure provide additional fresh developer for adjacent areas, thereby increasing development in those areas. Dark streamers ooze from areas of low exposure and density, while light streamers result from nearby areas of heavy exposure and high density.

If film is made to move in a single direction during processing, and if the agitation is inadequate, the leading edge of the film will receive greater development than the trailing edge. To test for this directional effect, place two sensitometric step-tablet exposures side by side on film, with the heavily exposed end leading on one, the heavily exposed end trailing on the other. The result of such a test is shown in Figure 4–64.

Films with perforated edges, such as motion picture films and 35 mm still-camera films, present a special problem. As the film passes through the developer, the sprocket holes cause increased local agitation and, therefore, increased density around the holes

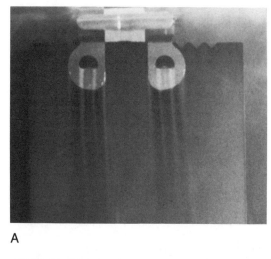

A

B

Figure 4–63   Bromide streaks.

(see Figure 4–65). If there is adequate agitation, however, the increased local agitation or turbulence has little effect. Good agitation of such films is particularly important when either a density sound track or telemetering information is contained along the side of the film.

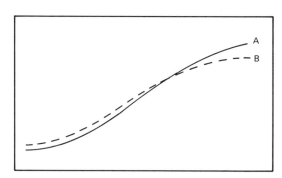

Figure 4–64   Directional effects. Curve A: Dense end of the sensitometric strip leads; opposite for curve B.

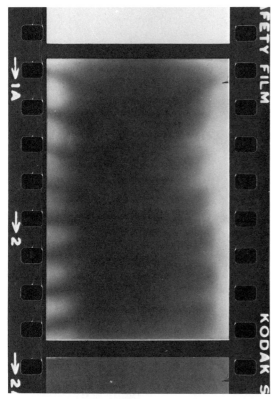

Figure 4–65   Sprocket hole effect.

Figure 4–66   X-ray marks.

X-RAY FOG

Photographic film and paper can be fogged by X-radiation, including the low-level X-radiation given off by detection equipment found in airports. In the U.S., such equipment is regulated by law to produce no more than a relatively low level of one milliroentgen of radiation. (A roentgen is a unit of X-radiation named in honor of Wilhelm Konrad Roentgen, who discovered the radiation in 1895 and designated it X because he was not sure what it was.)

Film or paper that has been fogged by X-rays may exhibit a faint sharp-edged "shadow" if any object is in contact with or close to the emulsion (see Figure 4–66). This is particularly noticeable in uniform areas. The "shadows" projected by the X-rays that are blocked by objects in their path will appear as *lighter* areas on a negative and *darker* areas on the print as well as on reversal-type materials.

The danger of X-ray damage to photographic film is slight for those who fly in the U.S. and other countries using low-level X-radiation. Several factors, however, can contribute to a visible fogging of film:

1. Film that has had repeated screening is more likely to exhibit fog since exposures are cumulative.
2. Some overseas airports use a higher level of radiation than the one milliroentgen used in the U.S.
3. Equipment may not be properly adjusted and give off more X-ray exposure than it should.
4. Some film is more sensitive to X-rays than others. For example, color films rated at 400 or higher are more susceptible to X-ray fog than similarly rated black-and-white films.

Lead-lined film pouches help protect against X-ray fog. The surest way to avoid any risk of X-ray fog, however, is not to have film or paper pass through detection equipment. In U.S. airports and those of some other countries one can choose to have carry-on luggage inspected by a security agent.

Higher speed films are more vulnerable to being fogged by X-rays, including those used for baggage inspection.

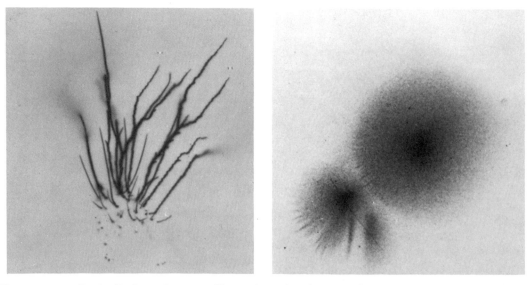

Figure 4–67   Static discharge between film and nearby object or objects. Left, negatively charged film; right, positively charged film. (© Eastman Kodak Company 1976.)

STATIC ELECTRICITY FOG

Film is more susceptible to being fogged by static electricity when the atmospheric humidity is low.

Photographic film passing through a film cartridge, or over rollers in a camera, processing machine, or rewinding setup, can acquire high electrostatic charges due to friction. When the accumulated static electricity is discharged, the air becomes ionized, producing light and heat. The result of this phenomenon is the appearance of *static marks* on film (see Figure 4–67). Static discharges are easily observed in a darkroom when the adhesive that secures roll film to a spool is quickly pulled from the film. To avoid such discharges, handle film carefully and wind slowly. Avoid winding film rapidly when the humidity is very low. People who have a tendency to produce static electricity when handling film can minimize this by wearing shoes with leather soles instead of rubber.

REVIEW QUESTIONS

1. In tone-reproduction studies, the term "input data" refers to information about the . . . (*p. 86*)
   A. subject
   B. image
   C. type of meter
   D. exposure
   E. film development

2. A subject has a highlight luminance of 200 c/sq.ft. and a shadow luminance of 0.50 c/sq.ft. The luminance ratio of the scene is . . . (*p. 86*)
   A. 20:1
   B. 90:1
   C. 150:1
   D. 200:1
   E. 400:1

3. A subject luminance ratio of 200:1 corresponds to a log luminance range of . . . (*p. 87*)
   A. 1.3
   B. 1.7
   C. 1.9
   D. 2.3
   E. 3.0

4. Assume that a gray scale has been photographed, and the print made from the negative matches the original gray scale exactly. The tone-reproduction curve for this situation would be . . . (*p. 88*)
   A. a straight line
   B. a C-shaped curve
   C. an S-shaped curve

5. Using current international symbols, photographic exposure is defined as . . . (*p. 89*)
   A. $E = I + T$
   B. $E = I \times T$
   C. $H = I \times T$
   D. $H = E + T$
   E. $H = E \times T$

6. Photographing a gray scale with a camera is not a completely satisfactory alternative to exposing film in a sensitometer because . . . (p. 90)
   A. camera shutters tend to be inaccurate
   B. camera flare affects image illuminances
   C. light falls off toward the corners of the film
   D. All of the above.
   E. None of the above. The camera is a satisfactory substitute for a sensitometer.

7. The part of sensitometers that represents the subject is the . . . (p. 90)
   A. light source
   B. shutter
   C. step tablet
   D. filter
   E. adjustable diaphragm

8. The most important reason for agitating film properly during development is to . . . (p. 91)
   A. shorten the time of development
   B. produce uniform development
   C. keep the film wet
   D. increase the effective film speed

9. If 80 units of light are incident on an area of a negative and 20 units are transmitted, the transmittance is . . . (p. 93)
   A. 0.20
   B. 0.25
   C. 0.80
   D. 2.50
   E. 4.0

10. Density is defined as the . . . (p. 94)
    A. antilog of the opacity
    B. log of the transmittance
    C. log of 1/opacity
    D. antilog of the transmittance
    E. log of the opacity

11. The density that corresponds to an opacity of 8 is . . . (p. 94)
    A. 0.4
    B. 0.6
    C. 0.8
    D. 0.9
    E. 1.0

12. Most commercial densitometers are designed to measure . . . (p. 95)
    A. specular density
    B. diffuse density
    C. doubly diffuse density

13. The maximum slope of film characteristic curves occurs in the . . . (p. 97)

A. toe region (only)
B. shoulder region (only)
C. straight line region (only)
D. straight line and shoulder regions
E. base-plus-fog region (only)

14. "Total negative contrast" is a synonym for . . . (p. 100)
    A. log exposure range
    B. luminance ratio
    C. curve slope
    D. density range
    E. Dmax plus Dmin

15. The useful log exposure range is the distance along the horizontal axis between the . . . (p. 101)
    A. two ends of the straight line
    B. two ends of the characteristic curve
    C. minimum and maximum useful points

16. Gamma is defined as . . . (p. 103)
    A. $(\Delta \text{Log } H) / (\Delta D)$
    B. $(\Delta D) / (\Delta \text{Log } H)$
    C. $(\Delta D) \times (\Delta \text{Log } H)$
    D. $(\Delta D) + (\Delta \text{Log } H)$

17. Gamma is considered to be a measure of . . . (p. 104)
    A. subject contrast
    B. negative contrast
    C. printing contrast
    D. development contrast
    E. exposure contrast

18. In the relationship SPEED = 1/H, an appropriate unit for H is . . . (p. 106)
    A. footcandle
    B. second
    C. f-number
    D. lux-second
    E. meter-second

19. The lowest point on film curves where good shadow detail is retained is considered to be where the . . . (p. 106)
    A. log exposure is 0.1 above base plus fog
    B. slope is 0.1 above base plus fog
    C. density is .01 above base plus fog
    D. None of the above.

20. The ISO (ASA) film speed for a film that requires .0064 lux-seconds exposure to produce the specified density is . . . (p. 107)
    A. 100
    B. 125
    C. 160
    D. 200
    E. 400

21. In comparison with conventional black-and-white films, the speeds of conventional black-and-white papers are . . . (p. 110)
    A. fast
    B. slow
22. For proper reproduction of the darkest detailed subject shadows in reflection prints, they should be located on the paper curve at a density of . . . (p. 112)
    A. 0.04 above base plus fog
    B. 0.10 above base plus fog
    C. 0.60 above base plus fog
    D. 90% of the maximum density
    E. the maximum density
23. The main difference between a number 1 grade paper and a number 4 grade paper is in the . . . (p. 113)
    A. range of log exposures they can accept
    B. minimum density obtainable
    C. maximum density obtainable
    D. location of the speed point on the curve
24. For printing with diffusion enlargers, the negative density range should . . . (p. 114)
    A. be 0.6
    B. be 0.85
    C. match the density range of the paper
    D. match the log exposure range of the paper
    E. match the contrast index of the paper
25. Enlarger flare light affects print contrast . . . (p. 116)
    A. more in the highlights than in the shadows
    B. more in the shadows than in the highlights
    C. uniformly in all areas
26. The exposure time range that produces minimum reciprocity failure with pictorial films is . . . (p. 121)

A. 1 sec. to 60 sec.
B. 1/10 sec. to 1 sec.
C. 1/1000 sec. to 1/10 sec.
D. 1/100,000 sec. to 1/1000 sec.
27. To obtain the correct density and contrast with films exposed for long exposure times at low illuminance levels, it is necessary to . . . (p. 122)
    A. increase the exposure and increase the development
    B. increase the exposure and decrease the development
    C. decrease the exposure and increase the development
    D. decrease the exposure and decrease the development
    E. increase the exposure only
28. Mackie lines are associated with . . . (p. 125)
    A. the Herschel effect
    B. reciprocity effects
    C. solarization
    D. the Sabattier effect
    E. the Mackintosh effect
29. Edge effects produce . . . (p. 126)
    A. an increase in image sharpness
    B. a decrease in image sharpness
30. The byproducts of development act in a manner similar to the developer ingredient identified as the . . . (p. 126)
    A. developing agent
    B. activator
    C. preservative
    D. restrainer
    E. solvent
31. The danger of film damage due to static electricity is greater when the humidity is . . . (p. 130)
    A. high
    B. low

5 Photographic Optics

Leslie Stroebel. *Six Eyes.*

IMAGE FORMATION WITH A PINHOLE

The pinhole, as an image-forming device, has played an important role in the evolution of the modern camera. The observation of images formed by a small opening in an otherwise darkened room goes back at least to Aristotle's time, about 350 B.C.; and, indeed, the pinhole camera still fascinates many of today's photography students because of the simplicity with which it forms an image. The darkened room, or camera obscura, evolved into a portable room that could be moved about and yet was large enough to accommodate a person. The portable room in turn shrank to a portable box with a small opening and tracing paper, used as a drawing aid. By about 1570 the pinhole was replaced by a simple lens that produced a brighter image, which was easier to trace. The name *camera obscura* survived all these changes.

Pinholes are able to form images because light travels in a straight line. Thus for each point on an object, a reflected ray of light passing through the pinhole can fall on only one spot on the ground glass or film. Since light rays from the top and bottom of the scene and from the two sides cross at the pinhole, the image is reversed vertically and horizontally so that lettering in a scene will appear correct on the ground glass if it is viewed upside down (see Figure 5–1) and on film images viewed through the base.

There are two basic variables with a pinhole camera—the pinhole-to-film or image

A pinhole aperture on a view camera operates somewhat like a zoom lens.

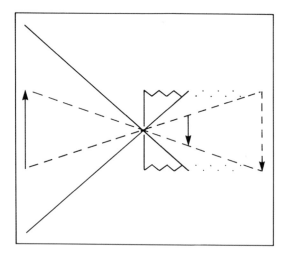

Figure 5–2   As the pinhole-to-film distance increases, the image size increases and the angle of view decreases.

distance and the diameter of the pinhole. If a pinhole is used on a view camera or other camera having a bellows, it is the equivalent of a zoom lens because the image size will increase in direct proportion to the pinhole-to-film distance. The angle of view, on the other hand, will decrease as the image distance increases (see Figure 5–2). Since a pinhole does not focus light as a lens does, changing the image distance has little effect on the sharpness of the image; but when the image is examined critically, it is found that there is an optimum pinhole-to-film distance for a pinhole of a given diameter.

Increasing the size of a pinhole from the optimum allows more light to pass through, therefore increasing the illuminance at the film plane and decreasing the required exposure time, but it also reduces the image sharpness. Decreasing the pinhole size, however, does not increase image sharpness. When the size is decreased beyond the optimum size for the specified pinhole-to-film distance, diffraction causes a decrease in sharpness (see Figure 5–3). The optimum pinhole size can be calculated with the formula

$$D = \frac{\sqrt{f}}{141}$$

where D is the diameter of the pinhole in inches and f is the pinhole-to-film distance

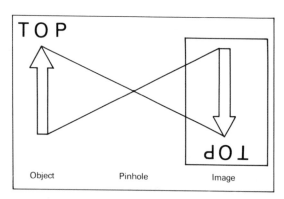

Figure 5–1   Reversal of the image vertically and horizontally by the crossing of the light rays at the pinhole produces correct reading of the lettering if viewed upside down from the rear.

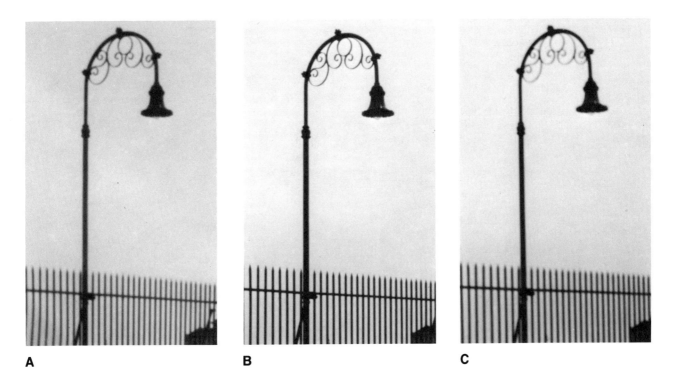

A   B   C

Figure 5–3 Three photographs made with pinholes of different sizes—one-half the optimum size (A), the optimum size (B), and two times the optimum size (C). The images represent small sections cropped from 8 × 10-inch photographs.

in inches (see Table 5–1).* Thus, for a pinhole-to-film distance of eight inches, the diameter of the optimum size pinhole is about 1/50 inch. A No. 10 needle will produce a pinhole of approximately this size.

If millimeters are used as the unit of measurement, the following formula should be used:

$$D = \frac{\sqrt{f}}{28}$$

A basic formula that takes into account the wavelength of the light is $D = \sqrt{2.5\lambda f}$, where λ is the average wavelength of the exposing radiation in millimeters. For white-light and panchromatic film a value of 500 nm (.00050 mm) is used. The optimum pinhole size is somewhat smaller for photographs made with ultraviolet radiation and somewhat larger for photographs made with infrared radiation. Doubling the wavelength or the pinhole-to-film distance will increase the size of the optimum pinhole by a factor of

Image definition varies with the size of the pinhole aperture and there is an optimum size for a given pinhole-to-film distance.

Table 5–1 Optimum pinhole diameters for different pinhole-film distances

(Distance) f	(Diameter) D	(F-Number) F-N
1 in.	1/140 in.	f/140
2 in.	1/100 in.	f/200
4 in.	1/70 in.	f/280
8 in.	1/50 in.	f/400
16 in.	1/35 in.	f/560

$$D = \frac{\sqrt{f}}{141} \text{ (in.)}$$

$$D = \frac{\sqrt{f}}{28} \text{ (mm.)}$$

$$F\text{-}N = \frac{f}{D}$$

*When photographing relatively close subjects, the value of f should be determined by substituting the object distance (u) and the image distance (v) in the formula $1/f = 1/u + 1/v$.

Figure 5–4   An anamorphic pinhole photograph (right) made with a vertical slit placed 1½ times as far from the film as a horizontal slit. The comparison photograph was made with a camera lens.

the square root of 2 or 1.4. Various references do not agree on the constant in the formula, and the 2.5 constant used above represents an average value.

Pinholes are usually made by pushing a needle through a piece of thin, opaque material, such as black paper, or drilling a hole in very thin metal. It is also possible to make a pinhole by crossing a vertical slit in one piece of thin material with a horizontal slit in another. The fact that this pinhole is square rather than round is of no great importance. Placing the vertical and horizontal slits at different distances from the film produces different scales of reproduction for the vertical and horizontal dimensions of objects being photographed, i.e., an anamorphic image. Since the vertical slit controls the horizontal image formation, placing the vertical slit at double the distance of the horizontal slit from the film will produce a horizontally elongated or stretched image with a 2:1 ratio of the two dimensions (see Figure 5–4).

Anamorphic images can be produced with a pinhole formed by vertical and horizontal slits at different distances from the film.

Positive lenses are thicker in the center than at the edges. Negative lenses will not form real images that can be recorded on film.

IMAGE FORMATION WITH A LENS

A positive lens produces an image by refracting light so that all the light rays falling on

the front surface of the lens from an object point converge to a point behind the lens (see Figure 5–5). If the object point is at infinity or a large distance, the light rays will enter the lens traveling parallel to each other, and the image point where they come to a focus is referred to as the *principal focal point*. The *focal length* can then be determined by measuring the distance from the principal focal point to the *back nodal point*—roughly the middle of a single-element lens (see Figure 5–6). Reversing the direction of light through the lens with a distant object on the right produces a second principal focal point and a second focal length to the left of the lens, as well as a second nodal point (see Figure 5–7). The two sets of terms are distinguished by the adjectives *object* (or *front*) and *image*

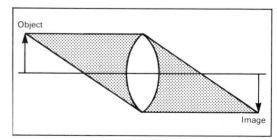

Figure 5–5   Image formation with a positive lens.

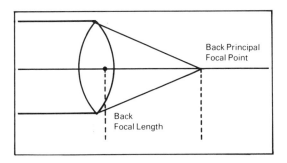

Figure 5–6   The back focal length is the distance between the back nodal point and the image of an infinitely distant object.

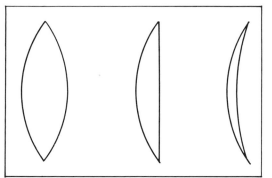

Figure 5–8   Three types of positive lenses—biconvex (left), plano-convex, and positive meniscus.

(or *back*)—e.g., front principal focal point and back principal focal point.

All positive lenses are thicker at the center than at the edges and must have one convex surface, but the other surface can be convex, flat, or concave (see Figure 5–8). The curved surfaces are usually spherical, like the outside (convex) or inside (concave) surface of a hollow ball. Actually, this is not the best shape for forming a sharp image, but it has been widely used because lenses having flat and spherical surfaces are easier to mass produce than those having other curved surfaces, such as parabolic. If the curvature of a spherical lens surface is extended to produce a complete sphere, the center of that sphere is then identified as the *center of curvature* of the lens surface. A straight line drawn through the two centers of curvature of the two lens surfaces is identified as the *lens axis* (see Figure 5–9). If one of the surfaces is flat, the lens axis is a straight line through the one center of curvature and perpendicular to the flat surface.

The *optical center* of a lens is a point on the lens axis where an off-axis undeviated ray of light crosses the lens axis. All rays of light that pass through the optical center are undeviated—i.e., they leave the lens traveling parallel to the direction of entry (see Figure 5–10). *Object distance* and *image distance* can be measured to the optical center when great precision is not needed; but when precision is required the object distance is measured from the object to the front (or object) nodal point, and the image distance is measured to the back (or image) nodal point.

The *front nodal point* can be located on a drawing by extending an entering undeviated ray of light in a straight line until it intersects the lens axis, and the *back nodal point* by extending the departing ray of light backward in a straight line until it intersects the lens axis, as shown in Figure 5–11. A convention (which is not always observed) is to place objects to the left of lenses in draw-

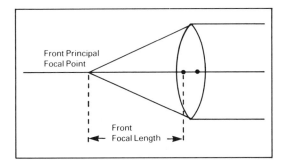

Figure 5–7   The front focal length is found by reversing the direction of light through the lens.

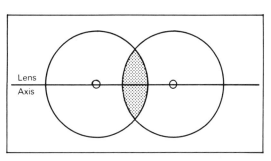

Figure 5–9   The lens axis is a straight line through the centers of curvature of the lens surfaces.

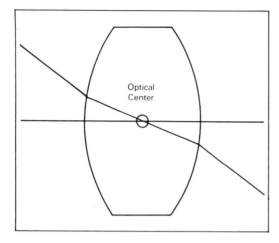

Figure 5–10 All rays of light passing through the optical center leave the lens traveling parallel to the direction of entry.

Figure 5–12 A simple nodal slide. The line on the side identifies the location of the pivot point on the bottom.

Nodal points are not marked on lenses by the manufacturers, but they can be located with a simple experiment.

ings and images to the right, so that the light travels in the same direction that our eyes move when reading.

Nodal points can also be located experimentally with actual lenses. The lens is first placed in a nodal slide, which is a device that enables a lens to be pivoted at various positions along its axis. For crude measurements, a simple trough with a pivot on the bottom is adequate (see Figure 5–12). Professional optical benches are used when greater precision is required (see Figure 5–13). The lens is first focused on a distant light source or the equivalent using a collimator with a

light source placed at an appropriate closer distance, with the front of the lens facing the light source. The lens is then pivoted from side to side while the image is observed. If the image does not move as the lens is pivoted, the lens is being pivoted about its back (image) nodal point. If the image does move, however, the lens is moved either toward or away from the light source, the image is brought into focus by moving the focusing screen or the microscope, and the lens is again pivoted. This procedure is repeated until the image remains stationary when the lens is pivoted. The front (object) nodal point is de-

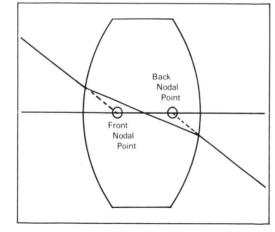

Figure 5–11 The nodal points can be located by extending the entering and departing parts of an unaeviated ray of light in straight lines until they intersect the lens axis.

Figure 5–13 A professional optical bench, which permits the aerial image formed by a lens to be examined with a microscope.

termined in the same way with the lens reversed so that the front of the lens is facing the image rather than the light source. A knowledge of the location of the nodal points can be useful, and it is unfortunate that the manufacturers of photographic lenses do not mark or otherwise indicate the location of the two nodal points on their lenses.

Three practical applications for knowledge of the location of the nodal points are:

1. When accuracy is required, the object distance must be measured from the object to the front nodal point, and the image distance from the image to the back nodal point. Accurate determination of the focal length requires measuring the distance from the sharp image of a distant object to the back nodal point. Measurements to the nodal points are also made when using lens formulas to determine image size and scale of reproduction (see Figure 5–14). With conventional lenses that are relatively thin, little error will be introduced by measuring to the physical center of the lens rather than to the appropriate nodal point. With thick lenses and some lenses of special design—such as telephoto, retrofocus, and zoom lenses—considerable error can be introduced by measuring to the center of the lens. With the latter types of lenses, the nodal points can be some distance from the physical center of the lens, and may even be in front of or behind the lens. When distances are measured for objects or images that are not on the lens axis, the correct procedure is to measure the distance parallel to the lens axis to the appropriate nodal plane rather than to the nodal point. The *nodal plane* is a plane that is perpendicular to the lens axis and

that includes the nodal point (see Figure 5–15).

2. If a view-camera lens can be mounted on the camera so that it tilts and swings about the rear nodal point, the image will remain in the same position on the ground glass as these adjustments are made to control the plane of sharp focus. Otherwise the image will move up or down as the lens is tilted and sideways as it is swung, requiring realignment of the camera to restore the original composition. Unfortunately, little consideration is given to this factor in the design of most view cameras and the fact that the nodal points are at different positions with various types of lenses complicates the task. Although a sliding adjustment, as on a nodal slide, is the ideal solution, some improvement may be obtained with certain lenses by using a recessed lens board, a lens cone, or by reversing the lens standard (on modular view cameras having that feature) to change the position of the lens in relation to the pivot point on the camera (see Figure 5–16).

3. The task of making graphical drawings to illustrate image formation with lenses is greatly simplified by the use of nodal points and planes rather than lens elements. Whereas lens designers must consider the effect that each surface of each element in a lens has on a large number of light rays, many problems involving image and object sizes and distances can be illustrated and solved by using the nodal planes to represent the lens in the

If a view-camera lens is tilted or swung about the image nodal point, the image will not move on the ground glass.

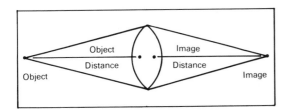

Figure 5–14 For an object point on the lens axis, object distance is measured to the front nodal point, and image distance is measured from the image to the back nodal point.

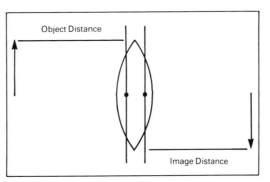

Figure 5–15 For an object point off the lens axis, object and image distances are measured in a direction that is parallel to the lens axis, to the corresponding nodal planes.

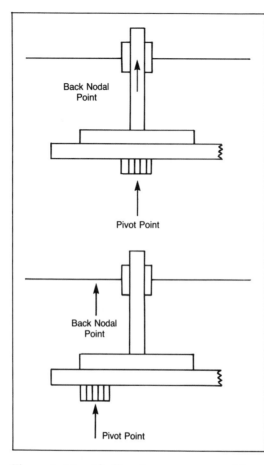

Figure 5–16 Ideally, view cameras would permit the lens pivot point to be adjusted so that all lenses could be rotated about the back nodal point.

drawing, regardless of the number of lens elements. This procedure will be used in the following section.

GRAPHICAL DRAWINGS

Graphical drawings are useful in that they not only illustrate image formation with lenses in simplified form, but they can be used as an alternative to mathematical formulas to solve problems involving image formation. The two drawings in Figure 5–17 show a comparison between the use of lens elements and the use of nodal planes. In the so-called thin-lens treatment, the two nodal planes are considered to be close enough to each other so that they can be combined into a single plane without significant loss of accuracy. If the drawing is to be used to solve a problem,

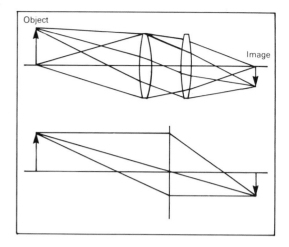

Figure 5–17 Graphical drawings to show image formation using the lens elements (top) and the simpler thin-lens procedure (bottom).

rather than as a schematic illustration of image formation, the drawing must be made either actual size or to a known scale. The original drawing in Figure 5–18 was actual size, but the reproduction is smaller. The steps involved in making a graphical drawing are:

1. Draw a straight horizontal line to represent the lens axis.
2. Draw a straight vertical line to represent the lens.
3. Place marks on the lens axis one focal length to the left and one focal length to the right of the lens. In this example, the focal length was 1 inch.
4. Draw the object at the correct distance from the lens. In this example, the object was 2 inches tall and was located 2 inches from the lens.
5. Draw the first ray of light from the top of the object straight through the optical center of the lens, i.e., the intersection of the lens axis and the nodal plane.
6. Draw the second ray, on a parallel to the lens axis, to the nodal plane, then through the back principal focal point.
7. Draw the third ray through the front principal focal point to the nodal plane, then parallel to the lens axis.
8. The intersection of the three rays represents the image of the top of the object. Draw a vertical line from that intersection to the lens axis to represent the entire (inverted) image of the object.

With a ruler, we could determine the correct size and position of the image from the original drawing. The image size was 2 inches and the image distance was 2 inches. From this we can generalize that placing an object two focal lengths in front of any lens produces a same-size image two focal lengths behind the lens.

This same drawing could be used as a one-quarter-scale drawing of a 4-inch focal length lens with an 8-inch-tall object located at an object distance of 8 inches. To determine the image size and image distance, the corresponding dimensions on the drawing are multiplied by 4 to compensate for the drawing's scale. Thus the image size is 2 inches × 4 = 8 inches, and the image distance is 2 inches × 4 = 8 inches.

Changing the distance between the object and the lens produces a change in the position where the image is formed. The relationship is an inverse one, so that as the object distance decreases the image distance increases. Since the two distances are interdependent and interchangeable, they are commonly referred to as conjugate distances. Image size also changes as the object and image distances change. Moving an object closer to the lens results in an increase in both the image distance and the image size. These relationships are illustrated in Figure 5–19.

The closest an object can be placed to a lens and still obtain a real image is theoretically slightly more than one focal length. Placing an object at exactly one focal length from the lens causes the light rays from an object point to leave the lens traveling parallel to each other so that we can think of an image only as being formed at infinity. In practice, the closest an object can be placed to a camera lens and still obtain a sharp image is determined by the maximum distance the film can be placed from the lens. Problems of this type can be solved with graphical drawings. If the maximum image distance is three inches for a camera equipped with a 2-inch focal length lens, an actual-size or scale drawing is made starting with the image located 3 inches to the right of the lens. Three rays of light are then drawn back through the lens, using the same rules as before, to determine the location of the object.

With lenses that are too thick to be considered as thin lenses, or where greater ac-

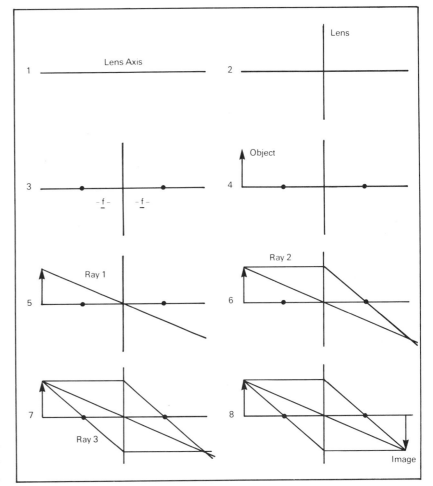

Figure 5–18 Making an actual-size or scale graphical drawing in eight steps.

Simple graphical drawings can be used to solve practical problems involving image size, scale of reproduction, and image and object distances.

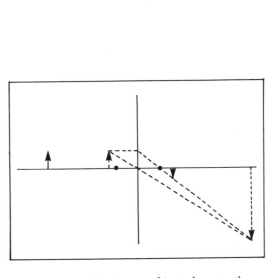

Figure 5–19 Moving an object closer to the lens results in an increase in both image distance and image size.

curacy is required, only small modifications of the thin-lens treatment are required. If it is known that the front and back nodal planes are separated by a distance of one inch in a certain lens, two vertical lines are drawn on the lens axis, to represent the two nodal planes, separated by the appropriate actual or scale distance. Then the three rays of light are drawn from an object point to the front nodal plane, as before, but they are drawn parallel to the lens axis between the two nodal planes before they converge to form the image (see Figure 5–20).

Angle of view can be determined with graphical drawings in addition to image and object distances and sizes. The term *angle of view* should not be confused with *angle of coverage*, a measure of the covering power of lenses that will be considered later in this chapter. Angle of view is a measure of how much of the scene will be recorded on the film as determined by the lens focal length and the film size. Angle of view is usually determined for the diagonal of the film, which is the longest dimension, although two angles of view are sometimes specified—one for the film's vertical dimension and one for the horizontal dimension. A horizontal line is drawn for the lens axis and a vertical line is drawn to represent the nodal planes, as with the thin-lens treatment above. A second vertical line is drawn one focal length (actual or scale distance) to the right of the nodal planes. The second vertical line represents the film diagonal, so it must be the correct actual or scale length.

The drawing in Figure 5–21 represents a 50 mm focal length lens on a 35 mm camera, where the diagonal of the negative image area

The angle of coverage of a lens must be at least as large as the angle of view of the lens-camera combination.

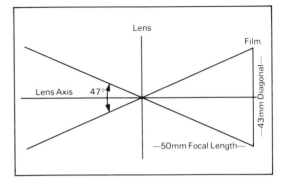

Figure 5–21　The angle of view of a lens-film combination can be determined by drawing a line with the length equal to the film diagonal at a distance of one focal length from the lens. The angle formed by the extreme rays of light can be measured with a protractor.

is approximately 43 mm. Lines drawn from the rear nodal point (i.e., the intersection of the nodal planes and the lens axis) to opposite corners of the film form an angle that can be measured with a protractor. No compensation is necessary for the drawing's scale, since there are 360° in a circle no matter how large the circle is drawn. The angle of view in this example is approximately 47°.

Two other focal length lenses, 15 mm and 135 mm, are represented in the drawing in Figure 5–22. The measured angles of view are approximately 110° and 18°. It should be apparent from these drawings that substituting a shorter focal length lens on a given camera will increase the angle of view, whereas substituting a smaller film, as in using a reducing back on a large-format camera, will decrease the angle of view. Angle of view is determined with the film placed one focal length behind the lens, which corresponds to focusing on infinity. When a camera is focused on nearer objects, the lens-to-film distance increases, and the *effective* angle of view decreases.

Graphical drawings are especially helpful for beginning photographers because they make it easy to visualize the relationships involved. With experience, it becomes more efficient to solve problems relating to distances and sizes using mathematical formulas. Most problems of this nature that are encountered by practicing photographers can be solved by using these four simple formulas:

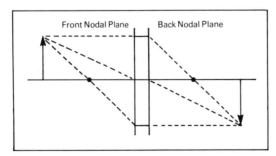

Figure 5–20　The nodal planes are separated by an appropriate distance for thick-lens graphical drawings.

1. $1/f = 1/u + 1/v$
2. $R = v/u$
3. $R = \dfrac{v - f}{f}$
4. $R = \dfrac{f}{u - f}$

where

f = focal length
u = object distance
v = image distance and
R = scale of reproduction,
    which is image size/object size, or I/O.

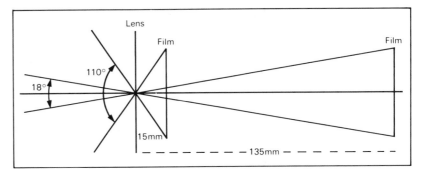

Figure 5–22 Angles of view for 15 mm and 135 mm focal length lenses with 35 mm film.

Thus I/O can be substituted for R in any of the formulas. Problems are typically solved by selecting the formula that contains two known factors plus the unknown. No two formulas contain the same three terms.

The focal length of an unmarked lens can be determined with formula No. 1 by forming a sharp image of an object, measuring the object and image distances, and solving the formula for f. Thus, if the object and image distances are both 8 inches, the focal length is 4 inches. The formula illustrates the inverse relationship between the conjugate distances u and v, whereby moving a camera closer to an object requires increasing the lens-to-film distance to keep the image in sharp focus.

It also illustrates that u and v are interchangeable, which means that sharp images can be formed with a lens in two different positions between an object and the film. For example, if a sharp image is formed when a lens is 8 inches from the object and 4 inches from the film, another (and larger) sharp image will be formed when the lens is placed 4 inches from the object and 8 inches from the film (see Figure 5–23). Exceptions to the statement that sharp images can be formed with a lens in two different positions are (a) when the object and image distances are the same, which produces an image that is the same size as the object; and (b) when an image distance is larger than the maximum bellows extension on the camera, so that the lens cannot be moved far enough away from the film to form the second sharp image.

A problem commonly encountered by photographers is determining how large a scale of reproduction (R) can be obtained with a certain camera-lens combination. If a view

camera has a maximum bellows extension (v) of 16 inches and is equipped with an 8-inch focal length (f) lens, formula No. 3 would be selected: $R = (v - f)/f = (16 - 8)/8 = 1$, where the image is the same size as the object. Although we would substitute a longer focal length lens to obtain a larger image of a distant object when the camera cannot be moved closer (see formula No. 4), a shorter focal length lens would be substituted to obtain a larger image in closeup work where the maximum bellows extension is the limiting factor. Replacing the 8-inch lens above with a 2-inch lens would increase the scale of reproduction from 1 to $(16 - 2)/2 = 7$.

The scale of reproduction for a photograph is equal to the image size divided by the object size, or, the image distance divided by the object distance.

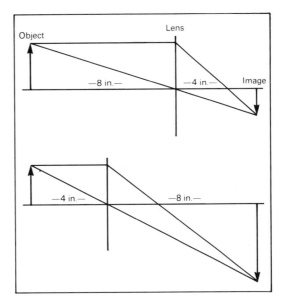

Figure 5–23 Since object and image distances are interchangeable, sharp images can be formed with a lens in two different positions between object and film.

PERSPECTIVE

Perspective refers to the appearance of depth when a three-dimensional scene is represented in a two-dimensional image such as a photograph, or when a scene is viewed directly. Photographers have a number of ways of creating the appearance of depth in photographs, including the following:

Depth of Field. Use of a limited depth of field, so that the images of objects in front of and behind the point focused on are unsharp, creates a stronger appearance of depth or perspective than when the entire scene appears sharp.

Lighting. Depth can be emphasized with lighting that produces a gradation of tones on curved surfaces, a separation of tones between the planes of box-shaped objects, and between objects and backgrounds; and that casts shadows of objects on the foreground or background.

Overlap. Arranging a scene so that a nearby object obscures part of a more distant object provides the viewer with a powerful clue as to the relative distances of the objects.

Aerial Haze. The scattering of light that occurs in the atmosphere makes distant objects appear lighter and less contrasty than nearby objects. Thick fog and smoke can create the illusion of depth with relatively small differences in distance.

Color. Red is referred to as an advancing color and blue as a receding color. In the absence of conflicting depth clues, a blue surface tends to appear farther away than a red surface at the same distance.

Stereophotography. Viewing two photographs taken from slightly different positions so that the left eye sees one and the right eye sees the other produces a realistic effect with strong depth clues similar to those produced with binocular vision and three-dimensional objects.

Holography. With a single photographic image, holograms present different images to the left and right eyes, as with stereo pairs, but they also produce motion parallax for objects at different distances when the viewer moves laterally.

Linear Perspective. Linear perspective is exemplified by the convergence of parallel subject lines, and the decrease in image size as the object distance increases. Linear per-

spective is so effective in representing depth in two-dimensional images that it is often the only type of depth clue provided by artists when making simple line drawings (see Figure 5–24).

LINEAR PERSPECTIVE

In photographs, changes in linear perspective are commonly associated with changes in camera position and focal length of camera lenses. If the definition of linear perspective were limited to the relative image size of objects at different distances (or the angle of convergence of parallel subject lines) the photographer's control would be limited to the choice of camera position (i.e., object distance). If changes in the appearance of linear perspective were included even when there is no change in the relative image sizes of objects at different distances, then lens focal length would have to be included as a control.

Two basic assumptions important to an understanding of the control of linear perspective are that image size is directly proportional

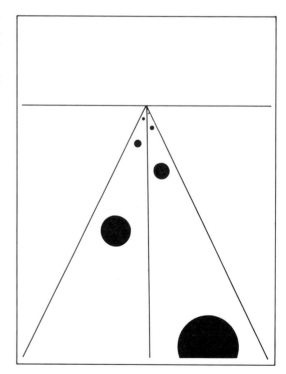

Figure 5–24   Depth is represented by the convergence of parallel subject lines and decreasing image size.

Photographers have many different ways of representing three-dimensional depth in two-dimensional photographs.

Image size is directly proportional to focal length and is inversely proportional to object distance.

Moving a camera farther away from a subject weakens the linear perspective.

to focal length, and that image size is inversely proportional to object distance.

OBJECT DISTANCE AND IMAGE SIZE

Placing two objects of equal size at distances of 1 foot and 2 feet from a camera lens produces images which vary in size in a 2:1 ratio. Since image size varies inversely with object distance, the smaller of the two images corresponds to the object at the larger distance. *Actual* image sizes could be determined for given object size, focal length and object distances with either graphical drawings or lens formulas, but for now we will be concerned only with *relative* image sizes.

Linear perspective is based on the ratio of image sizes for objects at different distances. Figure 5–25 shows two objects of equal size at distance ratios of 1:2, 1:3, and 1:4. The sizes of the resulting images are in ratios of 2:1, 3:1, and 4:1. Thus, with movable objects the relative image sizes and linear perspective can be controlled easily by changing the positions of the objects. Most of the time, however, we photograph objects that cannot be moved easily, and therefore we must resort to moving the camera or changing the focal length of the lens if we want to alter image size.

Starting with two objects at a distance ratio of 1:2 from the camera, moving the camera farther away to double the distance from the closer object does not double the distance to the farther object, and therefore the ratio of the image sizes will not remain the same. Since the ratio of object distances changes from 1:2 to 2:3 by moving the camera, the ratio of image sizes changes from 2:1 to 3:2 (or 1½:1) (see Figure 5–26). Moving the camera farther away not only reduces the size of both images but also makes them more nearly equal in size. The two images can never be made exactly equal in size, no matter how far the camera is moved away, but with very large object distances the differences in size can become insignificant. The linear perspective produced by moving the camera farther from the objects is referred to as a weaker perspective than that produced with the camera in the original position. Thus, weak perspective can be ascribed to a picture in which

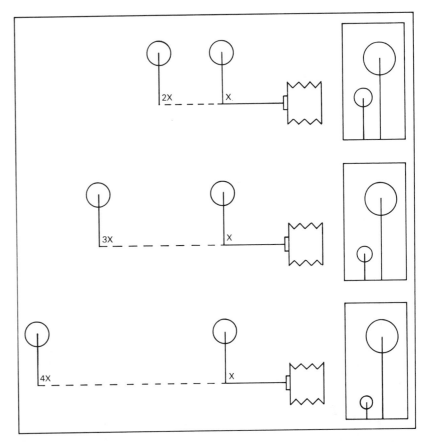

Figure 5–25   Image size is inversely proportional to object distance. The ratios of the object distances from top to bottom are 2:1, 3:1, and 4:1.

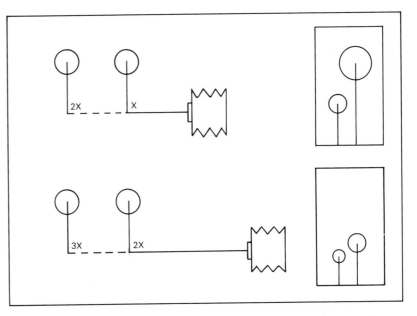

Figure 5–26   Doubling the distance from the camera to the near object changes the ratio of distances to the two objects from 1:2 to 2:3.

image size decreases more slowly with increasing object distance than expected. Another aspect of weak perspective is that space appears to be compressed, as though there were less distance between nearer and farther objects than actually exists (see Figure 5–27).

Conversely, moving a camera closer to two objects increases the image size of the nearer object more rapidly than that of the farther object, producing a stronger perspective. For example, with objects at a distance ratio of 1:2, moving the camera in to one-half the original distance to the near object doubles its image size but reduces the distance to the farther object from 2 to 1½, therefore increasing the image size of the farther object to only 1⅓ times its original size. Moving the camera closer to the subject produces a stronger linear perspective whereby image size decreases more rapidly with increasing object distance, and the space between the closer and farther objects appears to increase. Strong perspective is especially flattering to architectural photographs of small rooms because it makes the rooms appear more spacious.

A word of caution about the assumption that image size varies inversely with object distance: This relationship holds as long as the lens-to-film distance remains the same for the two objects. It will not hold when the back of a view camera is swung or tilted. Indeed, the purpose of these adjustments is to control the shape of the image. Neither will the relationship hold when separate photographs are made of each of the two objects and the camera is focused in turn on each object distance, changing the lens-to-film distance. This is of significance mostly with closeup photography. For example, with objects at distances of 8 inches and 16 inches from a camera equipped with a 4-inch focal length lens, the ratio of image sizes is 2:1 when the objects are photographed together but 3:1 when they are photographed separately and the camera is refocused. To solve this problem for other object distances, use the lens formula $R = f/(u - f)$.

Strong perspective makes a room appear larger in a photograph, but it is inappropriate for a formal portrait of a person.

Figure 5–27   Space appears to be compressed in the bottom photograph, made with a 1500 mm lens, compared to the top photograph, made with a 50 mm lens from a closer position.

FOCAL LENGTH AND IMAGE SIZE

Serious photographers have more than one focal length lens per camera in order to control image size and the corresponding angle of view. As stated above, image size is directly proportional to focal length. Thus, if an image of a building is 1/2-inch high in a photograph made with a 50 mm focal length lens on a 35 mm camera, substituting a 100 mm focal length lens will produce an image of the building that is 1 inch high. Since this is a direct rather than inverse relationship, doubling the focal length will double the image size. Also, since doubling the focal length will double the size of all parts of the image, it will not change the ratio of image sizes for objects at different distances (see Figure 5–28). This can be demonstrated convincingly

Figure 5–28   Photographs made from the same position with 135 mm (left), 55 mm (middle), and 28 mm (right) focal length lenses. Image size changes in proportion to focal length, but relative sizes for objects at different distances remain constant.

with a camera equipped with a zoom or variable focal length lens. As the focal length is changed, one can see the overall image change in size, but the relative sizes of images of objects at different distances remain the same. (Although the relative sizes remain the same, the linear perspective may appear to change, a phenomenon we will examine in the section on viewing distance.)

There is also an exception to the assumption that image size is directly proportional to focal length. The assumption holds when the image distance is approximately equal to the focal length—that is, when photographing objects at moderate to large distances. When photographing objects at close distances, the lens-to-film distance must be increased to focus the image, and the size relationship with different focal length lenses deviates from that predicted by the assumption. For example, with an object at a distance of 16 inches from a camera equipped first with an 8-inch focal length lens and then with a 4-inch focal length lens, the ratio of image sizes will be 3:1, rather than the expected 2:1. The same lens formula $R = f/(u - f)$ is used to solve this problem (see Figure 5–29).

CHANGING OBJECT DISTANCE AND FOCAL LENGTH

Photographers commonly change focal length and object distance simultaneously to control linear perspective and overall image size. For example, if the perspective appeared too strong and unflattering in a portrait made with a normal focal length lens, the photographer would substitute a longer focal length lens and move the camera farther from the subject to obtain about the same size image but one with weaker perspective. Because short focal length wide-angle lenses tend to be used with the camera placed close to the subject, and long focal length telephoto lenses tend to be used with the camera at relatively large distances, strong perspective is often associated with wide-angle lenses and weak perspective is similarly associated with telephoto lenses—although it is the camera position and

Linear perspective is determined by the camera-to-subject distance, not by the focal length of the lens.

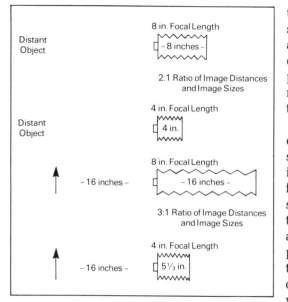

Figure 5–29   Image size is proportional to lens focal length with distant objects (top). With small object distances the ratio of the image sizes is larger than the ratio of the focal lengths (bottom).

Changing the viewing distance of a photograph can make the perspective appear stronger or weaker.

not the focal length or type of lens that produces the abnormal linear perspective.

The change in linear perspective with a change in object distance is seen most vividly when a corresponding change is made in the focal length to keep an important part of the scene the same size. In Figure 5–30, for example, the focal length of the lenses and the camera positions were adjusted to keep the images of the nearer object the same size, and the difference in linear perspective is revealed by the difference in size of the images of the farther object.

In situations where a certain linear perspective contributes significantly to the photograph's effectiveness, the correct procedure is first to select the camera position that produces the desired perspective, and then to select the focal length lens that produces the desired image size. For example, if the photographer wants to frame a building with a tree branch in the foreground, the camera must be placed in the position that produces the desired size and position relationships between the branch and the building. The lens is then selected that forms an image of an appropriate size. A zoom lens offers the advantage of providing any focal length be-

tween the limits. With fixed focal length lenses, if the desired focal length is not available, and changing the camera position would reduce the effectiveness due to the change in perspective, the best procedure is to use the next shorter focal length lens available and then enlarge and crop the image.

Cameras cannot always be placed at the distance selected on the basis of linear perspective. Whenever photographs are made indoors, there are physical limitations on how far away the camera can be placed from the subject. Fortunately, the strong perspective that results from using short focal length wide-angle lenses at the necessarily close camera positions enhances rather than detracts from the appearance of many architectural and other subjects. There are also many situations where the camera must be placed at a greater distance from the subject than would be desired. This, of course, applies to certain sports activities where cameras cannot be located so close that they interfere with the sporting event, block the view of spectators, or endanger the photographer.

Not all subjects are such that the perspective changes with object distance. Since two-dimensional objects have no depth, photographs of such objects reveal no change in the relative size of different parts of the image with changes in camera distance. Also, photographic copies of paintings, photographs, printed matter, etc., made from a close position with a short focal length wide-angle lens, and from a distant position with a long focal length telephoto lens, should be identical.

VIEWING DISTANCE

It would seem that the distance at which we view photographs should have no effect on linear perspective, since a 2:1 size ratio of images of objects at different distances will not be altered by changes in the viewing distance. In practice, however, changes in viewing distance can have a significant effect on the perspective, provided the photographs contain good depth clues. Photographs of two-dimensional objects appear to change little with respect to linear perspective when viewing distance is changed, whereas those containing dominant objects in the fore-

Figure 5–30   Photographs made with a 135 mm focal length lens (left) and a 28 mm focal length lens (right) with the camera moved closer to match the image size of the foreground object. The change in the relative sizes of the two objects is due to the change in the camera position.

ground and background or receding parallel lines can change dramatically.

We seldom experience unnatural-appearing linear perspective in real life. Abnormally strong or weak perspective tends to occur only when we look at photographs or other two-dimensional representations of three-dimensional objects or scenes. The reason perspective appears normal when we view a three-dimensional scene directly is that as we change the viewing distance, the perspective and the image size change simultaneously in an appropriate manner. Because we normally know whether we are close to or far away from the scene we are viewing, the corresponding large or small differences in apparent size of objects at different distances seems normal for the viewing distance. (It is possible to deceive viewers about the size or shape of objects or their distances, but we will discuss such illusions later.)

To illustrate how the situation changes when we view photographs rather than actual three-dimensional scenes, assume that two photographs are made of the same scene, one with a normal focal length lens and the other with a short focal length lens with the camera moved closer to match the image size of a foreground object (see Figure 5–30). Looking at the two photographs, viewers suppose that they are at the same distance from the two scenes because the foreground objects are the same size, but if the perspective appears normal in the first photograph, the stronger perspective in the second photograph will appear abnormal for what is assumed to be the same object distance. Viewers can make the perspective appear normal in the second photograph, however, by reducing the viewing distance. The so-called "correct" viewing distance is equal to the focal length of the camera lens (or, more precisely, the image distance) for contact prints, or the focal length multiplied by the magnification for enlarged prints.

The correct viewing distance for a contact print of an 8 × 10-inch negative exposed with a 12-inch focal length lens is 12 inches. Since we tend to view photographs from a distance about equal to the diagonal, the perspective would appear normal to most viewers. If the 12-inch lens were replaced with a 6-inch fo-

The correct viewing distance of a photograph for a normal appearing perspective is equal to the camera lens focal length multiplied by the negative-to-print magnification.

cal length lens, the print would have to be viewed from a distance of 6 inches for the perspective to appear normal. When the print is viewed from a comfortable distance of 12 inches, the perspective will appear too strong. Conversely, the perspective in a photograph made with a 24-inch focal length lens would appear too weak when viewed from a distance of 12 inches. It is fortunate that people tend to view photographs from standardized distances rather than adjusting the viewing distance to make the perspective appear normal, for that would deprive photographers of one of their most useful techniques for making dramatic and effective photographs.

THE WIDE-ANGLE EFFECT

Closely related to the way a change in viewing distance can change the perspective of photographs is the influence of viewing distance on the so-called wide-angle effect. The wide-angle effect is characterized by what appear to be distorted image shapes for three-dimensional objects near the corners of photographs. This effect is especially noticeable in group portraits where heads appear to be stretched out of shape in directions radiating away from the lens axis or the center of the photograph. Thus, heads near the sides seem to be too wide, those near the top and bottom seem to be too long, and those near the corners appear to be stretched diagonally (see Figure 5–31).

Such stretching occurs because rays of light from off-axis objects strike the film at oblique angles rather than at a right angle, as occurs at the lens axis. If the subject consists of balls or other spherical objects, the amount of stretching can be calculated in relation to the angle formed by the light rays that form the image and the lens axis. Thus, at an off-axis angle of 25° the image is stretched about 10%, and at 45° the image is stretched about 42%. (The image size of off-axis objects changes in proportion to the secant of the angle formed by the central image-forming ray of light with the lens axis. The reciprocal of the cosine of the angle may be substituted for the secant.) Normal focal length lenses, where the focal length is about equal to the diagonal of the film, have an angle of view of approximately 50°, or a half angle of 25°.

The wide-angle effect refers to elongated images of objects near the edges of photographs made with short focal length wide-angle lenses.

Figure 5–31   The image of a spherical light globe in this cropped section of a photograph taken with a short focal length lens is stretched approximately 60% in a direction away from the lens axis.

Why don't we notice the 10% stretching that occurs with normal focal length lenses? It is not because a 10% change in the shape of a circle is too small to be noticed, but rather that when the photograph is placed at the correct viewing distance, the eye is looking at the images at the edges at the same off-axis angle as the angle of the light rays that formed the image in the camera. Thus, the elliptical image of the spherical object is seen as a circle when the ellipse is viewed obliquely. The effect would be the same even with the more extreme stretching produced with short focal length wide-angle lenses if the photographs were viewed at the so-called correct viewing distance. Since the correct viewing distance is uncomfortably close for photographs made with short focal length lenses, people tend to view them from too great a distance, where the stretching is apparent.

One might question why the wide-angle effect does not occur when photographing circles drawn on paper as it does when photographing three-dimensional objects. The distinction is that if one looks at a row of balls from the position of the camera lens, the outline shape of all of the balls will appear circular, whether they are on or off axis, but off-axis circles drawn on paper will appear elliptical in shape because they are viewed obliquely. The compression that produces the ellipse when the circle is viewed from the lens is exactly compensated for by stretching when the image is formed because the light falls on the film at the same oblique angle at which it leaves the drawing (see Figure 5–32).

DEPTH OF FIELD

Camera lenses can be focused on only one object distance at a time. Theoretically, objects in front of and behind the object distance focused on will not be imaged sharply on the film. In practice, acceptably sharp focus is seldom limited to a single plane. Instead, objects somewhat closer and farther away appear sharp. Depth of field is defined as the range of object distances within which objects are imaged with acceptable sharpness. Depth of field is not limited to the plane focused on because the human eye has limited resolving power, so that a circle up to a certain size appears as a point. The largest circle that appears as a point is referred to as the *permissible circle of confusion*.

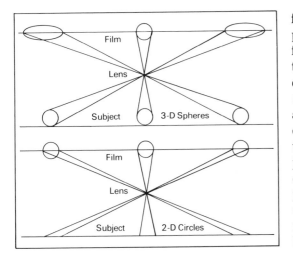

Figure 5–32 The wide-angle effect results in elongation of off-axis images of three-dimensional objects (top). Images of two-dimensional objects such as circles in a painting or photograph are not affected (bottom).

PERMISSIBLE CIRCLE OF CONFUSION

The size of the largest circle that appears as a point depends upon the viewing distance. For this reason permissible circles of confusion are generally specified for a viewing distance of 10 inches, and 1/100 inch is commonly cited as an appropriate value for the diameter. It is apparent that even at a fixed distance the size of the permissible circle of confusion will vary with such factors as differences in individual eyesight, the tonal contrast between the circle and the background, the level of illumination, and the viewers' criteria for sharpness. Nevertheless, when a lens manufacturer prepares a depth-of-field table or scale for a lens, these variables must be ignored. A single value is selected for the permissible circle of confusion that seems appropriate for the typical user of the lens. Although many photographic books that include the subject of depth of field accept 1/100 inch as being appropriate, it should not be surprising that there is less agreement among lens manufacturers. A study involving a small sample of cameras designed for advanced amateurs and professional photographers revealed that values ranging from 1/70 to 1/200 inch were used—approximately a 3:1 ratio. Two different methods will be considered for evaluating depth-of-field scales and tables for specific lenses.

One procedure is to photograph a flat surface that has good detail at an angle of approximately 45°, placing markers at the point focused on and at the near and far limits of the depth of field as indicated by the depth-of-field scale or table, as shown in Figure 5–33. The first photograph should be made at an intermediate f-number, with additional exposures bracketing the f-number one and two stops on both sides, with appropriate adjustments in the exposure time. A variation of this procedure is to use three movable objects in place of the flat surface, focusing on one and placing the other two at the near and far limits of the depth of field as indicated by the scale or table.

To judge the results, 6 × 8-inch or larger prints should be made without cropping and viewed from a distance equal to the diagonal of the prints. The diagonal of a 6 × 8-inch print is 10 inches, which is considered to be

Depth of field is the range of distances within which objects are imaged with acceptable sharpness.

When a camera is focused on the hyperfocal distance, everything should appear sharp in the photograph from infinity to one-half the hyperfocal distance.

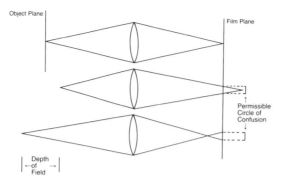

Figure 5-33A   Depth of field is the range of distances within which objects are imaged with acceptable sharpness. At the limits, object points are imaged as permissible circles of confusion.

Figure 5-33B   Depth-of-field scales and tables can be checked by photographing an object at an angle with markers placed at the point focused on and at the indicated near and far limits of the depth of field. A 6 × 8-inch or larger print should be made so that it can be viewed from the "correct" viewing distance.

the closest distance at which most people can comfortably view photographs or read. If the photograph made at the f-number specified by the depth-of-field scale or table has either too little or too much depth of field when viewed at the correct distance, the photograph that best meets the viewer's expectation should be identified from the bracketing series. A corresponding adjustment can be made when using the depth-of-field scale or table in the future.

The second procedure involves determining the diameter of the permissible circle of confusion used by the lens manufacturer in calculating the depth-of-field scale or table.

It is not necessary to expose any film with this procedure. Instead, substitute values for the terms on the right side of the formula $C = f^2/(N \times H)$ and solve for C, where C is the diameter of the permissible circle of confusion on the film, f is the focal length of the lens, N is any selected f-number, and H is the hyperfocal distance at that f-number.

Hyperfocal distance can be defined in either of two ways: (a) the closest distance that appears sharp when a lens is focused on infinity; or (b) the closest distance that can be focused on and have an object at infinity appear sharp. Although the two procedures are different, the results will be essentially the same. We should also note that when a lens is focused on the hyperfocal distance, the depth of field extends from inifinity to one half the hyperfocal distance (see Figure 5-34). If we select f/16 as the f-number with a 2 inch (50 mm) focal length lens, the hyperfocal distance can be determined either from a depth-of-field table or from a depth-of-field

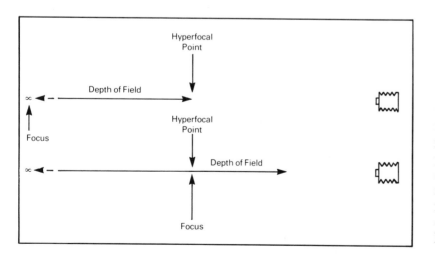

Figure 5-34   The hyperfocal distance is the closest distance that appears sharp when a lens is focused on infinity (top), or the closest distance that can be focused on and have an object at infinity appear sharp (bottom).

scale on the lens or camera by noting the near distance sharp at f/16 when the lens is focused on infinity. (If the near-limit marker on a DOF scale falls between two widely separated numbers, making accurate estimation difficult, set infinity opposite the far-limit marker and multiply the distance opposite the near-limit marker by two, as illustrated in Figure 5–35.)

Since the circle of confusion is commonly expressed as a fraction of an inch, the hyperfocal distance and the focal length must be in inches. The hyperfocal distance at f/16 for the lens illustrated is 24 feet or 288 inches. Substituting these values in the formula $C = f^2/(N \times H)$ produces $2^2/16 \times 288$ or 1/1,152 inch. This is the size of the permissible circle of confusion on the negative, but a 35 mm negative must be magnified six times to make a 6 × 8-inch print to be viewed at 10 inches. Thus, $6 \times 1/1,152 = 1/192$ inch, or approximately half as large a permissible circle of confusion as the 1/100-inch value commonly used.

It is important to note that the size of the permissible circle of confusion used by a lens manufacturer in computing a depth-of-field table or scale tells us nothing about the quality of the lens itself. The manufacturer can arbitrarily select any value, and in practice a size is selected that is deemed appropriate for the typical user of the lens. If a second depth-of-field scale is made for the lens in the preceding example based on a circle with a diameter of 1/100 inch rather than approximately 1/200 inch, the new scale would indicate that it is necessary to stop down only to f/8 in a situation where the original scale specified f/16. Lens definition, however, is determined by the quality of the image for the object focused on, not the near and far limits of the indicated depth of field.

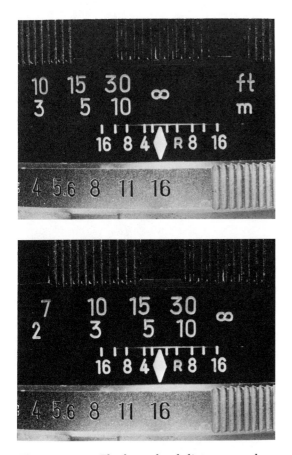

Figure 5–35 The hyperfocal distance can be determined from a depth-of-field scale either by focusing on infinity and noting the near distance sharp at the specified f-number (top) or by setting infinity opposite the far-distance sharp marker and multiplying the near distance sharp by two (bottom).

DEPTH-OF-FIELD CONTROLS

Photographers have three basic controls over depth of field: f-number, object distance, and focal length. Since viewing distance also affects the apparent sharpness of objects in front of and behind the object focused on, it is generally assumed that photographs will be viewed at a distance about equal to the diagonal of the picture. At this distance, depth of field will not be affected by making different-size prints from the same negative. For example, the circles of confusion at the near and far limits of the depth of field will be twice as large on a 16 × 20-inch print as on an 8 × 10-inch print from the same negative, but the larger print would be viewed from double the distance, making the two prints appear to have the same depth of field. If the larger print were viewed from the same distance as the smaller print, it would appear to have less depth of field. Cropping when enlarging will decrease the depth of field because the print size and viewing distance will not increase in proportion to the magnification.

Photographers can change the depth of field in photographs by changing f-number, the object distance, or the focal length of the camera lens.

DEPTH OF FIELD AND F-NUMBER

Depth of field is directly proportional to the f-number.

Fortunately, the relationship between f-number and depth of field is a simple one, whereby doubling the f-number doubles the depth of field. In other words, depth of field is directly proportional to the f-number, or $D_1/D_2 = N_1/N_2$. Thus, if a certain lens has a range of f-numbers from f/2 to f/22, the ratio of the depth of field at these settings would be D_1/D_2 = f/22/f/2 = 11/1. Changing the f-number is generally the most convenient method of controlling depth of field, but occasionally insufficient depth of field is obtained with a lens stopped down to the smallest diaphragm opening or too much depth of field is obtained with a lens wide open. In these circumstances other controls must be considered.

DEPTH OF FIELD AND OBJECT DISTANCE

Depth of field increases rapidly as the distance between the camera and the subject increases. For example, doubling the object distance makes the depth of field four times as large. The differences in depth of field with very small and very large object distances are dramatic. In photomacrography, where the camera is at a distance of two focal lengths or less from the subject, the depth of field at a large aperture sometimes appears to be confined to a single plane (see Figure 5–36).

At the other extreme, by focusing on the hyperfocal distance, depth of field extends from infinity to within a few feet of the camera with some lenses (see Figure 5–37). The mathematical relationship between depth of field and object distance (provided the object distance does not exceed the hyperfocal distance) is represented by the formula $D_1/D_2 = U_1^2/U_2^2$. For example, if two photographs are made with the camera 5 feet and 20 feet from the subject, the ratio of the depths of field will be

$$\frac{D_1}{D_2} = \frac{20^2}{5^2} \text{ or } \left(\frac{20}{5}\right)^2 = \frac{16}{1}$$

If, however, object distance is increased to obtain a larger depth of field when a camera

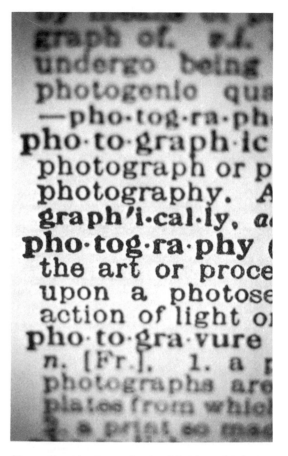

Figure 5–36   Since depth of field varies in proportion to the object distance squared, photographs made at a scale of reproduction of 1:1 and larger tend to have a shallow depth of field.

lens cannot be stopped down far enough, it is necessary to take into account the enlarging and cropping required to obtain the same image size as with the camera in the original position. There is still a net gain in depth of field in moving the camera farther from the subject, even though some of the increase is lost when the image is cropped in printing. The net gain is represented by the formula $D_1/D_2 = U_1/U_2$, which is the same as the preceding formula with the square sign removed.

DEPTH OF FIELD AND FOCAL LENGTH

There is an inverse relationship between focal length and depth of field, so that as focal length increases, depth of field decreases. Before specifying the relationship more exactly, however, it is necessary to distinguish between situations where the different focal length lenses are used on different format cameras, such as 35 mm and 8×10-inch, and where the lenses are used on the same camera. When a large-format camera and a small-format camera are each equipped with a normal focal length lens, i.e., one with a focal length about equal to the film diagonal, the depth of field will be inversely proportional to the focal length. For example, a 2-inch (50 mm) focal length lens on a 35 mm camera will produce about six times the depth of field of a 12-inch (305 mm) lens on an 8×10-inch camera. Even though enlarging negatives does not affect the depth of field, it is easier to compare the depth of field on two prints when they are the same size. Thus it would be appropriate to make an 8×10-inch enlargement from the 35 mm negative and a contact print from the 8×10-inch negative. The above relationship of depth of field and focal length is expressed as

$$\frac{D_1}{D_2} = \frac{f_2}{f_1}$$

Using the same two lenses on one camera produces a more dramatic difference in depth of field. It is now necessary to square the focal lengths so that $\dfrac{D_1}{D_2} = \dfrac{f_2^{\,2}}{f_1^{\,2}}$. Comparing the depth of field produced with 50 mm and 300 mm focal length lenses on a 35 mm camera, the ratio of the focal lengths is 1:6, and the ratio of the depths of field is 36:1. The great increase in depth of field with the shorter lens evaporates, however, if the camera is moved closer to the subject to obtain the same size image on the film as with the longer lens. In this example the camera would be placed at distances having a ratio of 1:6 to obtain the same image size with the 50 mm and 300 mm lenses, and we recall that depth of field increases with the distance squared, so that the $36\times$ increase obtained with the shorter lens would be exactly offset by the reduction in object distance.

There is still a net gain in using a shorter focal length lens on the same camera if the negative is enlarged and cropped to obtain the same image size and cropping on a print as that produced with a longer lens. This is essentially the same situation as using different format cameras each with a normal focal length lens.

Figure 5–37 Focusing on the hyperfocal distance produces a depth of field that extends from infinity to one-half the hyperfocal distance. (Photograph by John Johnson.)

DEPTH OF FOCUS

Depth of focus can be defined as the focusing latitude when photographing a two-dimensional subject. In other words, it is the distance the film plane can be moved in both directions from the optimum focus before the circles of confusion for the image of an object point match the permissible circle of confusion used to calculate depth of field. It is important to note that for depth-of-field calculations it is assumed the film occupies a single flat plane, and for depth-of-focus calculations it is assumed that the subject occupies a single flat plane (see Figure 5–38). If a three-dimensional object completely fills the depth-of-field space, there is only one position for the film, and there is in effect no depth of focus and no tolerance for focusing errors.

With a two-dimensional subject, the depth of focus can be found by multiplying the permissible circle of confusion by the f-number by 2. Thus, using 1/200 inch for the permissible circle of confusion on a 6×8-inch print

A 50 mm focal length camera lens produces four times as much depth of field as a 100 mm focal length lens, with all other conditions remaining the same.

Depth of focus refers to the focusing latitude when photographing a two-dimensional subject.

The tilt and swing adjustments on view cameras provide control over image shape and the angle of the plane of sharp focus.

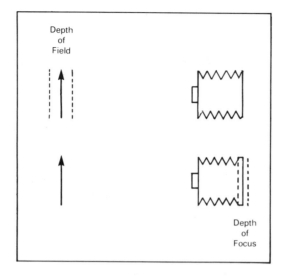

Figure 5–38   Depth-of-field calculations are based on the assumption that the two-dimensional film is in the position of optimum focus (top). Depth-of-focus calculations are based on the assumption that the subject is limited to a two-dimensional plane (bottom).

or 1/1,200 inch on a 35 mm negative, the depth of focus is $C \times N \times 2 = 1/1,200 \times f/2 \times 2 = 1/300$ inch. It can be seen from this formula that depth of focus varies in direct proportion to the f-number, as does depth of field.

Whereas depth of field decreases as a camera is moved closer to the subject, depth of focus increases. This is because as the object distance decreases, the lens-to-film distance must be increased to keep the image in sharp focus, and this increases the effective f-number. It is the effective f-number, not the marked f-number, that is used in the formula.

Whereas increasing focal length decreases depth of field, it has no effect on depth of focus. The explanation is that at the same f-number the diameter of the effective aperture and the lens-to-film distance both change in proportion to changes in focal length so that the shape of the cone of light falling on the film remains unchanged. Since focal length does not appear in the formula $C \times N \times 2$, it has no effect on depth of focus.

Although changing film size would not seem to affect depth of focus, using a smaller film reduces the correct viewing distance and, therefore, the permissible circle of confusion. Substituting a smaller value for C in the formula $C \times N \times 2$ reduces the depth of focus.

VIEW-CAMERA MOVEMENTS

The basic view-camera movements include (1) an adjustment of the distance between the lens and the film to permit focusing on objects over a wide range of distances and to accommodate lenses having a wide range of focal lengths, (2) vertical and horizontal perpendicular movements of the lens and film to provide control over the positioning of the image on the film without altering image shape or the angle of the plane of sharp focus, and (3) tilt and swing movements of the lens and film to provide control over image shape and the angle of the plane of sharp focus. Some view cameras also have revolving backs that allow the film to be rotated in the film plane to provide angular control of cropping.

PERPENDICULAR MOVEMENTS

The circular area within which satisfactory image definition can be obtained is called the *circle of good definition*. The circle of good definition is one measure of the covering power of a lens (see Figures 5–39 and 5–40). If the circle of good definition is somewhat larger than the film, it will be possible to select different parts of the image within the circle to record on the film. View cameras typically have rising–falling adjustments to move the lens and/or film up and down, and lateral shifts to move the lens and/or film from side to side.

Angle of coverage is a second measure of the covering power of a lens, and can be determined by drawing straight lines from op-

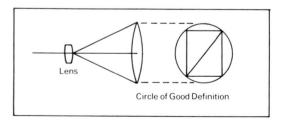

Figure 5–39   The diameter of the circle of good definition of a lens must be at least as large as the diagonal of the film.

posite sides of the circle of good definition to the back nodal point of the lens. Changes in object and image distances affect the size of the circle of good definition, but do not affect the angle of coverage (see Figure 5–41).

BACK MOVEMENTS AND IMAGE SHAPE

When the front of a long building or box-shaped object is photographed at an oblique angle, the near end is taller than the far end in the photograph, and the horizontal lines converge toward the far end. If the back of the camera is swung (i.e., rotated about a vertical axis) parallel to the front of the object, the near and far ends will be the same size and the horizontal lines will be parallel in the photograph regardless of the camera-to-object distance (see Figure 5–42). Conversely, the ratio of image sizes and the convergence of the horizontal lines can be increased by swinging the back of the camera in the other direction, away from being parallel to the front of the object.

Tilting a camera upward to photograph a tall building causes the vertical subject lines to converge in the photograph for the same reasons as do horizontal lines, since the top of the building is at a greater distance from the camera than the bottom. To make the image lines parallel in this situation, the back is tilted (rotated about a horizontal axis) until it is parallel to the vertical subject lines, and the convergence can be exaggerated by tilting the back in the opposite direction. It would

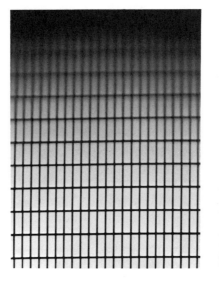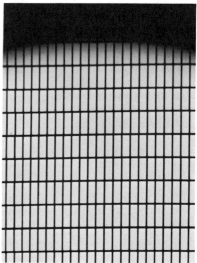

Figure 5–40 Edges of the circle of good definition and the circle of illumination of a camera lens at the maximum aperture (left) and the minimum aperture (right). Stopping down increases the size of the circle of good definition and the abruptness of the transition from the illuminated area to the nonilluminated area.

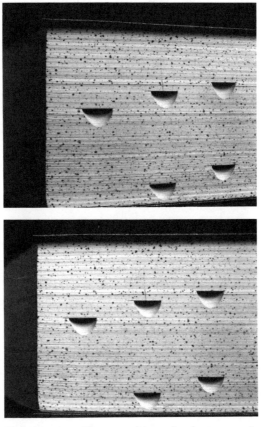

Figure 5–42 Photographing a book at an angle to show the front edge and one end causes the horizontal lines to converge with increasing distance (top). Swinging the back of the camera parallel to the front edge of the book eliminates the convergence of the horizontal lines in that plane (bottom).

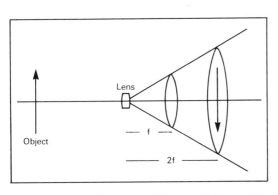

Figure 5–41 Changes in object and image distances do not affect the angle of coverage of a lens.

even be possible to make the top of the building appear to be larger than the bottom by overcorrecting, that is, by tilting the back to the vertical position and beyond.

Swinging and tilting the lens on a view camera will not alter the convergence of subject lines or the shape of the image.

Image shape is controlled by swinging and tilting the back of the camera. Image sharpness can be controlled by altering the angle of either the front or the back.

BACK MOVEMENTS AND IMAGE SHARPNESS

In situations where the swing and tilt adjustments on the camera back are not needed to control the convergence of parallel subject lines or to otherwise control image shape, they can be used to control the angle of the plane of sharp focus in object space. Whereas only the back adjustments can be used to control image shape, either the back or lens adjustments can be used to control the angle of the plane of sharp focus. When both image shape and sharpness must be controlled, the back adjustments are used for shape (since there is no other choice) and the lens adjustments are used for sharpness.

Figure 5–43 illustrates that the image of a distant object on the left comes to a focus closer behind the lens than the image of a nearby object on the right, as specified by the lens formula $1/f = 1/u + 1/v$, where the conjugate object and image distances vary inversely. To obtain a sharp image, the back is swung in a direction away from being parallel to the object plane containing the two object points—the opposite direction to that used to prevent convergence of parallel lines in the same object plane.

This relationship of the plane of the subject, the plane of the lensboard, and the plane of sharp focus in image space is known as the Scheimpflug rule (see Figure 5–44).

LENS MOVEMENTS AND IMAGE SHARPNESS

The Scheimpflug rule indicates that if the back of the camera is left in the zero position with a subject that is at an oblique angle, the lensboard can be swung (or tilted) to obtain convergence of the three planes at a common location. The lensboard is swung toward a position parallel to the subject plane to obtain a sharp image, whereas the back of the camera was swung in the opposite direction.

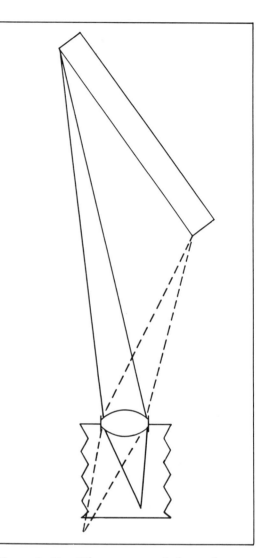

Figure 5–43   When a camera is focused on an intermediate distance, the image of the far end comes to a focus in front of the film, and the image of the near end comes to a focus behind the film.

LENS TYPES

Descriptive names applied to different types of camera lenses include *normal, telephoto, catadioptric, wide-angle, reversed telephoto (retrofocus), supplementary, convertible, zoom, macro, macro-zoom, process, soft focus,* and *anamorphic.* There are considerable variations among so-called normal lenses, especially in the characteristics of focal length,

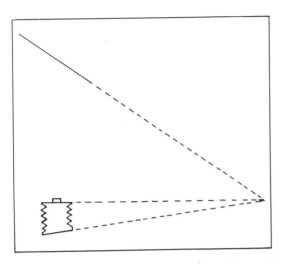

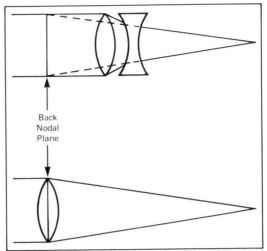

Figure 5–44 The convergence of the planes of the subject, the lensboard, and the film illustrates the Scheimpflug rule for controlling the angle of the plane of sharp focus.

Figure 5–45 The lens-to-film distance is shorter for a telephoto lens (top) than for a normal-type lens (bottom) of the same focal length.

speed, image quality, and price. The characteristic most common to normal-type lenses is an angular covering power of about 53°, which is just sufficient to cover the film when the focal length is equal to the film diagonal. The rule of thumb that recommends using a lens having a focal length about equal to the film diagonal is reasonable for normal-type lenses with most cameras, but longer focal length normal-type lenses should be used with view cameras to provide sufficient covering power to accommodate the camera movements. In the past many cameras were built with the lens permanently attached, the implication being that one lens should be able to satisfy all picture-taking needs. Most contemporary cameras are constructed so that other lenses can be substituted, enabling photographers to take advantage of the great variety of special-purpose lenses available.

TELEPHOTO LENSES

Telescopes and microscopes were invented to enable us to see distant objects as well as small objects more clearly. Photographers often want to record larger images of these distant and small objects than can be produced with lenses of normal focal length and design. We know that the image size of a distant object is directly proportional to the focal length. Thus, to obtain an image that is six times as large as that produced by a normal lens, the focal length must be increased by a factor of six, but the lens-to-film distance will also be increased six times unless the lens design is modified. The lens-to-film distance is shorter with telephoto lenses than with normal lenses of the same focal length. Compactness is the advantage of equipping a camera with a telephoto lens rather than a normal-type lens of the same focal length. Photographs made with the two lenses, however, would be the same with respect to image size, angle of view, linear perspective, and depth of field.

The basic design for a telephoto lens is a positive element in front of and separated from a negative element (see Figure 5–45). When a telephoto lens and a normal-type lens of the same focal length are focused on a distant object point, both images will come to a focus one focal length behind the respective back (image) nodal planes; but the lens-to-image distance will be smaller with the telephoto lens. The reason for the reduced lens-to-image distance is that the back nodal plane is located in front of the lens with telephoto lenses rather than near the center, as with normal lenses. It is easy to locate the position of the back nodal plane in a ray tracing of a telephoto lens focused on a distant object point, as in the preceding figure, by reversing

Telephoto lenses have shorter lens-to-film distances than normal-type lenses of the same focal length.

the converging rays of light in straight lines back through the lens until they meet the entering parallel rays of light. To determine the position of the back nodal point with an actual telephoto lens, the lens can be pivoted about various positions along the lens axis on a nodal slide until the image of a distant object remains stationary; or the lens and camera can be focused on infinity, whereby the back nodal point will be located exactly one focal length in front of the film plane. The position of the back nodal plane should be noted in relation to the front edge of the lens barrel, since this relationship will remain constant. If the back nodal plane is found to be located 2 inches forward of the front of a telephoto lens, 2 inches should be added to the distance from the front of the lens to the film any time the image distance (V) is used in a calculation, such as to determine scale of reproduction or the exposure correction for closeup photography. Although telephoto lenses are not generally used for closeup photography, the closeup range begins at an object distance of about 10 times the focal length, which can be a considerable distance with a long focal length telephoto lens.

CATADIOPTRIC LENSES

Catadioptric lenses contain mirrors and glass elements.

Notwithstanding the shorter lens-to-film distance obtained with telephoto lenses compared to normal lenses of the same focal length, the distance becomes inconveniently large with very long focal length telephoto lenses. Catadioptric lenses achieve a dramatic improvement in compactness through

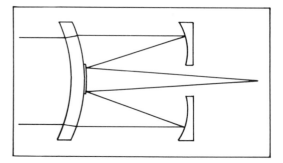

Figure 5–46 Catadioptric lenses make use of folded optics to obtain long focal lengths in relatively short lens barrels.

the use of folded optics. They combine glass elements and mirrors to form the image. The name catadioptric is derived from dioptrics (the optics of refracting elements) and catoptrics (the optics of reflecting surfaces). Figure 5–46 illustrates the principle of image formation with a catadioptric lens. A beam of light from a distant point passes through the glass element, except for the opaque circle in the center; it is reflected by the concave mirror and again by the smaller mirror on the back of the glass element, and it passes through the opening in the concave mirror to form the image on the film. The glass element and the opaque stop reduce aberrations inherent in the mirror system. Additional glass elements are commonly used between the small mirror and the film.

Location of the image nodal plane and the focal length of a catadioptric lens can be determined by the same methods described above for telephoto lenses. When the converging light rays that form the image on the film are reversed in straight lines on a ray tracing until they meet the entering rays of light, it can be seen that the image nodal plane is located a considerable distance in front of the lens, and the lens-to-film distance is small compared to the focal length (see Figure 5–47).

Catadioptric lenses are capable of producing images having excellent definition. There are also disadvantages with this type of lens. Due to the long focal length, the lens diameter would have to be very large to match the low f-numbers commonplace on lenses of normal design. Since a variable diaphragm cannot be used with this type of lens, exposure must be controlled with the shutter

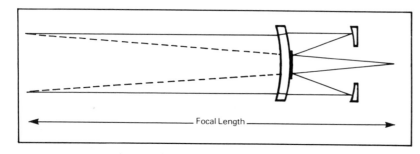

Focal Length

Figure 5–47 The image nodal plane and the focal length of a catadioptric lens can be determined by reversing the converging rays of light that form the image in straight lines until they meet the corresponding entering rays.

or by using neutral-density filters, and there is no control over depth of field. An additional disadvantage is that f-numbers calculated by dividing the focal length by the effective aperture do not take into consideration the light that is blocked by the mirrored spot in the center of the glass element.

WIDE-ANGLE LENSES

Two especially important reasons for substituting a shorter focal length lens on a camera equipped with a lens of normal focal length and design are: (a) the need to include a larger area of a scene in a photograph from a given camera position, and (b) the need to obtain a larger scale of reproduction when photographing small objects and the maximum lens-to-film distance capability of the camera is the limiting factor. In the latter situation a shorter focal length lens of normal design can be used satisfactorily because the diameter of the circle of good definition of the lens increases in proportion to the lens-to-film distance, which is necessarily larger for closeup photography and photomacrography. The same shorter focal length lens would not have sufficient covering power to photograph more distant scenes where the lens-to-film distance is about equal to the focal length.

A wide-angle lens can be defined as a lens having an angular covering power significantly larger than the approximately 53° angle of coverage provided by a normal-type lens, or as having a circle of good definition with a diameter considerably larger than the focal length when focused on infinity (see Figure 5–48). Wide-angle lenses are not restricted to short focal length lenses. It would be appropriate to use a wide-angle lens with a focal length equal to the film diagonal on a view camera where the extra covering power is needed to accommodate the view camera movements.

There is no distinctive basic design for wide-angle lenses comparable to the arrangement of positive and negative elements in telephoto lenses, except for that of the reversed telephoto wide-angle lenses. Early wide-angle lenses tended toward symmetry about the diaphragm, with few elements, and they usually had to be stopped down some-

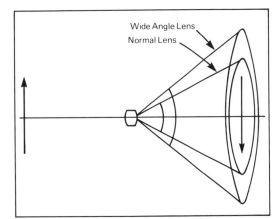

Figure 5–48   The covering power of a wide-angle lens compared to that of a normal-type lens of the same focal length. The images formed by the two lenses would be the same size.

what to obtain an image with satisfactory definition. Most but not all modern wide-angle lenses have a considerable number of elements, and they generally produce good definition even at the maximum aperture, with much less falloff of illumination toward the corners than the earlier lenses.

Wide-angle lenses of the fisheye type are capable of covering angles up to 180°, but only by recording off-axis straight subject lines as curved lines in the image. At this time, rectilinear wide-angle lenses are available that cover an angle of 110° with a 15 mm focal length on a 35 mm camera. There is no minimum angle of coverage that a lens must have to qualify as a wide-angle lens—the label is used at the discretion of the manufacturer. A 35 mm focal length wide-angle lens for a 35 mm camera, for example, only needs to have a 63° angle of coverage.

REVERSED TELEPHOTO WIDE-ANGLE LENSES

Problems may be encountered when using short focal length wide-angle lenses of conventional design due to the concomitant short lens-to-film distances. With view cameras, the short distance between the front and back standards can interfere with focusing or use of the swing, tilt, and other camera movements due to bellows bind. View-camera

The diameter of the circle of good definition of wide-angle lenses is considerably larger than the focal length.

Reversed telephoto wide-angle lenses are used on single-lens reflex cameras to avoid interference between the lens and the mirror.

manufacturers and users have found various ways of avoiding or minimizing these difficulties, as by using recessed lensboards with wide-angle lenses and substituting flexible bag bellows for the stiffer accordion type. With single-lens reflex cameras the placement of a short focal length lens close to the film plane can interfere with the operation of the mirror, requiring the mirror to be locked in the up position, which makes the viewing system inoperative. The shutter and viewing mechanisms in motion-picture cameras may also prevent short focal length wide-angle lenses from being placed as close to the film plane as required.

The lens designer's solution to the problems mentioned above is to reverse the arrangement of the elements in a telephoto lens, placing a negative element or group in front of and separated from a positive element or group. This design places the image nodal plane behind the lens (or near the back surface), which in effect moves the lens farther away from the film (see Figure 5–49). Lenses of this type are at different times referred to as *reversed telephoto*, *inverted telephoto*, and *retrofocus* wide-angle lenses. They have largely replaced the more traditional type of wide-angle lenses for small-format reflex cameras, but they have not yet invaded the large-format camera market.

SUPPLEMENTARY LENSES

Because camera lenses are expensive, photographers sometimes look for less costly alternatives to purchasing additional lenses when their general-purpose lens is not adequate. Supplementary lenses can be used to increase the versatility of a camera lens. Adding a positive supplementary lens produces the equivalent of a shorter focal length lens, and adding a negative supplementary lens produces the equivalent of a longer focal length lens. If the supplementary lens is positioned close to the front surface of the camera lens, the focal length of the combination can be computed with reasonable accuracy with the formula $1/f_c = 1/f + 1/f_s$, where f_c is the focal length of the combination, f is the focal length of the camera lens, and f_s is the focal length of the supplementary lens. For example, adding a positive 6-inch supplementary lens to a 2-inch (50 mm) camera lens produces a combined focal length of 1½ inches (38 mm). Adding a negative 6-inch supplementary lens produces a combined focal length of 3 inches (75 mm) (see Figure 5–50).

If the lenses are separated by a space (d), use the following formula:

$$1/f_c = \frac{1}{f} + \frac{1}{f_s} - \frac{d}{f \times f_x}$$

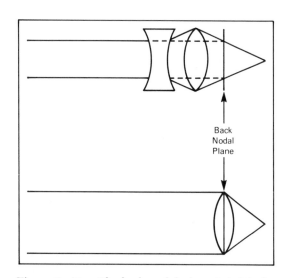

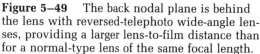

Figure 5–49 The back nodal plane is behind the lens with reversed-telephoto wide-angle lenses, providing a larger lens-to-film distance than for a normal-type lens of the same focal length.

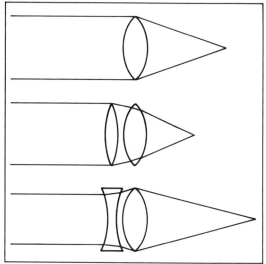

Figure 5–50 The effective focal length of a camera lens can be decreased by adding a positive supplementary lens (center), and increased by adding a negative supplementary lens (bottom).

Supplementary lenses are commonly calibrated in diopters, where the power in diopters equals the reciprocal of the focal length in meters, or $D = 1/f$. To convert from diopters to focal length, use the formula $f = 1/D$. For example, with a 2-diopter lens, $f = \frac{1}{2}$ meter or 500 mm. With a 4-diopter lens, $f = \frac{1}{4}$ meter or 250 mm. An advantage of using diopters is that the power of the combination of a camera lens and a supplementary lens is the sum of the individual diopters, or $D_c = D + D_s$.

Adding a positive supplementary lens in effect reduces the focal length of the camera lens, but does not convert it into a wide-angle lens. Therefore, the covering power of the combination may be insufficient to permit use with distant scenes. As long as the combined lenses are not moved closer to the film than the camera lens alone would be when focused on infinity, the covering power of the combination should be adequate. Figure 5–51 illustrates that when a camera is focused on infinity and a positive supplementary lens is added, the point of sharpest focus in object space moves from infinity to a distance of one focal length of the supplementary lens from the camera. Thus, to photograph a small object from a distance of 6 inches, sharp focus would be obtained with a 6-inch focal length positive supplementary lens on any camera focused on infinity, regardless of the focal length of the camera lens.

When photographing small objects, there can be an advantage in using a supplementary lens rather than increasing the lens-to-film distance with the camera lens alone (when the camera has sufficient focusing latitude). The aberrations in normal-type camera lenses are generally corrected for moderately large object distances, but the corrections do not hold with small object distances and the corresponding large image distances. There is the additional advantage that the camera exposure does not have to be increased when the camera is focused on infinity and a supplementary lens is added. If 1:1-scale reproduction photographs were made with the two procedures, four times the camera exposure would be required using the camera lens alone with the increased lens-to-film distance (see Figure 5–52).

Single-element supplementary lenses can be expected to introduce aberrations, but with the camera lens stopped down to a small opening the results may be quite satisfactory. Multiple-element supplementary lenses capable of producing high-quality images at large apertures exist, but they are relatively expensive. Commonly overlooked is the possibility of using a second camera lens as a highly corrected supplementary lens. The second lens should be turned around so that it is facing the camera to permit the light rays from the nearby object to enter and leave the lens at the same angles but in the reverse direction, as when it is used alone in the usual way with a distant object. This procedure retains the aberration corrections built into the lens. It is important to center the reversed second lens on the camera lens so that the two lens axes coincide.

Adding a positive supplementary lens to a camera lens produces the effect of reducing the focal length of the camera lens.

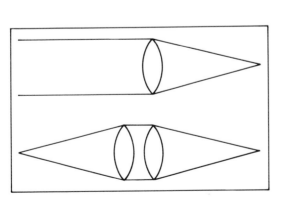

Figure 5–51 Positive supplementary lenses enable cameras to focus on near objects without increasing the lens-to-film distance.

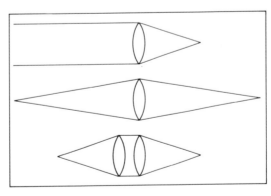

Figure 5–52 Making a 1:1-scale reproduction photograph by increasing the lens-to-film distance (center) requires a $4\times$ increase in the camera exposure. Using a supplementary lens (bottom) requires no increase in exposure.

SPECIAL SUPPLEMENTARY LENSES

Certain other multiple-element attachments are available to modify the image-forming capabilities of camera lenses. Some of these are not generally referred to as supplementary lenses but rather by terms such as *extender* (or *converter*), *afocal attachment*, and *monocular attachment*.

Extenders are negative lenses containing one or more elements that are used behind the camera lens to increase the focal length. They are commonly referred to as *tele-extenders*, as they are most effective when used with telephoto or other longer-than-normal focal length lenses and produce a telephoto effect with the addition of a negative lens behind the positive camera lens. A given tele-extender will increase the focal length of whatever camera lens it is used with by the same factor, such as 2×, and some are variable to produce different factors.

Afocal attachments combine positive and negative elements having appropriate focal lengths and separation between them so that rays of light entering the attachment from a distant object point leave traveling parallel, as in a Galilean telescope. Since the attachment does not form a real image, it has no focal length, hence the name *afocal*. Afocal attachments do alter the focal length of the camera lens, however, increasing it when the positive component is in front of the negative component, and decreasing it when the negative component is in front of the positive component, as illustrated in Figure 5–53. With the camera lens focused on infinity and the afocal attachment added, focus can be adjusted for different object distances by changing the distance between the positive and negative elements. Changing the ratio of the focal lengths of the positive and negative elements alters the effect of the attachment on the focal length of the combination and, therefore, image size. The afocal attachment has no effect on the f-number of the camera lens.

CONVERTIBLE LENSES

Some lenses, generally referred to as convertible lenses, are designed so that one or

Two different focal lengths can be obtained with convertible lenses by removing part of the lens.

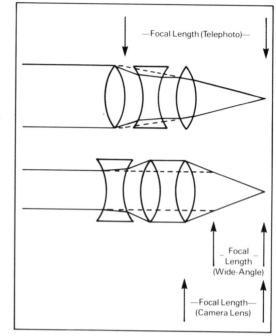

Figure 5–53   Afocal attachments change the effective focal length of camera lenses without changing the lens-to-film distance or the f-number. A two-element telephoto attachment is shown in front of a camera lens, at the top; a wide-angle attachment is shown at the bottom.

more elements can be removed to change the focal length. Removing a positive element or group of elements increases the focal length. Removing the part of a compound lens that is in front of or behind the diaphragm introduces other complications. Since the focal length and the lens-to-film distance are both increased with the removal of a positive component, the f-numbers will be affected and a separate set of markings must be provided. This differs from the addition of an afocal attachment, where the focal length is altered but the camera lens-to-film distance and therefore the f-numbers are unaffected.

Although multiple elements can be used in both components of a convertible lens to minimize aberrations, the photographer should not expect the same image quality when part of the lens is removed. If a component is removed from a convertible lens to obtain a longer focal length for the purpose of making portraits, for example, the loss of image sharpness at large diaphragm openings may be flattering rather than detrimental, and stopping the lens down will reduce any loss of sharpness. Disrupting the symmetry of the lens on both sides of the diaphragm by the removal of one component without introducing barrel or pincushion distortion presents the lens designer with a difficult problem.

A more recent variation of the convertible lens is the substitution of a different component. This procedure makes it possible to maintain a higher degree of aberration correction and to offer a greater variety of longer and shorter focal lengths at a lower price than for completely separate lenses with different focal lengths.

ZOOM LENSES

From the photographer's point of view, the ideal solution to the problem of having the right lens available for every picture-making situation is to have one versatile variable-focal length lens. Lens designers have made excellent progress toward the goal of a universal lens with the zoom design, but there is still far to go. With a zoom lens, the focal length can be altered continuously between limits, with the image remaining in focus. The basic principle involved in changing the focal length of a lens can be illustrated with a simple telephoto lens where the distance between the positive and negative elements is varied, as illustrated in Figure 5–54. This change in position of the negative element would change the focal length (and image size and angle of view), but the image would not remain in focus. Other basic problems include aberration correction and keeping the

relative aperture constant at all focal length settings. It should not be surprising that one of the early zoom lenses contained more than 20 elements, was large and expensive, and was not very successful in solving all of the basic problems. Better zoom lenses are now being mass produced with relatively few elements, due to design improvements.

Two methods have been used for the movement of elements in zoom lenses. One is to link the elements to be moved so that they move the same distance. This is called optical compensation because the mechanical movement is simple but the optical design is complex and requires more elements. The other method is called mechanical compensation and involves moving different elements by different amounts, requiring a complex mechanical design. For the problem of maintaining a constant f-number as the focal length is changed, an optical solution is to incorporate the concept of the afocal attachment at or near the front of the lens so that the aperture diameter and the distance between the diaphragm and the film can remain fixed. An alternative mechanical solution is to use a cam that adjusts the size of the diaphragm opening as the focal length is changed.

Similarly, there are mechanical and optical methods of keeping the image in focus. The mechanical method consists of changing the lens-to-film distance, as with conventional lenses. The optical method involves using a positive element in front of the afocal component that is used to keep the relative aperture constant. The range of maximum to minimum focal length varies with different zoom lenses from less than 3:1 to more than 20:1, with the larger ranges found only on lenses for motion-picture and television cameras.

MACRO-ZOOM LENSES

Zoom lenses generally were not made to focus on short object distances. In 1967 the first of a series of macro-zoom lenses was introduced. Macro-zoom lenses are designed to photograph small objects near the camera either by extending the conventional focusing range or by making a separate adjustment in the position of certain components for the

> A zoom lens should keep the image in focus and keep the image illuminance the same as the focal length is altered.

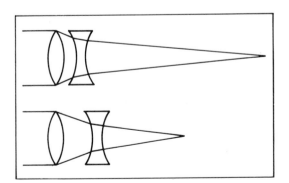

Figure 5–54   Changing the distance between the positive and negative elements of a telephoto lens changes the focal length.

so-called macro capability. Use of the term *macro* in this context is misleading, as most lenses of this type produce a maximum scale of reproduction no larger than 1:2. To date, none yield a scale larger than 1:1, which is considered the lower limit for the specialized area of photography called photomacrography.

MACRO LENSES

Macro lenses are designed to produce better image definition than conventional lenses when used at small object distances.

Lenses of normal design that produce images of excellent quality with objects at moderate to large distances may not perform well when used at small object distances. At scales of reproduction larger than 1:1, where the image distance is greater than the object distance, such lenses tend to produce sharper images when they are turned around so that the front of the lens faces the film. *Macro lenses* are small-format camera lenses especially designed to be used at small object distances. The important optical characteristic of macro lenses is the excellent image definition they produce under these conditions compared with normal-type lenses (see Figure 5–55). The lens designer's task of optimizing aberration correction for small object distances is made easier by removing the ad-

ditional requirement to make the lens fast. Thus most macro lenses are two or three stops slower than comparable normal-type lenses. The implication that macro lenses are not suitable for photographing objects at larger distances is not entirely valid, however. Because of the slower maximum speed, the aberration corrections are not as sensitive to changes in object distance, and some photographers prefer to use a macro lens for general-purpose photography when a faster lens is not needed.

SOFT-FOCUS LENSES

Spherical aberration is commonly used to produce the soft focus with soft-focus lenses.

Photographers have for the most part demanded lenses that produce sharp images, but for some purposes a certain amount of unsharpness is considered more appropriate. Soft-focus lenses are sometimes labeled portrait lenses because they have been used so widely for studio portraits, but they have also been used extensively by other photographers, including pictorialists and even photographers doing advertising illustration, when certain mood effects are desired.

The soft-focus effect is generally achieved by undercorrecting for spherical aberration in designing the lens. Since spherical aberration is reduced as the lens is stopped down, the photographer can control the degree of unsharpness by the choice of f-number. To the discerning viewer, the effect produced with a soft focus lens on the camera is not at all similar to that produced by defocusing the enlarger or diffusing the image while exposing the print with a sharp negative. Rays of light near the axis of a soft-focus lens form a sharp image on the film, which is surrounded by an unsharp image formed by the marginal rays of light so that highlight areas in the photograph appear to be surrounded by halos (see Figure 5–56). If the same lens were used on an enlarger, the shadows in prints would be surrounded by dark out-of-focus images, except when making reversal prints from transparencies.

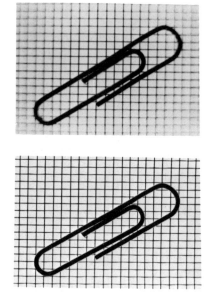

Figure 5–55   Photographs of a small object made with a normal-type camera lens (top) and a macro lens (bottom), both at the maximum aperture.

Motion picture anamorphic camera lenses squeeze a wide angle of the scene onto standard width film. The projector anamorphic lens unsqueezes the film image onto a wide projection screen.

ANAMORPHIC LENSES

An anamorphic lens produces images having different scales of reproduction in each of two perpendicular directions, usually the vertical and horizontal directions. It is common practice to think of one of the two dimensions of the image as being normal and the other as being either stretched or squeezed. Even before photography some artists were experimenting with anamorphic drawings where, for example, it was necessary to view the image at an extreme angle for it to appear normal. Lens designers have long been familiar with the concept of anamorphic lenses but there was little demand for such lenses before the introduction of wide-screen motion pictures.

Motion-picture cameras equipped with anamorphic lenses have a wider-than-normal horizontal angle of view, but the extra width is squeezed by the lens to fit the conventional film format, which typically has an aspect ratio of 1.33:1. The projector, in turn, is equipped with an anamorphic lens that will stretch or unsqueeze the horizontal dimension to produce a picture with a higher aspect ratio (1.8:1, for example) and images that appear normal (see Figure 5–57).

Figure 5–56 Photographs made with a normal-type camera lens (left) and a single-element lens that was uncorrected for spherical aberration (right).

ENLARGER LENSES

The requirements for enlarger lenses are similar to those for a camera lens intended for copying, with aberrations minimized for small object distances; and in practice the degree of correction can be expected to vary with the price for a given focal length. In the past, most enlarging lenses were designed for normal covering power with the expectation that the photographer would select a focal length about equal to the diagonal of the film format. Recent years have seen the increasing use of shorter focal length lenses with larger angles of coverage to increase the range of scales of reproduction, and the range of image sizes that can be obtained with a given enlarger at the upper and lower limits of elevation.

The introduction of variable focal length enlarger lenses followed years behind widespread use of variable focal length lenses for motion-picture cameras, television cameras, small-format still cameras, and slide projectors.

Figure 5–57 The water-filled parts of the glasses function as anamorphic lenses to stretch the images horizontally. Two stretched images of the penny on the left were rotated 90° and placed on the glass on the right where the bottom one was stretched back to a circular shape by the water. With wide-screen cinematic photography, the anamorphic camera lens squeezes a wide angle of the scene onto normal-width film, and the anamorphic projector lens stretches or unsqueezes the image to fit a wide projection screen.

Four types of lens shortcomings are related to (1) image definition, (2) image shape, (3) image illuminance, and (4) image color.

LENS SHORTCOMINGS

The geometrical drawings used to illustrate image formation in preceding sections imply that the lenses form perfect images. Such is not the case, of course. Actually, there is no need for such perfect images in the field of pictorial photography, where photographs tend to be viewed from a distance about equal to the diagonal, and, as noted earlier, circles of confusion up to a certain maximum size are perceived as points. This tolerant attitude does not apply to photographs viewed through magnifiers or microscopes to extract as much information as possible, or even pictorial photographs that are enlarged and cropped in printing. Lens shortcomings can be subdivided into four categories: those that affect image definition, image shape, image illuminance, and image color.

DIFFRACTION

The diffraction-limited resolving power of a lens can be calculated by dividing a constant by the f-number.

Diffraction is the only lens aberration that affects the definition of images formed with a pinhole. According to the principles of geometrical optics, which ignore the wave nature of light, image definition will increase indefinitely as a pinhole is made smaller. In practice there is an optimum size, and image definition decreases due to diffraction as the pinhole is made smaller. The narrow beam of light passing through the pinhole from an object point spreads out in somewhat the same manner as water coming out of a nozzle on a hose, and the smaller the pinhole the more apparent the effect.

Similarly, the definition of an image formed by an otherwise perfect lens would be limited by diffraction. Some lenses are referred to as diffraction-limited because under the specified conditions they are that good. Using resolution as a measure of image definition, the diffraction-limited resolving power can be approximated with the formula $R = 1,800/N$ where R is the resolving power in lines per millimeter, 1,800 is a constant for an average wavelength of light of 550 nm, and N is the f-number. Thus a lens having minimum and maximum f-numbers of f/2 and f/22 would have corresponding diffraction-limited resolving powers of 900 and 82 lines/mm. If the resolution is to be based on points

Practical lens tests done by examining photographs made with the lens are systems tests rather than lens tests.

rather than lines, the formula is changed to $R = 1,500/N$ (see Table 5–2).

LENS TESTING

Photographers who inquire about testing a lens are commonly advised to do so by making photographs of the same type that they expect to be making with the lens in the future, and to leave the more analytical testing to lens designers, manufacturers, and others who have the training and the sophisticated equipment necessary for the job. It makes sense that a photographer should not worry about shortcomings in the imaging capabilities of a lens if the effects are not apparent in photographs made with the lens. On the other hand, many photographers use their normal lens in a variety of picture-making situations, and certain lens shortcomings may appear in one photograph and not in another. Thus the subject matter and the conditions for even a practical test of this type must be carefully controlled.

To test for image definition, the following are required:

1. The subject must conform to a flat surface that is perpendicular to the lens axis and parallel to the film plane.
2. The subject must exhibit good detail with local contrast as high as is likely to be encountered, not of a single hue such as red bricks or green grass.
3. The subject must be large enough to cover the angle of view of the camera-lens combination at an average object distance.
4. The subject must also be photographed at the closest object distance likely to be used in the future.
5. Photographs must be made at the maximum, minimum, and at least one intermediate diaphragm opening.
6. Care must be taken to be certain that the optimum image focus is at the film plane, which may require bracketing the focus.
7. Care must be taken to avoid camera movement, noting that mirror action in single-lens reflex cameras can cause blurring of the image, especially with small object distances or long focal length lenses. Outdoors, wind can cause camera movement.

If the same subject is to be used to test for image shape, it must have straight parallel lines that will be imaged near the edges of the film to reveal pincushion or barrel distortion (see Figure 5–58). If it is also to be used to test for uniformity of illumination at the film plane, it must contain areas of uniform luminance from center to edge. Tests for flare and ghost images require other subject attributes. We can conclude that a practical test cannot be overly simple if it is to provide much information about the lens. In any event, evaluation of the results is easier and more meaningful if a parallel test is done on a lens of known quality for comparison.

Practical tests such as these are not really lens tests, but rather they are systems tests, which include subject, lens, camera, film, exposure, development, the enlarger, and various printing factors. The advantage that such tests have of being realistic must be weighed against the disadvantage that tests of a system reduce the effect of variations of one component, such as the lens, even if all the other factors remain exactly the same. For example, if two lenses having resolving powers of 200 and 100 lines/mm are used with a film having a resolving power of 100 lines/mm, the resolving powers of the lens-film combinations would be 67 and 50, using the formula $1/R = 1/R_L + 1/R_F$. Thus, only dramatic differences between lenses will be detected easily in the photographs.

The influence of other factors in the system can be eliminated by examining the optical image directly, either on a finely textured ground glass or by removing the ground glass and examining the aerial image. A good-quality magnifier or low-power microscope should be used. An artificial star, made by placing a light bulb behind a small hole in thin, opaque material, is commonly used to check image definition by noting how the image deviates from an ideal point image. Since the ideal is seldom approximated closely, it is better to evaluate a lens by comparison with a lens of known quality than by judging it alone. By placing the artificial star at an appropriate distance on the lens axis, the effect of stopping down on spherical and longitudinal chromatic aberrations can be seen. The chromatic aberration can be removed by placing a green filter behind the hole to study spherical aberration alone.

Table 5–2 Diffraction-limited resolving power vs. f-numbers. The underlined rows identify the f-numbers that would produce the same diffraction-limited resolving power of 14 lines/mm on 8×10-inch prints made from five different size negatives

F-Number	l/mm	Film	Adjusted for 8×10 print
$R = \dfrac{1800}{\text{F-Number}}$			
f/1	1800		
f/1.4	1286		
f/2	900		
f/2.8	643		
f/4	450		
f/5.6	321		
f/8	225		
f/11	160		
f/16	112	$1 \times 1\frac{1}{2}$	$112/8 = 14$
f/22	80		
f/32	56	$2\frac{1}{4}$	$56/4 = 14$
f/45	40		
f/64	28	4×5	$28/2 = 14$
f/90	20		
f/128	14	8×10	$14/1 = 14$
f/180	10		
f/256	7	16×20	$7/0.5 = 14$
f/360	5		

Note: 1500 used for points, 1800 for lines.

$R = \dfrac{10^6}{\lambda \times \text{f-number}}$ where $\lambda = 550$ nm

Figure 5–58 An example of barrel distortion produced by a 7.5 mm fisheye lens with a 180° angle of view. Note that the distortion does not affect radial subject lines—straight subject lines that intersect the lens axis—such as the horizon line.

Moving the star laterally so that the image appears near a corner will reveal off-axis defects, including coma and lateral chromatic aberration.

Resolving power has been a widely used but controversial method of checking image definition. The testing procedure is simple. The lens is focused on a row of resolution targets that contain alternating light and dark stripes and that are placed at a specified distance from the lens (see Figure 5–59). The separate targets are arranged so that the images fall on a diagonal line on the ground glass or film plane with the center target on the lens axis, and oriented so that the mutually perpendicular sets of stripes constitute radial and tangential lines. The aerial images are examined through a magnifier or microscope of appropriate power, and the smallest set of stripes that can be seen as separate "lines" is noted for each target. Resolving power is the maximum number of light-dark line pairs per millimeter that can be resolved.

Critics of resolving power note that different observers may not agree on which is the smallest set of stripes that can be resolved, and that in comparing two lenses, photographs made with the lens having the higher resolving power sometimes appear to be less sharp. In defense of resolving power, it has been found that observers can be quite consistent even if they don't agree with other observers. Consistency makes their judgments valid on a comparative basis such as between on-axis and off-axis images, images at two different f-numbers, and images formed

with two different lenses. It is appropriate to make comparisons between lenses on the basis of resolving power as long as it is understood that resolving power relates to the ability of the lens to image fine detail rather than the overall quality of the image.

Electronic methods have largely replaced visual methods of testing lenses in industry. The equipment required is complex and expensive but it is capable of providing comprehensive and objective data quickly. The results are commonly presented in the form of modulation transfer function curves in which the input spatial frequency is plotted as cycles per millimeter on the horizontal axis against output contrast as percent modulation on the vertical axis. Representative curves for three different lenses, two having imaging shortcomings, are shown in Figure 5–60, where lens B has higher contrast in the high-frequency (fine detail) areas but lower contrast in the low-frequency (coarse detail) areas than lens C. The accompanying pictures illustrate that a photograph in which fine detail is better resolved may appear less sharp because of the lower contrast in the larger image areas.

TESTING FLARE

Optical images formed with lenses are always less contrasty than the scenes being photographed due to the effects of flare light. Antireflection lens coatings are effective in reducing the proportion of light reflected from lens surfaces and increasing the proportion of light transmitted, but they reduce flare light rather than eliminate it. In practice, the flare light that falls on the film in a camera is a combination of lens flare and camera flare, and the amount of flare light can vary greatly with a given lens, depending upon the distribution of light and dark tones in the scene, the lighting, the interior design of the camera, and whether or not a lens shade is used. Thus, flare tests can be conducted with the lens in a laboratory where a standard test target is used and the effects of the camera are eliminated, or they can be conducted with the lens on a camera with a representative scene or a variety of scenes.

Flare light can be specified in various ways: as a flare factor, as a percentage of the max-

Resolving power is not a reliable indicator of the appearance of sharpness of photographic images.

Camera flare light reduces the contrast of photographic images, especially in shadow areas.

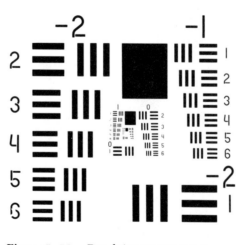

Figure 5–59 Resolving-power target.

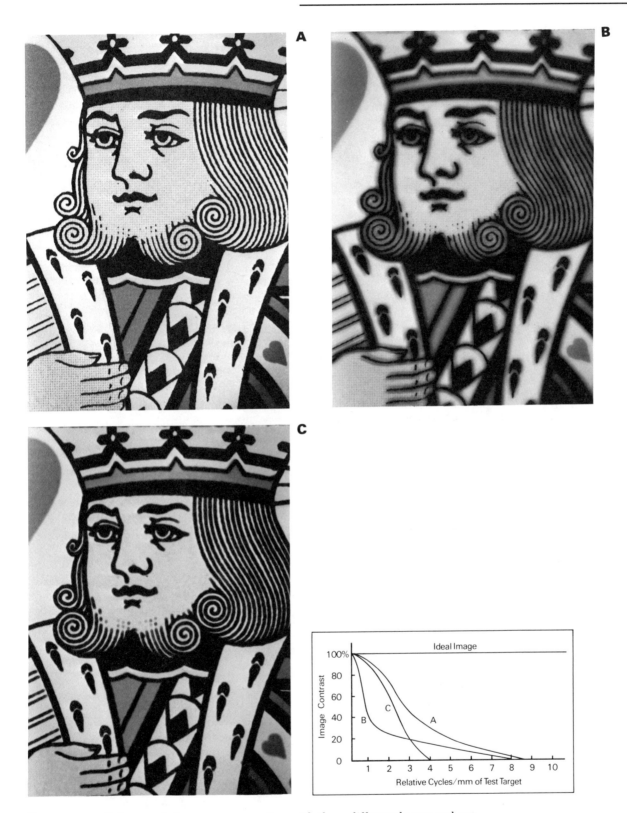

Figure 5–60   Prints made from a copy negative with three different lenses and representative modulation transfer function curves. Lens A was a high-quality enlarging lens. The lens B image shows a large loss of contrast in the intermediate frequencies, while the lens C image shows the largest loss in the high frequencies. Lens B would score higher on a resolving-power test even though photographs made with lens C tend to appear sharper due to the higher contrast of the intermediate-size tonal areas.

Figure 5–61   A black hole in a neutral test card surrounded by a gray scale, used to determine the flare factor in a given situation.

imum image illuminance, and graphically as a flare curve. The flare factor is defined as the scene luminance ratio divided by the image illuminance ratio. The flare factor can most easily be determined with a large-format camera and an in-camera meter of the type where a probe can be positioned to take readings in selected small areas such as a highlight and a shadow. If the image illuminance ratio is 80:1 with a scene having a luminance ratio of 160:1, the flare factor is 160/80 = 2. Flare factors are typically between 2 and 3, but they can be much higher.

An opaque black card placed in front of a large transparency illuminator in an otherwise darkened room can be used to determine the percentage of flare light by dividing an illuminance reading of the image of the black card by an illuminance reading of the image of the transparency illuminator. A white surface and a black hole can be placed in any scene and used in the same way, where the black hole is a small opening in an otherwise light-tight enclosure painted black on the inside (see Figure 5–61). Standard test targets and simple procedures are available for the routine determination of the percentage of flare with process cameras in the

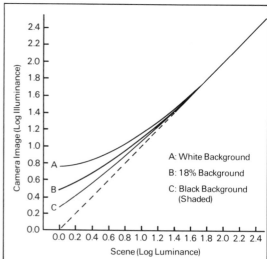

Figure 5–62   Flare curves prepared by taking meter readings at the film plane of the image of a gray scale with white, gray, and black backgrounds. The straight broken line represents the complete absence of flare light.

graphic arts field where flare levels above 1.5% are considered excessive. With a normal scene having a luminance ratio of 160:1, a flare level of 1.5% corresponds to a flare factor of 2.4.

Flare curves can be prepared in various ways. One method is to place a step tablet or gray scale, with an appropriate surround, in front of a large-format camera, measure the relative illuminance of each step of the image with an in-camera meter, and plot log relative illuminance on the vertical axis vs. log relative luminance (i.e., 1/density) of the original on the horizontal axis (see Figure 5–62).

REVIEW QUESTIONS

1. If a pinhole aperture is used on a view camera in place of the lens, increasing the pinhole-to-ground-glass distance results in . . . *(p. 134)*
   A. an increase in the image size
   B. a decrease in the image size
   C. no change in the image size
2. Diffraction causes the greatest degradation of the pinhole image when . . . *(p. 134)*
   A. the pinhole-to-ground-glass distance is less than the film diagonal

B. the pinhole-to-ground-glass distance is larger than the film diagonal

C. the pinhole is too large

D. the pinhole is too small

3. Anamorphic pinhole images can be obtained by using . . . (p. 136)

A. two pinholes, side by side

B. a rectangular pinhole aperture

C. crossed slits at different distances from the film

D. anamorphic film

4. The focal length of a lens is the distance from the . . . (p. 137)

A. film to the lens

B. principal focal point to the back nodal point

C. front nodal point to the back nodal point

D. subject to the lens

5. The image distance that produces an in-focus image for an object located 300 mm from a 200 mm lens is . . . (p. 143)

A. 200 mm

B. 300 mm

C. 400 mm

D. 600 mm

E. 800 mm

6. Image size is . . . (p. 144)

A. directly proportional to focal length

B. directly proportional to focal length squared

C. inversely proportional to focal length

D. inversely proportional to focal length squared

7. Two objects of equal size are located at distances of 10 feet and 20 feet from a camera. If the image of the nearer object measures 1 inch in the photograph, the image of the farther object will measure . . . (p. 145)

A. 1/4 inch

B. 1/2 inch

C. 1 inch

D. 2 inches

E. 4 inches

8. When a camera is moved farther away from a subject, the perspective . . . (p. 145)

A. becomes stronger

B. becomes weaker

C. is unchanged

9. The image of a distant building formed with a 90-mm focal length lens on a 4 × 5-inch view camera is 1 inch tall. To obtain an image that is 3 inches tall with the camera in the same position, it would be necessary to substitute a lens with a focal length of . . . (p. 146)

A. 30 mm

B. 60 mm

C. 135 mm

D. 180 mm

E. 270 mm

10. The recommended procedure for selecting the lens and the camera position for a photograph when perspective is an important factor is to select the . . . (p. 148)

A. lens first, then the camera position

B. camera position first, then the lens

11. The "correct" viewing distance for an 8 × 10-inch print made from a 35 mm negative exposed with a 50 mm focal length lens is approximately . . . (p. 149)

A. 50 mm

B. 8 inches

C. 10 inches

D. 12.8 inches

E. 16 inches

12. The heads near the left and right edges of a group portrait appear to be abnormally broad. This unnatural shape is associated with . . . (p. 150)

A. barrel distortion

B. pincushion distortion

C. curvature of field

D. the wide-angle effect

E. the balloon effect

13. The largest circle that will appear as a point on a print viewed at the standardized viewing distance is generally considered to have a diameter of approximately . . . (p. 151)

A. 1/50 inch

B. 1/100 inch

C. 1/200 inch

D. 1/400 inch

E. 1/800 inch

14. A definition of hyperfocal distance is . . . (p. 152)

A. the nearest distance that appears sharp when the camera is focused on infinity

B. the distance from the lens to the film when the camera is focused on infinity

C. one-half the distance to infinity when the lens is one focal length from the film

D. the distance between the nearest and farthest objects that appear sharp

15. A lens has a maximum aperture of f/2.0 and a minimum aperture of f/16. At f/16, the depth of field will be . . . (p. 154)
    A. 4 times the DOF at f/2
    B. 8 times the DOF at f/2
    C. 16 times the DOF at f/2
    D. 32 times the DOF at f/2
    E. 64 times the DOF at f/2

16. The object distance at which the largest depth of field is obtained for a given lens and f-number is . . . (p. 154)
    A. infinity
    B. 1/2 the hyperfocal distance
    C. the hyperfocal distance
    D. 2 times the hyperfocal distance

17. Two photographs are made with a 4 × 5-inch camera, the first with a 200 mm focal length lens and the second with a 400 mm focal length lens. If both photographs are made at the same f-number and with the camera in the same position, the depth of field in the first photograph will be . . . (p. 155)
    A. 2 times that in the second
    B. 4 times that in the second
    C. 8 times that in the second
    D. the same as that in the second

18. Image shape is controlled by tilting or swinging . . . (p. 157)
    A. the lens (only)
    B. the back (only)
    C. either the lens or the back

19. To control both shape and sharpness of the image of an object, the photographer should control . . . (p. 158)
    A. sharpness with the lens and shape with the back
    B. sharpness with the back and shape with the lens
    C. both sharpness and shape with the back
    D. both sharpness and shape with the lens
    E. None of the above.

20. The plane of sharp focus is controlled by tilting or swinging . . . (p. 158)
    A. the lens (only)
    B. the back (only)
    C. either the lens or the back

21. A 50 mm focal length lens produces a 1/2 mm diameter image of a full moon. To obtain an image with a diameter of 24 mm to completely fill the width of a 35 mm negative would require a lens having a focal length of . . . (p. 159)

    A. 240 mm
    B. 480 mm
    C. 2400 mm
    D. 4800 mm
    E. Insufficient information provided.

22. A simple telephoto lens consists of . . . (p. 159)
    A. a positive element in front of and separated from a negative element
    B. a negative element in front of and separated from a positive element
    C. a positive element and a negative element with an adjustable space between them
    D. None of the above.

23. A valid reason for using a 90 mm focal length wide-angle lens in preference to a 210 mm focal length lens of normal design on a view camera would be to obtain . . . (p. 161)
    A. a larger angle of view
    B. a smaller angle of view
    C. a shallower depth of field
    D. a larger circle of good definition

24. A problem that would be encountered in using a 28 mm focal length wide-angle lens of conventional design on a 35 mm SLR camera is inability to . . . (p. 162)
    A. focus on infinity
    B. focus close up
    C. use the focal plane shutter
    D. use the reflex viewfinder
    E. stop the lens down beyond f/8

25. Adding a +12-inch focal length supplementary lens to a 6-inch focal length camera lens produces a combined focal length of approximately . . . (p. 162)
    A. 2 inch
    B. 3 inch
    C. 4 inch
    D. 9 inch
    E. None of the above.

26. A 35 mm camera that is equipped with a 50 mm focal length lens is focused on infinity. A 50 mm positive supplementary lens is added to the camera lens. The camera is now in sharp focus for an object distance of . . . (p. 163)
    A. 35 mm
    B. 50 mm
    C. 100 mm
    D. None of the above.

27. An advantage of a macro lens over a lens of normal design of the same focal length is . . . (p. 166)

A. lower cost

B. larger depth of field

C. higher speed (smaller f-number)

D. better definition with small object distances

28. The only aberration that affects images formed with pinhole apertures is . . . (p. 168)

A. spherical aberration

B. diffraction

C. coma

D. curvature of field

E. distortion

29. The diffraction-limited resolving power for a 50 mm focal length lens on a 35 mm camera, at a diaphragm opening of f/2, is . . . (p. 168)

A. 1800 lines/mm

B. 900 lines/mm

C. 450 lines/mm

D. 225 lines/mm

30. In lens testing, resolving power is associated with . . . (p. 170)

A. graininess

B. sharpness

C. detail

D. overall quality

6 Chemistry for Photographers

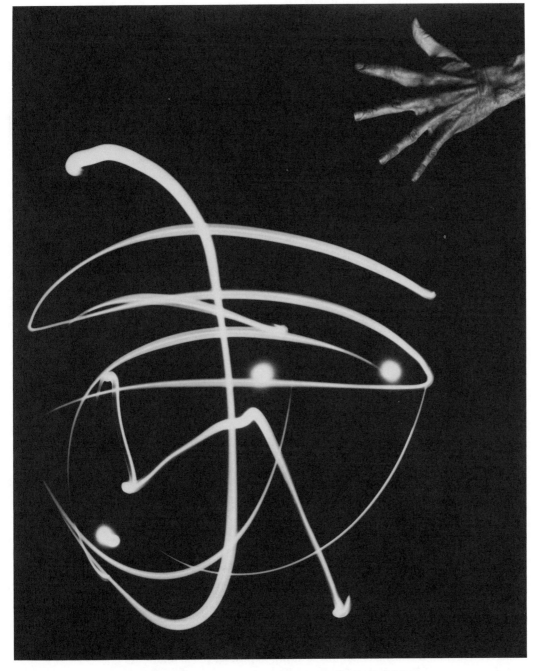

Barbara Morgan. *Pure Energy and Neurotic Man.* Copyright © 1980 by Barbara Morgan.

BASIC CHEMISTRY

The periodic table is a logical arrangement of the elements according to their chemical and physical properties.

Atoms are the smallest bits of elements that can combine chemically.

In order to understand photographic chemistry, it is necessary to review the basic concepts of general chemistry. The universe is made up of a number of elements. Ninety-two of these occur naturally, but elements heavier than uranium have been created by means of nuclear bombardment and radioactive decay. These elements have distinct characteristics and can be divided into two groups—metals and nonmetals. Each element is composed of atoms, which are the practical minimum particle that has the characteristic of an element. Scientists break the atom into smaller particles or subdivisions, but it is not necessary to go deeply into this to understand photographic chemistry.

THE PERIODIC TABLE

In 1869 Mendeleyev arranged the then-known atoms into a table, according to their chemical and physical properties. At first there were 89 elements in the table. A few years ago this number was extended to 103 (see Figure 6–1). More recently, it has been increased to 109, and more elements will likely be produced in the future. When recent information concerning the atomic structure became available, it gave further support to the arrangement of the table. The atomic number is equal to the positive charge of the nucleus in the atom, that is, the number of protons in the nucleus. The first column of the table contains the *alkalis*—lithium, sodium, potassium, rubidium, cesium, and francium. Near the other side of the table can be found the *halogen* family, the *halides*—fluorine, chlorine, bromine, iodine, and astatine. Other families of elements are similarly grouped in the table.

ATOMS

The smallest atom is that of hydrogen, which consists of a single positive nucleus surrounded by one electron. All of the other atoms have a nucleus of higher positive charge, with corresponding electrons orbiting around the nucleus, and which have a total negative charge equal to the positive charge of the nucleus. Thus a stable atom would be neutral. The *atomic number* is the number by which the charge on the hydrogen nucleus must be multiplied to equal the nuclear charge of the atom. These numbers determine the chemical behavior of the elements, and also support the arrangement of the atomic table (see Figure 6–2).

The various elements are designated by their symbols; that is, the symbol for hydro-

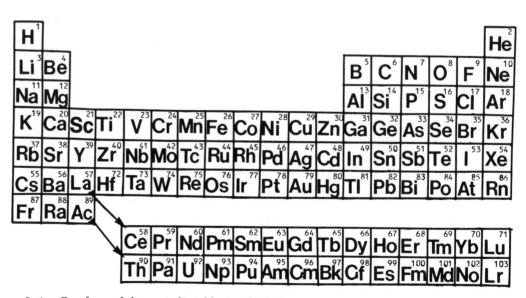

Figure 6–1   One form of the periodic table in which the atoms are arranged according to their chemical and physical properties. As the atomic numbers increase, so do the atomic weights.

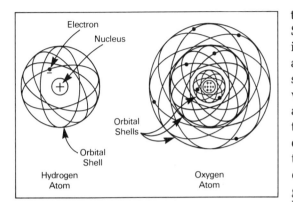

Figure 6–2 The hydrogen atom has 1 electron in orbit around a nucleus with a single positive charge. The oxygen atom has 8 electrons orbiting in shells around its nucleus, with an equal number of positive charges.

gen is H, He for helium, Na for sodium (for *natrium*, its Latin name), K for potassium (*kalium*), and so on. Strictly speaking, a single atom can be called a molecule—i.e., a "monatomic molecule"—or two or more atoms can be combined, such as O_2 and O_3 to become molecules of oxygen and ozone that are made up of more than one atom of oxygen; or molecules can be composed of different atoms. In the latter case they are called *chemical compounds*. Potassium bromide is a compound commonly used in photography.

Some compounds are made up of more than two elements, in which case two or more of each of the elements assume the charge characteristic of a single element. For example, ammonium chloride (NH_4Cl) in solution ionizes into NH_4^+ and Cl^-. The NH_4 is known as a *radical*.

Atoms ordinarily exist with a balance of charges; that is, the electrons orbiting the nucleus have an electron charge that balances the positive charge of the proton nucleus. Under certain conditions, a molecule can dissociate into separate atoms, each having a separate positive or negative electronic charge. Sodium chloride (NaCl), for example, when heated to make it molten, can exist with sodium atoms (Na^+) and chlorine atoms (Cl^-); or a solution of sodium chloride in water dissociates into atoms having $^+$ and $^-$ charges, which are identified as *ions*.

Chemical compounds have different physical characteristics than those of the atoms that are combined to make the compounds. Such attributes as melting points, solubility in water or other solvents, crystalline form and color, and specific gravity or density are some of the characteristics that differentiate various compounds. *Inorganic compounds* are those containing metals, and do not contain the element carbon, with the exception of the oxides of carbon—compounds containing the carbonate (CO_3) radical—carbon disulfide and a few other compounds. *Organic compounds* are those containing carbon—with the exceptions noted above—usually along with hydrogen, oxygen, sulfur, nitrogen, iodine, phosphorous, and other elements. Organic compounds are often not very soluble in water, and they are combined with sodium or similar alkali to form a salt that is more soluble, with the compound retaining most of its desirable photographic characteristics.

VALENCE

Depending on the free charges in their outer electron orbits, elements combine to form compounds in various whole-number proportions. Thus a sodium atom (ion) with a valence of one (Na^+) and a chlorine atom (ion) with a valence of one (Cl^-) combine to form NaCl, sodium chloride. Gold, with a valence of one or three, takes a valence of three (Au^{+++}) and each atom requires three chlorine atoms to form gold chloride ($AuCl_3$). Nitrogen can have a valence of three or five. Oxygen has a valence of two, so two atoms of nitrogen can unite with three atoms of oxygen to form N_2O_3 (nitrous anhydride). When nitrous anhydride combines with a molecule of water (H_2O), it forms two molecules of nitrous acid ($2HNO_2$).

ATOMIC WEIGHT

The *atomic weight* is the relative weight of the atom—with that of oxygen taken as 16—or very near to the weight of the nucleus (protons + neutrons) of the atom. The *molecular weight* of a compound is the total of the weights of the atoms making up the molecule. To form a given quantity of a compound without any of the atomic or molecular com-

Ions are atoms that have a positive or negative charge.

Atoms can combine to form compounds. The number of each type of atom required depends on its valence.

Atomic weights are the relative weights of atoms with the weight of oxygen taken as 16.

ponents left over, the reaction is carried out with amounts that are proportional to their atomic or molecular weights. When 23 grams of sodium (atomic weight—23) are combined with 35.5 grams of chlorine (atomic weight—35.5), 58.5 grams of sodium chloride (molecular weight—58.5) are formed. A *molar solution* is one that has the same number of grams as the molecular weight dissolved in one liter of water. Such solutions would have equal numbers of molecules of chemicals dissolved in them, and equal numbers are then available for reaction. A molar solution of sodium chloride would contain 58.5 grams per liter.

CHEMICAL REACTIONS

When two compounds in solution are mixed by adding separate solutions of the compounds together, a chemical reaction may take place. If the new compounds remain in solution, they assume an equilibrium of the ions, and the solution may take on some of the characteristics of either of the compounds in solution alone. If one of the products of the reaction is a gas, it is given off and removed from the reaction. If one of the products is a solid that is less soluble than either of the compounds entering into the reaction, a precipitate is formed, which is removed from the reaction. In either case, the reaction proceeds in the direction of the compound that is being removed, until the quantity that can remain in solution under the conditions is reached. If the reaction products (precipitate or gas) are removed, the reaction can continue as long as the unreacted chemicals are added.

For example, a solution of silver nitrate added to a solution of sodium chloride will form relatively insoluble silver chloride and soluble sodium nitrate. (Chemical reactions can be indicated by means of chemical equations.) Silver nitrate added to sodium chloride can be indicated by the following (see Figure 6–3):

$$AgNO_3 + NaCl \longrightarrow AgCl + NaNO_3$$
$$\downarrow$$
(precipitate)

or more precisely,

> **When oxidation occurs, reduction also occurs.**

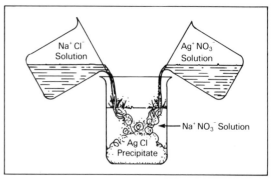

Figure 6–3   When clear, colorless solutions of sodium chloride and silver nitrate are added together a white insoluble precipitate of silver chloride is formed, along with a clear solution of sodium nitrate.

$$Ag^+ NO_3^- + Na^+ Cl^- \longrightarrow$$
$$AgCl + Na^+ NO_3^-$$
$$\downarrow$$

The silver nitrate and sodium chloride in solution have dissociated into their ions or radicals. When the silver and chlorine ions combine, they no longer exist as ions (are not dissociated), but form an insoluble *precipitate* that is removed from the reaction. Na^+ and NO_3^- ions remain in the solution.

OXIDATION/REDUCTION

Chemical reactions can take place without being in solution; that is, between solids, liquids, and gases. When something burns, it means that oxygen has combined with it, giving off heat at the same time. This is referred to as *oxidation*. Whenever oxygen unites with any other material, oxidation takes place. Oxygen is called the *oxidizing agent*. The material with which the oxygen combines is referred to as the *reducing agent*. Thus, *reduction* also occurs when a material takes on oxygen—one is the reciprocal of the other.

The oxidation/reduction concept can be applied to elements other than oxygen. Any element that gains electrons more rapidly than another element is said to have higher oxidizing power. Conversely, any element that loses electrons more readily than another element is said to have higher reducing power. When silver metal is subjected to chlorine gas, the two elements combine to form silver

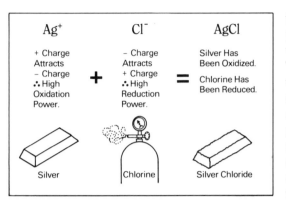

Figure 6–4 Oxidation occurs when a positively charged atom attracts a negatively charged atom, and reduction occurs when a negatively charged atom attracts a positively charged atom. In this reaction silver has been oxidized and chlorine has been reduced to form silver chloride.

chloride (see Figure 6–4). This is an oxidation/reduction reaction where the silver is oxidized (and the chlorine is reduced), even though no oxygen is involved.

When treated with a reducing agent, such as a solution of ferrous sulfate, the silver chloride will be converted to metallic silver to form ferric sulfate and ferric chloride (see Figure 6–5). The iron, in changing from the ferrous form with a valence of two (Fe^{++}) to the ferric form with a valence of three (Fe^{+++}), provides the electron that neutralizes the $^+$ of the silver:

$$3\ Ag^+\ Cl^-\ +\ 3\ Fe^{++}\ SO_4^{--} \longrightarrow$$
$$Fe_2^{+++}\ (SO_4)_3^{---}$$
$$+\ 3\ Ag°\ +\ Fe^{+++}\ Cl_3^{---}$$

If the silver chloride occurs in a photographic emulsion, a mild reducing solution,

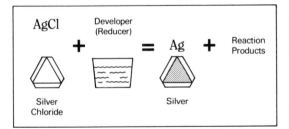

Figure 6–5 Silver chloride in a photographic emulsion is acted on by a reducer, the developer, to form metallic silver along with the other products of the reaction.

the developer, will selectively reduce (by providing electrons) only that part of the silver chloride that has been exposed to light. The "latent image" is converted into a real, visible silver image. Oxidation/reduction concepts can become rather involved. (If it seems inappropriate for a reducing agent to *provide* electrons rather than *accept* them, it should be noted that the oxidation/reduction concept was formulated at a time when it was commonly thought that electrons had positive charges.)

ACID/BASE

An *acid* is a compound containing hydrogen and another element or radical. Hydrochloric acid, for example, is a compound containing hydrogen and chlorine, HCl. This is a strong acid. HCl can exist as a gas, but as an acid it is a gas in a water solution. Sulfuric acid is made up of hydrogen and the SO_4 radical (SO_3 [gas] + H_2O [water] = H_2SO_4 [sulfuric acid]). Like hydrochloric acid, this is also a strong acid, which means that it is highly ionized or dissociated in solution; that is, H^+ SO_4^-, and H^+ Cl^-. Acetic acid, an organic compound, CH_3COOH, is a relatively weak acid by comparison. Acids are formed when oxides of nonmetallic elements are dissolved in water, and bases are formed when oxides of the metallic elements are dissolved in water. Essentially, a base is a compound containing a hydroxyl (OH^-) radical and another element or radical. Calcium oxide, CaO, when dissolved in water, yields calcium hydroxide, $Ca(OH)_2$, a *base*. A base is also known as an *alkali*. Practically all developers are basic in composition. When placed in water, metallic sodium produces a violent reaction in which one of the hydrogen atoms of water is given off as a gas, which ignites from the heat of the reaction and burns, leaving a strong alkali, NaOH (Na^+ + OH^-), as shown in Figure 6–6. Ammonium hydroxide, NH_4OH, is also a base, and has the characteristic "ammonia" odor. (Ammonia with moisture, ammonium hydroxide, is the developing agent for diazo papers.)

Acids are sour to the taste (such as acetic acid, the chief component of vinegar). Bases are characterized by their "slippery" feeling—a weak solution of lye, NaOH, feels very

When an acid and an alkali combine, water is formed along with a salt, gas, or precipitate.

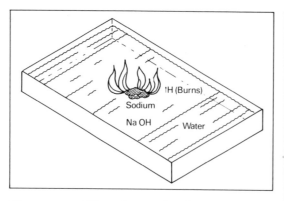

Figure 6–6 When the metal sodium is placed in water a violent reaction takes place in which one of the hydrogen atoms of water is released. The hydrogen burns while the remaining single atom each of oxygen and hydrogen combine with the sodium to form strongly alkaline sodium hydroxide.

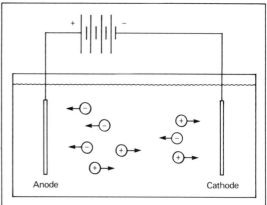

Figure 6–7 A solution of sodium chloride contains negatively charged chlorine ions and positively charged sodium ions. Electrodes placed in the solution and connected to a battery will cause the negatively charged ions to move to the anode where they lose their charge and become chlorine gas. The positively charged sodium ions will move to the cathode to form sodium metal, which reacts with the water to form sodium hydroxide and hydrogen gas.

slippery, as do most developer solutions that contain an alkali. When an acid solution is combined with a solution of a base, in equal molecular amounts, a *salt* is formed. For example, if hydrochloric acid is added to sodium hydroxide, sodium chloride, a salt, and water are formed. The water of formation is simply added to the water containing the acid and the base:

$$HCl + NaOH \longrightarrow NaCl + H_2O$$

$$H^+ Cl^- + Na^+ OH^- \longrightarrow$$
$$Na^+ Cl^- + H_2^+ O^-$$

IONIZATION

Pure water is made up of equal parts of H^+ (hydrogen) and OH^- (hydroxyl) radicals. When a salt is dissolved in water, it dissociates into the ions of the elements or radicals making up the salt. Sodium chloride, for example, ionizes into Na^+ and Cl^- ions. The presence of these free ions permits the passage of an electrical current through the solution, and it becomes known as an *electrolyte* (see Figure 6–7). A solution of sugar, an organic compound that does not dissociate into ions, therefore does not carry an electric current.

pH

It is the custom for chemists to specify the hydrogen ion concentration in terms of pH.

pH is the negative logarithm of the hydrogen ion concentration. A strong acid with a hydrogen ion concentration of 10^{-2} (i.e., 1/100) has a pH of 2, and a strong alkali with a hydrogen ion concentration of 10^{-12} (i.e., 1/1000000000000) has a pH of 12. Pure water or a neutral salt solution that has an equal number of positive and negative ions has a pH of 7. The higher negative exponents, representing lower hydrogen ion concentrations, and thus higher hydroxyl concentrations, range from 7 through 14, and are the alkaline side of the scale. (D-72 developer for paper has sodium carbonate as the alkali, and a pH in the vicinity of 10.) Conversely, lower negative exponents, representing higher hydrogen ion concentrations, are on the acid side of the scale, and range from 1 through 7. (An acetic acid stop bath, for neutralizing the alkali of the developer after development, might have a pH in the vicinity of 3.5.) Or this can be restated as follows:

$$pH = \frac{1}{\log (H^+)}$$
$$OR \qquad = -\log (H^+)$$

The atoms of salts dissociate when dissolved in water, forming negative and positive ions.

Acids have pH values below 7 and alkalis have pH values above 7.

or pH is the negative logarithm of the hydrogen ion concentration.

pH METERS

Since pH is an indication of hydrogen ion concentration, it is also related to electrical potential. That is, if suitable electrodes are placed in such an ionized solution, it acts as a battery or galvanic cell, and produces a voltage, E. The relationship between pH and the voltage (or electrical potential) is rather complex, and therefore several qualifications have to be made when using this potential to measure pH. Since we are interested in the potential—without the passage of current, which would have the effect of immediately lowering the voltage—the voltage is usually measured in comparison to a standard cell incorporated in the meter. Some modern meters, however, with very little internal resistance, are direct reading. The relationship between the measured potential and pH is as follows:

$$\text{pH} = \frac{(E - E_o)}{0.0591}$$

where E_o is a constant.

pH meters are calibrated directly in pH values, and it is not necessary to read a voltage and then convert it to pH. The measurements are directly affected by temperature; this has to be taken into consideration and adjustments made if it is different from 25 C.

pH INDICATORS

Many dyes change their color under the influence of a particular pH condition. These are referred to as *pH indicators*. Papers that have been impregnated with the dyes are available for indicating in various pH ranges (see Figure 6–8). These are generally more broad in their measure than pH meters, and there is a limit to the number of useful indicators that are available. Also, some judgment may be required in the interpretation of the colors produced. However, they are useful under many circumstances.

The performance of various photographic chemical solutions is influenced to a great

Figure 6–8 Dye-treated pH indicator paper. A small strip of the paper is taken from the package and dipped in the solution to be tested. The dye in the paper changes color according to the pH of the solution. The pH value is then assessed by comparing it with the color references on the package. (Photograph by Ira Current.)

extent by pH, and the control of this factor in one way or another is important. For example, the rate of development tends to increase as the pH, or alkalinity, of the developer is increased. Bleaches and other solutions used in color photography require control of pH to make sure that the bleaching action is adequate while at the same time dyes formed by the process are not adversely affected. The acid strength of stop baths and fixers has to be maintained to ensure that their neutralizing effects are adequate without producing other problems. Rate of fixing, hardening, and stability of the fixing solution are influenced by pH.

A pH meter is a voltmeter calibrated in terms of pH.

Some dyes change color in response to a particular pH condition and can therefore be used to measure pH.

COMPLEXES

A *complex compound* is one that is made up of two or more compounds or compounds and ions. In the fixing process used in photography, for example, silver complexes are formed when the unexposed and undevel-

During the fixation process, different silver complexes are formed in sequence with increasing degrees of solubility.

oped silver halide salt is treated with the fixing agent, sodium thiosulfate, $Na_2S_2O_3$. The fixing reactions pass through several complexes of varying solubility before one that is easily soluble is reached. The following represent several complexes that are thought to occur:

1. $2\ AgCl + Na_2S_2O_3 \longrightarrow$

    $Ag_2S_2O_3 + NaCl$
    (insoluble)

2. $Ag_2S_2O_3 + Na_2S_2O_3 \longrightarrow$

    $2\ NaAgS_2O_3$
    (slightly soluble)

3. $4NaAgS_2O_3 + Na_2S_2O_3 \longrightarrow$

    $Na_6Ag_4(S_2O_3)_5$
    (slightly soluble)

4. $Na_6Ag_4(S_2O_3)_5 + Na_2S_2O_3 \longrightarrow$

    $Na_5Ag_3(S_2O_3)_4$
    (soluble)

5. $2Na_5Ag_3(S_2O_3)_4 + Na_2S_2O_3 \longrightarrow$

    $3Na_4Ag_2(S_2O_3)_3$
    (soluble)

When the silver halide has been converted to complexes 4 and 5, they can be washed out of the emulsion and/or paper to increase the image permanence. Complexes 1, 2, and 3, while they are transparent and give the film or paper the appearance of being fixed, are insoluble or only slightly soluble, and thus are not removed by washing.

SOLUTIONS

When a substance, a solute, is dissolved in a solvent, a solution is formed.

The weight of a given volume of a substance compared to that of an equal volume of water is known as specific gravity.

Most of the chemical processing of photographs is accomplished with various *solutions*. A solution consists of the *solvent*—water in most photographic applications, with various chemical compounds dissolved in it. A chemical in solution is referred to as the *solute*. Various compounds have varying degrees of *solubility*; that is, capability of being dissolved in water, the solvent, until no more will dissolve. Such a solution is known as a *saturated solution*. The solubility of most chemicals varies with the temperature of the

solution; a greater amount of the chemical can be dissolved in water at higher temperatures. When a solution has as much solute dissolved in it as is possible at a higher temperature, and the solution is cooled, the solute is thrown out of solution, usually in the form of crystals. A *supersaturated* solution is one that contains more solute dissolved in it than would normally be possible at that temperature. The addition of a small crystal of the solute, or some other disturbance, will cause the excess to be crystallized out rapidly. If a chemical compound in solution is mixed with another one also in solution, the product of the chemical reaction may be insoluble, and will be thrown out as a *precipitate*.

Other liquids can also be considered to be solvents, and indeed in chemistry there are many systems in which solvents other than water are used.

Rate of solution also depends on the size of the particles being dissolved, as shown in Figure 6–9. Small particles or crystals have a much higher ratio of surface to volume, hence the solvent can act over a larger area in a given time and thus the particles go into solution faster. (Extremely fine powders may provide excess surface and permit such factors as hydrolysis—reactions with water—to

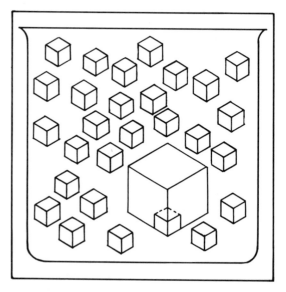

Figure 6–9   Volume of large cube (hence weight) is equal to that of all the smaller cubes, but surface area of large cube is one third that of the smaller cubes.

decrease solubility. The chemistry of solubility is complex and involves rates of diffusion, degree of dissociation, and many other factors.) Chemical compounds may either give off heat—*exothermic*—or absorb heat—*endothermic* when they are dissolved. Anhydrous sodium thiosulfate ($Na_2S_2O_3$) gives off heat when it goes into solution; but the compound of crystallization with water ($Na_2S_2O_3 \cdot 5H_2O$) absorbs heat when it goes into solution; that is, it is endothermic and the solution becomes cooler as the thiosulfate is dissolved. Monohydrated sodium carbonate ($Na_2CO_3 \cdot H_2O$) gives off heat when it is dissolved, and is thus exothermic; but sodium carbonate with 10 molecules of water of crystallization ($Na_2CO_3 \cdot 10H_2O$) absorbs heat, cools the solution, and is endothermic.

SPECIFIC GRAVITY

Specific gravity is the weight of a given volume of a substance, such as a solution, compared to that of an equal volume of water at a given temperature (which has to be taken into account for precise measurements). Specific gravity can be measured with a hydrometer, a graduated weighted tube that floats to a greater or lesser degree depending on the specific gravity of the solution in which it is floated. Specific gravity is read from a scale inside the tube.

REVIEW QUESTIONS

1. An element that is not a member of the halogen family is . . . *(p. 178)*
   A. chlorine
   B. fluorine
   C. helium
   D. iodine
   E. bromine
2. Atoms normally have . . . *(p. 178)*
   A. a positive electrical charge
   B. a negative electrical charge
   C. a neutral electrical charge

3. Since the chemical formula for water is H_2O, if it is known that hydrogen has a valence of 1, it can be assumed that oxygen has a valence of . . . *(p. 179)*
   A. 0
   B. 1
   C. 2
   D. 3
   E. 4
4. The light-sensitive compound silver bromide (AgBr) could be produced with a chemical reaction between . . . *(p. 180)*
   A. silver nitrate and sodium chloride
   B. silver chloride and sodium nitrate
   C. silver nitrate and silver bromide
   D. silver nitrate and sodium bromide
5. Developers for black-and-white films and papers are . . . *(p. 181)*
   A. reducing agents
   B. oxidizing agents
   C. nonparticipatory agents
6. One could identify a tray of developer by touch in a darkroom sink that also contained trays of stop bath, fixer, and water because the developer would feel . . . *(p. 181)*
   A. warm
   B. slippery
   C. sticky
   D. stingy
   E. gritty
7. An ion is an atom that has . . . *(p. 182)*
   A. no electrical charge
   B. a positive charge
   C. a negative charge
   D. either a positive or a negative charge
8. A photographic processing liquid that has a pH of 9 is probably . . . *(p. 182)*
   A. a developer
   B. a stop bath
   C. a fixing bath
   D. a hypo clearing agent bath
   E. the wash water
9. Indicator papers identify pH by means of a change in . . . *(p. 183)*
   A. lightness
   B. hue
   C. conductivity

7 Photographic Emulsions, Films, and Papers

Weston Kemp. *Stonehenge, Druids at Summer Solstice.* Copyright © 1988 by Weston Kemp.

PHOTOGRAPHIC EMULSIONS

Silver is used in photographic emulsions because its compounds are more sensitive to light than any other photochemical material.

A photographic emulsion consists of silver halide crystals suspended in gelatin. The emulsion is usually coated on a support that may be clear, as in the case of negative films, or opaque, as for photographic prints. The emulsion has the most significant controlling effect on the photographic and physical properties of the final product.

To visualize the emulsion, think of a chocolate candy bar with peanuts imbedded in it as shown in Figure 7–1. The peanuts are suspended in the chocolate medium. Similarly, the silver halide particles are suspended in gelatin. There are many different recipes for making photographic emulsions. Choice of gelatins and different combinations of salts (sodium, potassium, ammonium, or calcium and chloride, bromide, or iodide, for example), along with other chemical ingredients impart characteristics to the emulsion such as speed, spectral sensitivity, contrast, resolution, graininess, and physical properties. The emulsion is applied to the film or paper in a "liquid" state, and thickness of the emulsion layer is adjusted by controlling viscosity and speed of coating. The gelatin is then chilled (just as a gelatin dessert is when placed in the refrigerator), and then slowly dried to remove most of the water.

The silver halides respond to light to produce a latent image that is later developed to produce a visible silver image. The gelatin also acts as a "binder," serving to protect the silver halide from abrasion and other mechanical and chemical influences; in some cases the gelatin can even serve as the support for the image. Through "impurities" and other characteristics, the gelatin contributes to the photographic and chemical performance of the silver halides in the emulsion. Silver is used because its compounds are more sensitive to light than any other photochemical material.

FILM SUPPORTS

Cellulose nitrate, which is chemically similar to guncotton, was not the safest support for photographic films.

In the early 1900s flexible film supports were introduced. These early supports were essentially cellulose nitrate, and were produced in a manner not unlike the making of a wet collodion plate. In the beginning, a solution of the cellulose nitrate in solvents was flowed from a hopper on to a long glass table, and allowed to dry (see Figure 7–2). Then the film was stripped off of the glass to provide a flexible support that could be coated with a gelatin silver halide emulsion similar to that used today.

As can be imagined, cellulose nitrate, which is also chemically similar to guncotton, an inflammable compound, was not the

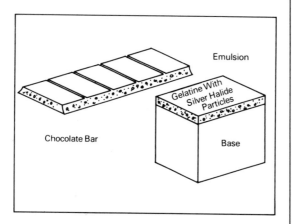

Figure 7–1 The photographic emulsion can be compared to the chocolate nut bar. The nuts are suspended and separated by the chocolate as the silver halide grains are in the emulsion.

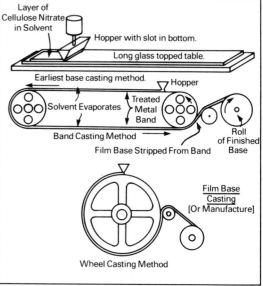

Figure 7–2 Film base casting (or manufacture).

safest support for photographic films. It had reasonably good physical characteristics, but anywhere that large collections of negatives existed there was a pronounced danger of fire. In fact, there were many fires in motion picture theater booths where the film could be ignited accidentally by the carbon arc light, particularly if the film was delayed for some reason in the projector film gate. There were also some disastrous hospital fires where accumulations of X-ray film made with nitrate base became ignited. Nitrate bases can spontaneously ignite as the result of deterioration, and archives of this material have to be carefully watched. Considerable effort has gone into preserving early nitrate-based images by photographically transferring them to more stable modern bases.

SAFETY BASE FILMS

The Eastman Kodak Company developed a nonflammable cellulose acetate safety film very early in the 1900s, but it was not accepted by the motion picture industry because of cost; and some of the other physical characteristics, such as wear on repeated showings, were not as good as those of the nitrate film. With more experience and further research and development, improved versions of "acetate" films were produced, and they are still the best support for commercial motion picture film.

Nitrate base of greater thickness was also used for early sheet films replacing glass plates. These nitrate films were supplanted by acetate bases when this type of material became available. While acetate bases had better dimensional stability during processing and storage than did nitrate bases, they were still far from ideal for some applications. They were less flexible, more brittle, and did not wear as well as the nitrate films.

POLYESTER FILM BASES

In recent years many of the products previously manufactured with acetate base have been produced on polyester bases, which are made in an entirely different manner. These are manufactured by extruding the heated polyester material through a slot, stretching

it in a longitudinal direction at the same time that it is stretched by means of grippers in a widthwise direction while still heated and soft, as shown in Figure 7–3. This orients the molecules of plastic, yielding a base material that is free from stresses and strains, and that has very good dimensional stability and certain other characteristics.

One problem with this type of base is the need of a suitable subbing so that the emulsion will adhere properly during the life of the material. While the techniques of subbing for good adhesion have been worked out satisfactorily for the old solvent cast base materials, there have been some lingering problems with the newer polyester materials, leaving some doubt as to their suitability for "archival" storage of images. Another early problem with the polyester films for motion picture applications was that of splicing, but this has been solved by new heat and tape splicing techniques (in those applications where the stronger base might be required).

Such polyester films can act like fiber optics and pipe light in a transverse direction, thus not serving well as a self leader when the film itself is intended to protect the image-forming part of the film from ambient light when on a daylight loading spool. In many applications, light piping of films is eliminated by introducing dyes or pigments into the base itself, as shown in Figure 7–4. Some light is absorbed in the viewing direction, where the density of the base may be on the order of 0.1 to 0.2, and the thickness of the base may be around 0.002 to 0.004 inch. In

Polyester film base materials have high strength, good dimensional stability, and other desirable characteristics, but . . .

they also tend to generate more static electricity, scratch more easily, and pipe light in from the edges.

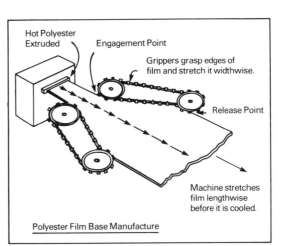

Hot Polyester Extruded

Engagement Point

Grippers grasp edges of film and stretch it widthwise.

Release Point

Machine stretches film lengthwise before it is cooled.

Polyester Film Base Manufacture

Figure 7–3   Polyester film base manufacture.

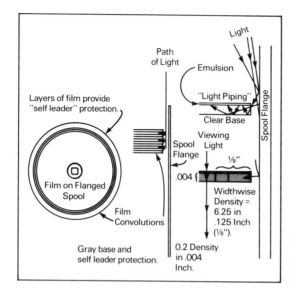

Figure 7–4   Daylight loading spools depend on the protection provided by several convolutions of film (4 to 7 feet in length). The necessary light stopping density is provided by the combination of emulsion, antihalation protection, and base density. Light entering between the film and the spool flange may be decreased by incorporating a pigment or dye in the base. The density in the viewing direction is relatively low, but adds up rapidly in the widthwise dimension.

The resin coating on the paper base of RC papers protects it from processing solutions and therefore shortens processing time.

the crosswise direction the density adds up very rapidly, and thus absorbs most of the light traveling in this direction.

Polyester films also have a greater tendency to generate static electricity—they are good insulators—and this makes them more susceptible to picking up dust and lint from the atmosphere. They are also more susceptible to scratching, and readily show scratches on viewing the photographic images if the base is not protected by a gelatin NC (non-curl) or antihalation coating. In the beginning, their very high tensile strength prevented them from breaking when trouble occurred in processing machines and similar equipment, and therefore the machinery was damaged rather than the film. The solvent cast acetate types of film are still largely used for commercial and amateur motion-picture applications, and for most roll films.

FILM STRUCTURE

A photographic film can be very simple in structure—a coating of a gelatin–silver hal-

ide emulsion on a transparent base, such as positive film, used for printing black-and-white motion pictures. Negative films used for pictorial photography are generally more complex in their structure. In many films two or more coatings of emulsion are used to derive the sensitometric characteristics that yield good pictorial tone rendition. On top of this there is generally an overcoating or surface layer that controls many of the physical characteristics of the film.

PAPER BASES

Most photographic prints are made on some form of paper base. Traditionally this type of base was manufactured with pulp made from old rags, but in more recent years practically all of it has been made from a high-quality wood pulp (alpha cellulose). These are referred to as *fiber-base* papers. Many photographs are made on paper that has been coated with a plastic or resin (such as polyethylene) on both sides. These are referred to as *RC* (resin-coated) papers. The resin surface coating of the stock protects it from the chemical solutions and water of processing, so that processing, washing, and drying times can be much shorter than with the fiber-base papers. Because of the problems of emulsion adhesion to polyethylene coatings, fiber-base papers are still preferred for producing photographs with archival permanence.

Several paper surfaces are produced by embossing the stock with an engraved roller under pressure to impart a distinctive textured surface pattern. These patterns can be irregular, but some are geometric in nature, for example, "silk" and "linen" surfaces. Any of the above can be coated with one or more layers of barium sulfate (baryta), and/or other white pigments (sometimes before embossing) before coating with photographic emulsion. This imparts a better, brighter color to the base, and also limits the penetration of the emulsion into the base, which gives a more uniform coating and a more even image, with more uniform blacks. The nature of the baryta coating also affects the reflectivity of the coating's surface. The baryta coating is omitted from photographic papers that may be folded in use, to prevent the emulsion layer from cracking.

BRIGHTENERS

It is common to incorporate optical brighteners in both nonbaryta-coated and baryta-coated paper surfaces, producing the effect of intensifying the paper's brightness. These are similar to laundry brighteners, and the effect is produced by using a dye or other substances that fluoresce on exposure to ultraviolet energy (available to some extent from most light sources) to convert the invisible ultraviolet radiation to light.

PAPER WEIGHT AND THICKNESS

Photographic papers are manufactured in various "weights." These are referred to in such terms as single weight, document weight, light weight, medium weight, and double weight (see Figure 7–5). Paper stock is customarily manufactured, controlled, and sold in terms of weight, but photographers are more aware of the different thicknesses of photographic paper bases. Hence, international standards designate papers in terms of their thicknesses. American National Standard PH1.1-1974, *Thickness of Photographic Paper, Designation For,* lists nine groups of paper thicknesses, with ranges for each group both in English and in metric units, and gives common trade designations (in terms of weights) for each of the groups.

Other related American National Standards publications include *Dimensions for Photographic Roll Paper, Dimensions for Photographic Sheet Paper for General Use, Requirements for Spooling Photographic Paper for Recording Instruments,* and *Methods for Determining the Dimensional Change Characteristics of Photographic Films and Papers.*

THE GELATIN COLLOID

Gelatin is a *colloid.* A colloid is a particulate material that can be suspended in water or other solute without settling out. The sizes of the particles range from approximately 1 to 1,000 nm, or intermediate between visibly suspended particles and invisible molecules. (One nanometer is one billionth of a meter,

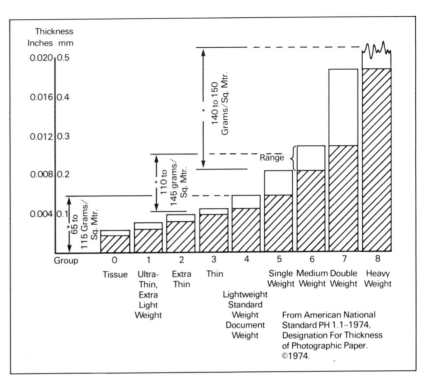

Figure 7–5   Paper thicknesses.

or one millionth of a millimeter.) Other colloids have been used for photography, and colloids that are a more similar substitute for gelatin have been tried, but gelatin has been, is, and will continue, for the near future at least, to be the best material for the preparation of photographic emulsions. (Albumen and collodion are examples of other colloids that have been important in photography.)

PROPERTIES OF GELATIN

The gelatin colloid used to make photographic emulsions has several important properties:

1. Gelatin disperses the light-sensitive silver halide crystals and prevents them from adhering to one another, or coagulating, and thus forming in effect larger crystals or "grains."
2. When wet, gelatin can be changed to a "solid" gel or liquid reversibly by changing the temperature.
3. When dry, gelatin is reasonably stable, and thus protects the silver halide grains.

A standard on photographic papers lists nine different categories of paper thicknesses.

An important feature of gelatin is that it has a large molecular structure.

Photographic gelatin has many desirable characteristics in addition to being transparent.

4. Gelatin serves as a receptor for halogen atoms in some aspects of latent image formation.
5. Gelatin has no effect on the silver halide, other than the protective function, although impurities found in many gelatins contribute to the photographic result (sensitivity, fog, etc.), sometimes beneficially, at other times in a manner that degrades the image.
6. Gelatin permits the processing solutions to penetrate and chemically react with the silver halide grains in development, fixing, etc.
7. Gelatin can be produced uniformly and inexpensively, and stored for long periods prior to use in manufacture.
8. Gelatin is transparent.

To help ensure uniformity, numerous batches of gelatin of a given type are blended so that when a new batch is added, it contributes only a relatively small amount to the overall characteristics of the blend. (American table wines are often blends of various batches of grapes for the same reason.)

CHEMICAL NATURE OF GELATIN

Gelatin tolerates both acid and alkaline solutions, prolonged washing, and changes in temperature.

Basic emulsion making consists of precipitating a silver halide in a gelatin solution.

Gelatin is an organic compound or, more precisely, a group of compounds (they are derived from "organic" materials—living animals—and can be burned) largely made up of carbon, hydrogen, oxygen, and nitrogen atoms having a composition of approximately 50, 7, 25, and 18 parts of these atoms, respectively. It has a very complex molecular structure made up of various amino acids. The average molecular weight of gelatin is about 27,000 or some multiple of this. (Molecular-weight values determined in a variety of different ways have ranged from 768, corresponding to a formula $C_{32}H_{52}O_{12}N_{10}$, to 96,000.) Since the amino acid molecules contain both acid carboxyl groups and basic amino groups, they are *amphoteric*; that is, they can act either as an acid or a base. Since the acidic and basic characteristics are not strong, it acts as a buffer: Large additions of either an acid or a base do not have a large effect on the hydrogen ion concentration (pH).

PHYSICAL PROPERTIES OF GELATIN

Dry gelatin contains about 10% water, and is a tough material with great mechanical strength. To prepare an emulsion, dry gelatin is soaked in water, which penetrates into the gelatin structure and causes it to swell many times its original dimension. When thus wet, it is soft and easily damaged. When the soaked gelatin is heated to about 40 C (100 F), it melts and can be further diluted with water indefinitely. If the concentration of gelatin in water is greater than 1%, it will "set" when cooled, just as dessert gelatin does, to become a gel that can be dried with dry air without remelting. Before drying, the set gelatin can be remelted by raising the temperature, and reset by cooling, repeatedly; but the setting and melting temperatures do not coincide.

EMULSION MAKING

Throughout the history of photography, the theory of the emulsion-making process has been difficult to understand. The characteristics of the final emulsion are governed by many factors, including the choice of soluble halides and gelatin, the method of silver halide precipitation, the cooking or ripening processes, and the choice of chemical additives to the emulsion. Since many of these

Figure 7–6   Processed gelatin ready for use in photographic emulsion. The flakes produced in the gelatin extraction process have been ground into a powder that has a pale amber color.

processes cannot be patented, photographic manufacturers take great pains to maintain the secrecy of their emulsion-making techniques. The following basic considerations, however, are well known. The basic emulsion-making steps for a typical black-and-white film emulsion are: precipitation, ripening, washing, digestion, additions, and coating.

A photographic emulsion is formed by treating a soluble silver salt with a soluble halide or halides in the presence of gelatin in solution; for example,

$$AgNO_3 + KCl \longrightarrow AgCl + KNO_3$$

Silver nitrate plus potassium chloride yields silver chloride plus potassium nitrate.

The light-sensitive silver halide, in this example silver chloride, exists in the form of crystals or "grains" that are one micrometer (one thousandth of a millimeter) in diameter or less. Mixtures of halides (chloride, bromide, iodide) are commonly in the form of mixed crystals containing two or three halides in each crystal (see Figure 7–7). The gelatin acts as the protective colloid in that it prevents the crystals from coalescing, and also controls the size and distribution of the crystals to some extent. Silver halides are primarily sensitive to ultraviolet radiation and blue light. Small amounts of compounds that react with the silver halide to increase sensitivity may be present in the gelatin, or they may be added separately to inert gelatin. Dyes may also be added to extend sensitivity to other regions of the spectrum than the ultraviolet and blue.

EMULSION CHARACTERISTICS

The photographic properties of an emulsion depend on a number of factors including its silver halide composition; the shape, average size, and size distribution of the crystals, and the presence of substances that affect sensitivity. All of this is governed by the amount and kind of gelatin(s) in the original solution, choice of halide compounds, the way in which they are mixed together, the way they are treated following this "precipitation," what other substances are added, and the coating procedure. These factors control photographic properties such as the speed, characteristic-curve shape, spectral sensitivity, and exposure latitude of the emulsion; and image characteristics such as graininess, sharpness, and resolving power.

GRAIN SIZE, SENSITIVITY, AND CONTRAST

The grains of the emulsion can be considered as individual units as far as exposure and development are concerned. A latent image formed in one of the grains does not ordinarily spread to the other grains unless they are touching. For a given exposure there is a greater probability that a large grain will absorb a quantum of light than will a small grain. If the grains were uniform in size, and in a single layer (one grain thick), the probability that individual grains would be exposed and made developable would depend on the random distribution of photons reaching the grains. Since such grains are uniform in size, density after development will vary with the fraction absorbing at least the number of photons required to become developable. When several photons are required, most of the grains will receive enough at the same time; and thus the characteristic curve will have a steep slope, indicating high contrast. The toe of the curve represents the region of exposure where only a small fraction of the grains receive enough photons to become developable.

Controls such as choice of silver halides, type of gelatin, method of mixing, additives, and cooking time make it possible to produce a great variety of film and paper emulsions.

The size and size distribution of silver halide grains have an important effect on the sensitivity and contrast of photographic emulsions.

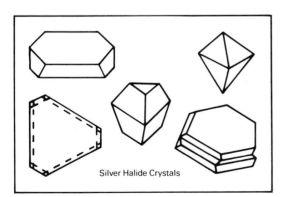

Silver Halide Crystals

Figure 7–7   Silver halide crystals.

If the grains are large, they provide a greater area for receiving photons, are thus more likely to be exposed, have a greater number of silver ions available for reduction, and after development they have a greater light-absorbing capability. This has the effect of giving the larger grains a greater amplification effect than smaller grains.

Intermediate-size grains would produce an amplification effect between those of the largest and smallest sizes (see Figure 7–8). Emulsions with a wide variety in grain sizes will have inherently lower contrast than emulsions having equal grain sizes, and will have greater sensitivity due to the availability of larger grains. The resulting characteristic curve will have a lower slope than in the case where the grains are all nearly the same size.

Silver bromide, chloride, and iodide have different ranges of spectral sensitivity in the short wavelength part of the spectrum.

GRAIN COMPOSITION

The composition of the silver halide grains plays an important part in the emulsion. The presence of iodide in small amounts enhances the sensitivity of silver bromide grains. The composition of the grains formed by precipitation of a silver salt and two or three halogen salts depends on the solubility characteristics of the precipitated halides. The three light-sensitive silver halides used in photography are relatively insoluble compared to the salts from which they are formed.

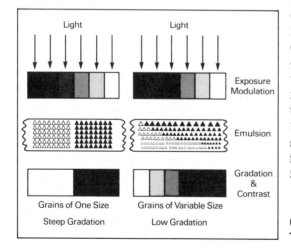

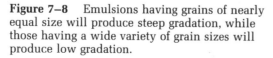

Figure 7–8   Emulsions having grains of nearly equal size will produce steep gradation, while those having a wide variety of grain sizes will produce low gradation.

The overall solubility of the silver halides in emulsions is increased by virtue of the small size of the crystals, 1 micrometer or less, which increases the surface-to-weight ratio significantly. This difference in solubility according to grain size accounts for the increase in average grain size during ripening, in which the smallest grains become smaller and the dissolved silver halide comes out of solution as an addition to the larger grains.

SPECTRAL SENSITIVITY

The fine division of halides in emulsions provides a greater amount of surface for adsorption of materials to the silver halide grains. Adsorption is the adherence of atoms, ions, or molecules to the surface of another substance. First, silver halide grains adsorb additional halide ions, with the greatest adsorption occurring with iodide and the least with chloride ions. In addition, the grains adsorb gelatin, sensitizing dyes, and other compounds. These all contribute to the overall sensitivity of the emulsion. The silver halide grains themselves have varying sensitivities to light of different wavelengths, depending on the halide. Silver chloride, which is colorless, is sensitive at the shortest wavelengths, mostly ultraviolet energy, extending up to about 420 nm. Silver bromide, which is more yellowish in appearance, has sensitivity extending to about 500 nm. Additions of iodide, which has a strong yellow color, in amounts ranging from 0.1% to 1.0% of the chloride crystals, extends the sensitivity to 450–475 nm. A similar extension is achieved when about 40% bromide is added to chloride. Bromide with about 3% iodide extends the sensitivity to about 525 nm (see Figure 7–9). Even without further extensions of sensitization, such as with dyes, the safelight filter must not have any transmission below 550 nm.

CRYSTAL STRUCTURE

Defects in the crystals of silver halides are an important aspect of light-sensitive emulsions. Silver halide crystals are usually composed of ions in a cubic lattice so that each

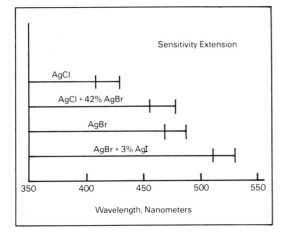

Figure 7–9   In silver halide emulsions, grains of silver chloride are sensitive to the shortest wavelengths of light, silver bromide to considerably longer wavelengths, and combinations of bromide and iodide to the longest wavelengths. (Chateau et al., *The Theory of the Photographic Process,* 1966, p. 6.)

halide ion is surrounded by six silver ions, and each silver ion is surrounded by six halide ions, as shown in Figure 7–10. This arrangement can exist within the interior of the crystal, but at the outer surface there has to be one of each with only five of the other surrounding it. Even in the cubic system, crystals can be formed that are octahedral in shape, or in the form of hexagonal plates (see Figure 7–7). Different photographic effects are produced by the surfaces presented by these different crystal shapes.

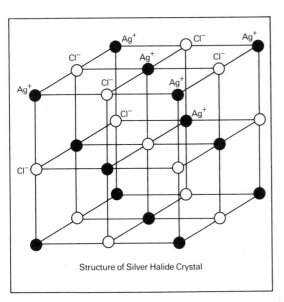

Structure of Silver Halide Crystal

Figure 7–10   Structure of silver halide crystal.

DEFECTS

The defects in the crystal lattice can be divided into extended imperfections and point defects. Whereas the faces of the crystal have only five ions around each one of opposite charge, at the corners there can be only three, and at the edges only four. Adsorption is greater at these positions and reactions may begin here. These positions are supplemented by other dislocations that can occur in the crystal's formation, and all are important in the photoconductive and photochemical processes. The most important type of point defect is interstitial silver ions; a minute fraction of the silver ions escape from their positions in the crystal lattice and can move through the spaces in the lattice, as shown in Figure 7–11.

Defects in the crystal act as locations for the formation of sub-image centers during latent image formation. Silver and halogen ions are not free to move, except to adjacent positions in the lattice, but electrons are free to move throughout the crystal. These motions are manifest as electrolytic conductivity, which plays an important part in the photolytic process. One kind of defect arises from a vacancy in the crystal lattice, such as would occur if a silver ion were removed to an in-

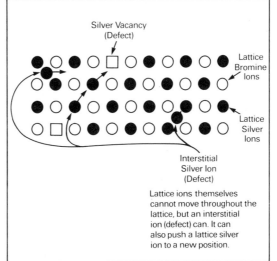

Lattice ions themselves cannot move throughout the lattice, but an interstitial ion (defect) can. It can also push a lattice silver ion to a new position.

Figure 7–11   The point defect of interstitial silver ions in a silver halide crystal. Lattice ions themselves cannot move throughout the lattice, but an interstitial ion (defect) can. It can also push a lattice silver ion to a new position.

The shape of silver halide crystals and the presence of defects in the crystals affect the photographic characteristics of the emulsion.

terstitial position. This would leave a silver-ion vacancy that would have a negative charge. The interstitial silver ion itself would have a positive charge. An extra electron would produce a negative charge, and a positive charge would occur when an electron is removed from the valence band of the crystal.

EMULSION PRECIPITATION

The rate of mixing emulsion ingredients affects the speed and contrast of the emulsion.

A typical emulsion-making process starts with a relatively dilute solution of gelatin in water (about 1%). Soluble halide salts (one or more) are added to this solution (potassium, sodium, or ammonium chloride, bromide, or iodide), as shown in Figure 7–12. Then, at a selected temperature, a soluble silver salt, such as silver nitrate, is added at a controlled rate and with controlled stirring. The silver halide or halides are precipitated out, and the crystals are first formed in a strong solution of soluble halide, in which the silver halide is much more soluble than in pure water. Under these conditions the crystals first formed grow rapidly and continue to grow throughout the precipitation.

Even a relatively dilute gelatin solution provides protection to the silver halide crystals formed. If it were not for the gelatin, the particles of the precipitate would coalesce and rapidly fall to the bottom of the reacting vessel, as shown in Figure 7–13. The growth

After the precipitation and ripening steps, excess soluble salts are removed by washing to prevent crystallization and excessive growth of the silver halide grains.

Figure 7–13   Silver bromide has been precipitated from silver nitrate and potassium bromide in the presence of gelatin in solution, in the beaker on the left. The silver bromide is kept in suspension, and the grains, "protected" by the gelatin, are essentially the same size as they were at the time of precipitation. In the beaker on the right, without gelatin, the precipitate has settled to the bottom, and the crystals have grown considerably larger in a relatively short time.

of crystals to grains of the desired size with good structure would be difficult.

WASHED EMULSIONS

If the emulsion is to be coated on a porous support such as a baryta-coated or otherwise uncoated paper base, the excess alkali nitrate (after the precipitation reaction) can be absorbed by the base. If the base or support is a film, or a paper base with a water-impermeable surface such as resin coated (RC), these excess salts have to be removed by washing to prevent them from crystallizing out on the surface (see Figure 7–14). It is also desirable in the emulsion-making process to remove excess salts to prevent further growth of the silver halide grain before continuing the process. This is accomplished by coagulating the gelatin, washing the emulsion, then redispersing (dissolving) the gelatin and halides. The gelatin at its isoelectric point can be coagulated by adding a salt such as a sulfate. There are other methods of accomplishing this coagulation.

Historically the emulsion, with added gelatin, was chilled and set in a manner similar to setting food dessert gelatin. This set emulsion was then extruded into noodles, or

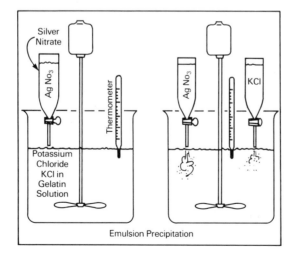

Figure 7–12   Emulsion precipitation.

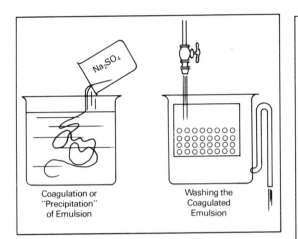

Figure 7–14   The gelatin emulsion is coagulated while at its isoelectric point by the addition of a salt solution such as sodium sulfate. It is then washed to remove the salts including those that were formed during the emulsion-making process. The washed grains can then be redispersed by raising the pH, along with the addition of more gelatin.

sometimes cut into cubes (the noodling procedure was established by F.C. Wratten—of Wratten Filters—around 1878), and washed with chilled water until conductivity or pH tests showed that enough of the soluble salts had been removed.

MULTIPLE COATINGS

Some black-and-white films owe their sensitivity and tone-reproduction characteristics to the coating of two different emulsions, one on top of another. A slow, relatively short-scale, or high gamma emulsion is coated first, then a faster, coarse-grained, long-scale emulsion is coated on top of it to provide a characteristic curve with increasing slope in the upper midtone or lower highlight regions (see Figure 7–15). This provides better speed and scale characteristics than would be achieved by blending the emulsions. Color films, of course, are made up of at least three emulsions, one for recording each of the primary colors. One or more of these three, in turn, may be made up of multiple coatings to provide required tonal response, such as required by internegative films and those for other special applications.

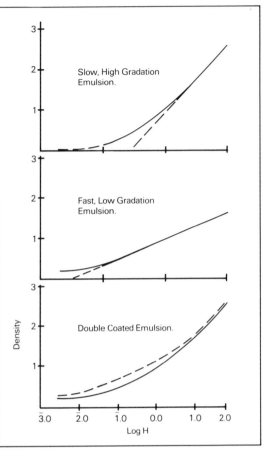

Figure 7–15   By coating a fast, low gradation emulsion over a slow, high gradation emulsion, the combination produces a sensitometric curve with increasing slope in the upper mid-tone and lower highlight regions.

EFFECTS OF DEVELOPMENT ON DEVELOPERS

The reaction products of development include soluble salts of the halides, along with modified developer components. The soluble halides act as restrainers, and thus curtail development. The kind and quantity of restrainer included in the developer formula is intended to minimize, as far as practicable, the effects of these reaction products of development.

DYE SENSITIZATION

The basic silver halide emulsion is sensitive only to the blue and ultraviolet regions of the

Multiple emulsion coatings are used on some black-and-white films to modify the tone-reproduction characteristics of the film.

spectrum (about 180 nm to 520 nm) that it absorbs. The human visual response is highest in the green region, with a peak in the vicinity of 550 nm. Thus, photographs made with an unsensitized emulsion will have different black-and-white rendering of the brightnesses of colors than that perceived with the human visual system.

In 1873, the photographic scientist H.W. Vogel found that emulsions could be made to respond to wavelengths of light in the green region of the spectrum by adding a pink or red dye. The emulsion became sensitized to the color absorbed by the dye. When an emulsion sensitized to green by means of a red dye is used in the camera, with a yellow filter over the lens to absorb some of the blue light (to which the film remains sensitive), it gives a response in terms of black-and-white rendering that is somewhat closer to that observed by the eye; and it is thus described as an *orthochromatic* or "correct" color rendering. The term *isochromatic* also has been used to identify this characteristic.

Dye-sensitized emulsions become sensitized to the colors of light that are absorbed by the dyes.

Later, a green or cyan dye that absorbs red light extended the emulsion's sensitivity. Thus the emulsion is sensitized to red light by the addition of a cyan (red-absorbing) dye. Combined with the added green sensitivity, the emulsion responds to all of the colors of the visible spectrum, which includes blue, green, and red, and it is termed *panchromatic* (see Figure 7–16). Spectral sensitization, while used in films mostly to provide appropriate tone reproduction of subject colors, also increases the film speed with white light, and some higher-speed films have a greater-than-normal sensitivity to red light.

These sensitizing dyes are sometimes retained by the film after processing, especially if a rapid processing technique is used, giving black-and-white negatives a pink cast. Normal development, fixation, and washing usually remove nearly all of the sensitizing and antihalation dyes from the negatives.

It is possible to dissolve all of the silver halide grains in an exposed film with hypo without destroying the latent image.

While most photographic papers are manufactured with emulsions that have been made to meet the sensitometric requirements without further modification, some papers require a final speed adjustment that can be achieved by addition of sensitizing dyes to the emulsion. This extends the sensitivity to some of the longer wavelengths of light. This technique can be satisfactory until a change

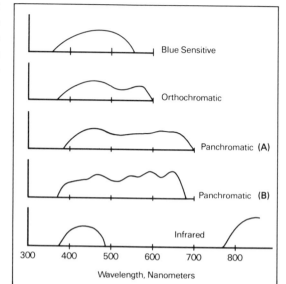

Figure 7–16   Traces from spectrograms for typical blue sensitive, orthochromatic, panchromatic, and infrared films. Panchromatic (A) has extended red sensitivity to nearly 700 nanometers; Panchromatic (B) is more representative of that used for pictorial photography.

in the color of the light sources used for printing or recording reveals a greater-than-expected shift in speed from one emulsion coating to another. Some variable-contrast papers have mixed emulsions of different contrast with spectral sensitivities that can be used to control the contrast by changes in filtration. A typical variable-contrast enlarging paper produces high-contrast images when exposed with blue light (with a magenta filter), low-contrast images when exposed with green light (with a yellow filter), and medium-contrast images when exposed with white light.

FORMATION OF THE LATENT IMAGE

After the sensitized material is exposed in a camera, examination of the surface would not reveal an image. It is said to be a *latent image*. Some of the chemical characteristics of this image are as follows:

1. It is weakened or destroyed by oxidizing agents such as chromic acid, which also oxidize metallic silver.

2. This oxidizing reaction does not destroy the sensitivity of the emulsion, which after washing and drying can be used to expose a new image, although spectral sensitivity and speed may be degraded.
3. It is not soluble in silver halide solvents. (If the exposed image is bathed in sodium thiosulfate solution and the remaining silver and halide ions are removed by subsequent washing, physical development can then be used to develop the latent image. Silver metal from silver ions in the developing solution plates out on image areas where exposure occurred.)
4. The reduction of silver ions to silver metal by the developer is catalyzed by the presence of silver atoms. The latent image also increases the reduction of silver ions.

From the above it appears that the latent image and silver have the same reactions. This indicates that exposure sets into operation a mechanism within the crystal that in the end produces silver atoms which distinguish exposed silver halide grains from those that are unexposed.

Thus, when the crystal of silver halide, made up of silver and halogen ions, is exposed to light, it becomes capable of being reduced by a developer to metallic silver, which along with all the other exposed crystals forms the image of the photograph. Among several theories that have been proposed for the mechanism of latent image formation, the Gurney-Mott hypothesis is a prominent one. It consists of two distinct steps in an exact order, but concluded in a short time interval. Using silver bromide as an example, they are:

1. The radiation of the silver bromide crystal produces electrons that are raised to a higher energy level associated with the conductance band. The electrons move through the crystal by photoconductance until they are trapped by the sensitivity specks. The specks, or traps, then possess a negative electrical potential. This concludes the primary process, and its final effect is to initiate the secondary process.
2. The secondary process involves the movement of the interstitial silver ions that are attracted to the negatively charged specks. The positive silver ions are neutralized by

the negative charges, and the production of silver atoms is completed.

The hypothesis does not explain the fate of the halogen. It is possible that the halogen atoms may either recombine with an electron, may attack the silver atoms produced, or react with a halogen acceptor such as gelatin or sensitizers that reduce the atoms to halogen ions (see Figure 7–17).

Absorption of a quantum of light by the crystal excites an electron so that it is free to move through the lattice and may combine with an interstitial silver ion, thus yielding an atom of silver. This is most likely to occur at a nucleus produced by chemical sensitization, called a sensitivity center, which traps the electron and holds it until an interstitial silver ion arrives. There is a strong tendency for this atom of silver to give up an electron and thus return to the ionic state. It is also possible for the electron to recombine with the positively charged "hole" left when it was released from a halide ion. If exposure is sufficient, two silver atoms together form a sub-latent image that is not capable of development. However, a greater time will be required for the two atoms to give off electrons and return to the ionic state than is the case for a single atom. When sufficient exposure has been received to bring about four atoms of silver at the site, the grain may become developable. Further exposure of the grain to light adds to the number of silver atoms at the original site, and thus increases

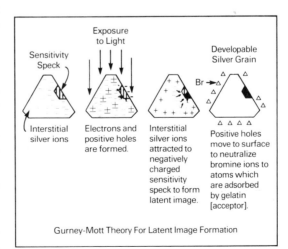

Gurney-Mott Theory For Latent Image Formation

Figure 7–17   Gurney-Mott hypothesis for latent-image formation.

the developability of the grain. About 10–20% of the grains in a fast negative emulsion are rendered developable when four atoms of silver have been formed at the site, but the average number of atoms required for developability is considerably higher.

The latent image is usually formed on the surface of the crystal, but a latent image can also be formed in the crystal's interior on exposure to light. This internal image is usually protected from the developer, but it can be developed in a solution containing a silver halide solvent. In nearly all emulsions the contribution of the internal latent image to the developed image is negligible. A grain with an internal image usually has a surface image. For purposes of experimental research, the surface image can be destroyed to demonstrate the existence of the internal latent image.

The positive holes, formed by the loss of electrons to form silver atoms, move to the crystal surface to neutralize halogen ions to atoms that are adsorbed by the gelatin. The gelatin in this context is referred to as a halogen or bromine acceptor.

While most of the latent image in ordinary photographic materials is formed on the surface of the crystal, the interior latent image plays an important part in some photographic effects. In the Gurney-Mott hypothesis the electron "traps" are considered to be distributed throughout the crystal but are more effective on the crystal's surface.

PHOTOGRAPHIC EFFECTS

The mechanism of latent-image formation is closely related to six photographic effects. These six are: Reciprocity, Intermittency, Herschel, Clayden, Solarization, and Sabattier.

Reciprocity Effects

Photographic exposure is the amount of light falling on the emulsion (Exposure = Illuminance × Time). Reciprocal combinations of illumination and time will give the same exposure, but not necessarily the same density. This decrease in density with certain combinations of illuminance and time is

called *reciprocity law failure*. When exposures are made at low light levels, the efficiency with which the sub-latent images are formed on the crystal is low because of the tendency of the silver atoms to give up an electron and return to the ionic state. Further exposure allows a greater number of silver atoms to accumulate around those few sub-images that survive, but these sub-images are relatively stable.

When exposures are made at high illuminance, with correspondingly short exposure times, the electrons are released so rapidly that the relatively slower-moving silver ions cannot neutralize them fast enough for the centers to grow to the predicted size. A greater number of sub-image centers are spread over the crystal surface, and sometimes into the crystal's interior. There is more competition for the additional silver atoms formed, and thus a smaller number of them grow large enough to become permanent developable latent-image centers.

Reciprocity law failure is caused by the relatively low efficiency of the formation of sub-image centers at low intensities with long exposure times; and by the high efficiency of the formation of sub-image centers at high intensities, with corresponding wide distribution of competing sub-image centers. Emulsions for pictorial use are formulated so that the most efficient compromise between exposures made at low intensities and those made at high intensities generally occurs when the exposure times are in the vicinity of 1/10 to 1/100 second. The reciprocity law failure of an emulsion is the same for all wavelengths of light when compared on the basis of equal densities and equal times (see Figure 7–18).

The reciprocity law is valid for the production of electrons in the primary process of the Gurney-Mott hypothesis, but does not apply to the secondary and other processes that are necessary for the production of the final image.

The reciprocity failure of a photographic material is dependent on the temperature of the material during exposure. Tests show that the failure virtually disappears at temperatures of −186 C, with an accompanying loss of sensitivity. Temperature variation has different effects on low and high intensity reciprocity law failure.

It is thought that a latent image must contain at least four atoms of silver to be stable and developable.

With long exposures at low light levels, silver atoms in the latent image can change back to silver ions.

Exposures made with X-rays and gamma rays show no reciprocity failure, due to the high velocity of the electrons first liberated, releasing large numbers of electrons on collision with ions in the grain. Print-out papers also do not show reciprocity failure.

Intermittency Effect

The intermittency effect is closely associated with reciprocity failure. Intermittent exposure means exposures in discrete installments rather than in one continuous installment. If the intermittency rate is low, the intermittent exposure will produce the same photographic effect as a continuous exposure of equal total energy. As the frequency is increased at a given level of illumination, the photographic effect is decreased until with a further increase in frequency the loss in photographic effect becomes constant. The point at which this occurs is a critical value that varies with the illuminance level. If the intermittent exposure is made at a sufficiently high illuminance level, the photographic effect will be greater than that of a continuous exposure of equal energy. In relation to the U-shaped reciprocity-law-failure curve, the intermittency effect at the critical interruption frequency is equivalent to moving to the left on the curve. In the case of low illuminances on the left half of the curve, the move is upward, indicating that *more* exposure is required to produce a specified density. In the case of high illuminances on the right half of the curve, the move to the left is downward, indicating that *less* exposure is required to produce a specified density (see Figure 7–18).

Solarization

When a sensitometric curve is considered, solarization is the reversal or decrease in density with additional exposure increase beyond that required to produce maximum density on the film (see Figure 7–18). The maximum solarization effect is produced with moderate developing times, while extended development reduces, or even eliminates, the effect. In addition, the presence of silver ha-

lide solvents, such as sodium thiosulfate and sodium sulfite, in the developer inhibits or removes the solarization. Developers that do not contain silver halide solvents usually produce the effect.

If halogen acceptors are present during exposure, solarization may be diminished or even eliminated. Thus, solarization is considered to be the result of rehalogenation of the photolytic silver formed at the sensitivity specks as the result of exposure. Normal exposure produces halogen at a rate that allows the halogen to react with acceptors such as gelatin. If the exposure is great, the production of the halogen proceeds at a rate beyond the capability of the acceptor, and thus may react with the latent-image silver to reform silver halide. This surface coating of silver halide, although it contains a latent image beneath it, will shield the latent image from the developer. This is sufficient to lower the number of developable crystals, and the result is a lower density. If the developer contains a silver halide solvent, it will remove the surface silver halide and thus expose the latent image for development.

A series of intermittent exposures may not produce the same density as a single exposure even though the total amounts of light are the same.

With a solarized image, heavily-exposed areas do not develop because rehalogenization produces a protective coating of silver halide on the latent images.

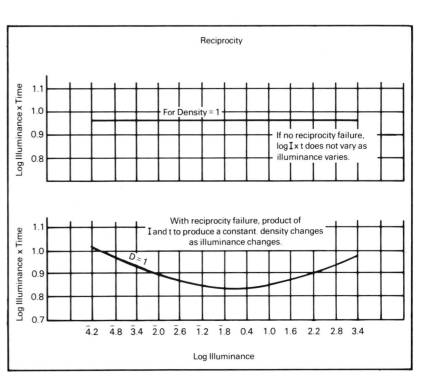

Figure 7–18   Reciprocity.

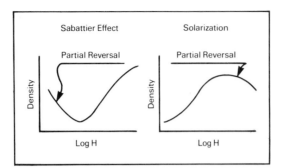

Figure 7–19 The Sabattier effect is the result of arrested development, wash, re-exposure, and further development. Solarization is the result of extended exposure and moderate development in a developer without silver halide solvents. Both effects produce a "reversal" image.

Sabattier Effect

The Sabattier effect is produced by developing an exposed photographic emulsion for a short time, washing it, and then allowing the emulsion to be exposed a second time. This is followed by further development, fixing, and washing (see Figure 7–19).The effect is sometimes confused with solarization, in that the final result is a partially reversed image. It appears to be caused by two mechanisms: (1) the image produced by the first development screens or acts as a negative and thus allows the exposure of the remaining silver halide to be modulated to produce a positive image during the second development; and (2) the byproducts of the first development act as a restrainer in the developed areas. The migration of used and fresh developer across image boundaries may produce Mackie lines, a line of increased density just inside the denser area and a line of decreased density just inside the thinner area.

LATENT-IMAGE STABILITY

The fact that amateur photographers sometimes allow months and even years to elapse between the time the first exposure is made on a roll of film and the time the film is processed attests to the stability of the latent image. Preservation has been exceptional in cases where exposed film has been frozen in ice, such as in the Arctic region, but images have been obtained with exposed film that has been stored for as long as 25 years under household conditions.

Latent images have produced usable images when development has been delayed for as long as 25 years.

These examples are not intended to suggest that the latent image is permanent and does not change with time. Since the latent image consists of a cluster of silver atoms, the loss of only a few atoms may render a silver halide grain undevelopable. An atom of silver can combine with an atom of bromine, for example, to form a molecule of silver bromide—reversing the effect of exposure. After long periods of time the image, when developed, may have less contrast and be less distinct due to the spontaneous development of a larger number of unexposed silver halide grains. As this type of development fog increases, a point is reached where the developed latent image can no longer be detected. A small amount of development fog, however, can produce the effect of increasing the speed of the photographic emulsion and the density of the developed latent image in the same manner as latensification (see Figure 7–20).

For critical work, detectable changes in developed latent images sometimes occur in remarkably short times. Exposing a large number of black-and-white prints identically, and then developing part of the batch one day and the balance the following day, has been reported to produce a significant difference in density. Improvements in emulsion technology in recent years have reduced

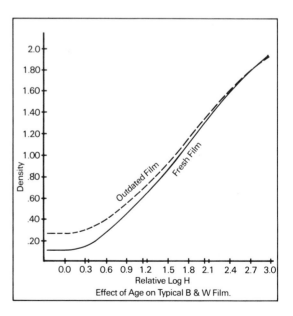

Figure 7–20 Effect of age on a typical black-and-white film.

the decay rate of the latent image in black-and-white printing papers, however.

Changes in the latent image of color printing papers are more serious because they can affect the color balance of the print in addition to the density and contrast. For this reason, recommendations are made to store color paper for a fixed time after exposure prior to processing to make the latent image change relatively constant. If the exposed paper is to be held for more than the hour or two normally built into the schedule, recommendations are that it be stored at 0 F (−18 C) or below, and then for a period of no more than 72 hours. Film manufacturers recommend that color films be processed as quickly as possible after exposure or that they be stored at a low temperature, noting that storage in a closed automobile on a hot day for only a few hours can have a serious effect on the developed image. They also warn against allowing color films to come into contact with various fumes, including those of chemical solvents and mothballs.

SOME ALTERNATIVE SYSTEMS

The familiar black-and-white and color films and papers that use gelatin as a suspension vehicle for the light-sensitive silver salts are but one of many ways to make a light-sensitive photographic system. Most of the older processes, such as calotype, daguerreotype, cyanotype, albumen, wet collodion, ambrotype, carbon, carbro, gum bichromate, platinum, and kallitype, use something other than silver salts and/or gelatin. None have the light-amplification ability of silver and therefore they all have very slow speeds. The required long exposure times or high light levels limit their use in a camera, so they are relegated to photographic printing or certain types of recording where high sensitivity is not required.

Many of these older processes are also being rediscovered by photographers who are exploring their delicate tones and esthetic qualities. Some processes utilize silver and/or gelatin only in an intermediate step. Many of the present-day color processes are examples of the latter where the final images consist of dyes. Newer processes such as electrophotography and television use electrical and magnetic fields for image recording and reproduction. Light-sensitive microchips containing many light-sensitive picture elements have sufficient sensitivity to be used as discs in cameras (see Chapter 11).

CALOTYPE PROCESS

Another early process utilizing silver but not involving gelatin was the *calotype* process, invented by William Henry Fox Talbot, and patented in 1841. Talbot had obtained negative silver images on silver chloride paper as early as 1839, with what was later called the Talbotype process. The negative and positive images formed the basis of modern photography. Paper was sensitized with silver iodide, silver nitrate, and gallic acid, and was developed in gallic acid; this paper was used for both the camera negative, and for printing a positive from the negative.

According to a modern version of the process, the paper is first iodized by coating it with an approximately 7% solution of silver nitrate in distilled water, as shown in Figure 7–21. After the paper is dried it is floated on a solution containing about 7% potassium iodide and 1% sodium chloride, and dried again. It is then sensitized by flowing onto the surface a solution containing about 10% silver nitrate, 7% acetic acid, and 1% gallic acid, producing light-sensitive silver halide. The sensitized paper is dried and kept in the dark until ready for camera exposure. The exposed paper is developed in a solution containing about 0.8% gallic acid and 2.5% silver nitrate, producing a silver image. The developed negative is fixed in a solution of about 30% sodium thiosulfate (hypo) solution, washed, and dried. Prints can be made by exposing this negative on a conventional modern printing paper, or on another sheet of calotype paper.

DAGUERREOTYPE

The process announced by Louis Jacques Mande Daguerre on August 19, 1839, introduced practical photography to the world. No gelatin was involved. A sheet of copper or brass was plated on one side with silver,

History has seen many photographic systems other than the silver halide-gelatin processes in use today—some of which still have significance.

The negative-positive calotype process, patented in 1841, used silver halide but did not use gelatin.

The Daguerreotype process used a silver surface that was treated with iodine vapor to form a light-sensitive silver halide.

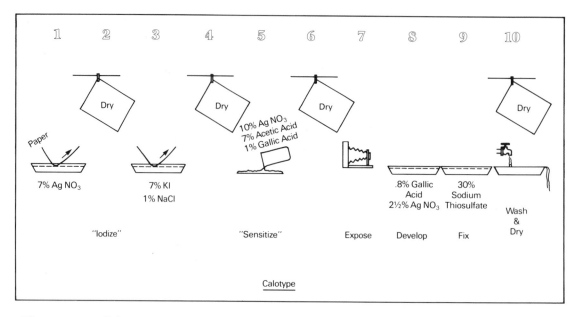

Figure 7-21 Calotype.

The cyanotype process is a nonsilver printout process that requires only washing after exposure.

which was buffed to a high mirrorlike polish. It was then sensitized by placing it in a light-tight box containing iodine crystals, which gave off iodine vapor that reacted with the silver to form a coating of silver iodide. Exposures were made in a camera, and since the image was viewed on the surface of the plate and not through it, it was reversed from left to right, a condition that was corrected in practice by placing a mirror at 45° to the lens in front of the camera.

After exposure, the plate was placed in a box containing a dish of mercury that was heated to about 75 C (165 F). The mercury adhered only to the exposed parts of the plate, giving a whitish amalgam of silver and mercury. The plate was fixed in a solution of sodium thiosulfate (hypo), washed, and dried. When the Daguerreotype was viewed so that the unexposed, undeveloped areas reflected the dark surroundings of a room, a positive image was seen. The speed of the plate was increased by adding bromine to the iodine vapor, and the image strength could be improved by toning with gold chloride after fixing. Since the silver image was readily tarnished or damaged by handling, it was protected by placing it in a decorative cutout frame and covered with glass.

CYANOTYPE

The *cyanotype* (or blueprinting) process, invented by Sir John Herschel in 1842, is a nonsilver, nongelatin process. The ferric iron salt used in coating the paper is reduced by light to the ferrous state, which is then precipitated to Prussian Blue (ferric ferrocyanide, $Fe_4[Fe(CN)_6]_3$), by the action of the potassium ferricyanide, the second component of the coating solution. The image is made up of varying densities of the Prussian Blue. A good grade of sized paper should be used; the paper is sensitized by coating (brushing, floating, swabbing) with a solution consisting of about 12.5% ferric ammonium citrate and 7.5% potassium ferricyanide (usually prepared by mixing separate solutions of these two compounds). Ferric ammonium oxalate can be substituted for the ferric ammonium citrate for increased speed, and about 1% of potassium dichromate can be added for increased contrast.

After exposure a pale image can be seen, and processing normally consists of washing the prints in plain water to remove the unexposed soluble ferric salt, followed by drying. Prussian Blue is a fairly stable compound but it is soluble in alkalis, and the image is affected by impurities in the atmosphere that sometimes cause the image to take on a "metallic" lustre in the denser areas. The cyanotype process can be used for printing continuous-tone photographs by contact from large negatives. It was also once used on a large scale for copying drawings from original tracings but has since been supplanted by the diazo process. (There are several other cyanotype formulas, but they all result in a Prussian Blue image).

Cyanotypes can be converted to purplish-black images by bathing in a 10% sodium carbonate solution, which bleaches the image; then redeveloping in about 1% tannic acid solution. Cyanotype images can be inked over with waterproof ink, and the blue bleached out with about 5% oxalic acid, followed by washing and drying, to produce a pen-and-ink drawing. A modification of the process (pellet process) produces a faint image that can be developed to a reversal image (positive to positive) upon development with potassium ferrocyanide. Another version, the pointevin process, is a positive-to-positive one that produces purplish-black lines on a light background.

WET COLLODION

The wet collodion process is a silver process, but it uses nitrocellulose as the binder for the halide crystals. It represents an important phase in the development of photography because it produces negative images on a transparent base, and was the principal method of making negatives from 1851, when it was introduced by Frederick Scott Archer, until the 1870s, when gelatin dry plates were introduced.

The cellulose nitrate is prepared by immersing cotton in a mixture of nitric and sulfuric acids, followed by washing in water. When dry, the nitrated cotton is dissolved in a mixture of ether and alcohol to produce collodion. (Prepared collodion can be obtained from chemical supply houses.) For photography, a small amount of sodium or potassium iodide or bromide is dissolved in the collodion. Some formulas have also made use of cadmium bromide and other halides. The prepared collodion is then flowed onto a clean glass plate until it is covered, and the excess collodion drained back into the bottle. The alcohol and ether evaporate and leave a tacky coating on the plate. Flow characteristics, which vary with atmospheric conditions, are adjusted by altering the relative amounts of ether and alcohol in the collodion. The tacky plate is immersed in the dark in a tank containing silver nitrate, about 65 grams per liter, for about one minute. The plate is then loaded into a special holder, while still wet, and inserted into the camera for exposure.

Following exposure the plate is developed before it has a chance to dry, by flowing a ferrous sulfate solution, or pyro solution, over the plate, and allowing the "puddle" to remain there during development. Some operators drain and replace the developer during development. However, it is thought that some physical development takes place, that is, some of the silver is plated back onto the image from the excess on the plate.

After development, the plates are fixed in a solution of potassium cyanide, or sodium cyanide, about 65 grams per liter, or with a sodium thiosulfate (hypo) fixing bath. The collodion image could be stripped off the glass to provide a film negative and a reusable glass plate, but this was not done often in the field. The wet collodion process continued to be used in the graphic arts industry until well into the twentieth century. The images were routinely stripped off the glass support, and recemented onto another support, or "flat"— a number of images placed together on a single support. The term stripping is still applied to the removal and repositioning of images into a new layout.

GUM BICHROMATE

The gum bichromate process depends on the hardening effect produced by light on a bichromated colloid; in this case the colloid is gum arabic. Before the turn of the century the process enjoyed great popularity because of its adaptability to printing controls, and the wide choice of colored pigments that could be used. While it is considered obsolete today, it is enjoying revived popularity as a

The wet collodion process was the first to produce a negative image on a transparent glass support, but the sensitized plate had to be exposed before it had time to dry.

With the gum bichromate process, light produces a hardened image that remains after the unhardened gum is removed by washing.

means of artistic expression. There are various formulas for preparing the sensitizer, depending on the tonal rendering desired, but a fairly standard formula consists of mixing equal parts of a 10% solution of potassium bichromate with a 30% solution of gum arabic. The potassium bichromate can be dissolved in hot water, but the gum arabic requires soaking overnight or longer at room temperature. A small amount of thymol or other preservative can be added to improve the keeping qualities of the gum solution. Ammonium bichromate is sometimes used in place of the potassium salt to produce increased sensitivity. Any of a wide variety of water-miscible colorants can be added to the mixture.

The mixture should be applied to paper or another support, which is temporarily attached to stiff cardboard, using a brush. Experience will result in a technique that will produce a uniform coating. The coated material is dried in a dark room. Sensitivity is low until the material is nearly dry. The speed is similar to that of POP (printing-out-paper), but it is necessary to run an exposure test, as the speed depends on variations in the colorant.

The prints are made by contact, using sunlight, or some other source rich in actinic ultraviolet radiation, using a printing frame. A typical exposure is about five minutes. Development consists of dissolving the unhardened coating. The print is placed face-down in a tray of cold water for a time that may vary from 15 minutes to hours. The process can be speeded up with a gentle spray after the initial soaking. The image should not be touched until it is completely dry. During the drying stage, the print should be attached to a stiff support to prevent curling.

ELECTROPHOTOGRAPHY

There are many different electrophotographic processes (which use electricity to form images), some of which have not yet been fully developed. Electrostatic photography (xerography), which had its beginnings with the inventions of Chester Carlson in 1938, is the most advanced electrophotographic system in use today. The basic principle of image formation is that certain materials, such as selenium (early experiments were conducted with sulfur) and zinc oxide, will take on an electrical charge when passed near a source of high voltage, a corona charge. This is usually a positive charge in systems utilizing selenium, but the zinc oxide (in a resin binder) requires a negative charge. The action of light is to eliminate the charge, resulting in an image that can be made visible by dusting the plate with pigmented "toner," which adheres to the charged (unexposed) areas. The toner also can be carried in a liquid.

With xerography (Xerography is a trade name, but *xerography* is now listed in dictionaries as a generic term), a drum plated or coated with selenium rotates first through an electric field (corona discharge) where it picks up a uniform positive electrical charge, as shown in planographic form in Figure 7–22. The image of the original is projected onto the charged drum, leaving an image consisting of positive ions in those areas that were not exposed. The drum is dusted with the toner, which adheres to the charged image. The drum then picks up a sheet of paper and both are passed through an electrical field that causes the toner to transfer from the drum to the paper. The paper is passed through heat to fix the pigmented toner permanently to the paper. If the exposure is made while the drum is rotating, a scanning system is used to synchronize the image with the rotation of the drum, but an alternative system uses a stationary charged surface with electronic flash illumination.

Good line copies are readily made with these processes, but one problem in their development has been to produce good continuous-tone reproductions. There have been efforts in this direction to devise methods that control the corona charging rates, among other things, which are meeting with continued success. Since the toner can be made in various colors, and since successive images can be made on the same sheet, it is possible to produce multicolored copies. Positive images are produced, since the toner adheres to the areas that are not exposed to light.

THERMOGRAPHY

Thermography is another nonsilver, nongelatin process. Images are formed by heat, usu-

With the xerographic process, light reduces the electrical charge in the exposed areas of a charged plate. The remaining charged areas attract a pigment that forms the final image.

Thermographic materials form images by reacting to heat radiation in any of several ways, including decomposition, softening, and the production of a transferrable image.

ally by radiation from an infrared lamp that is modulated by the image on the original document. The denser areas absorb infrared radiation and become hotter than the surrounding less-dense areas. The heat is transferred to the heat-sensitive material, which is placed in contact with the original. The most common changes are a physical softening in the heated areas, or a chemical decomposition. Since printing and typing inks generally do not strongly absorb infrared, it is necessary first to make an electrostatic copy of such originals and then make a thermographic copy from that copy—not a practical procedure if a single copy is needed.

DIAZO

The *diazo* family of nonsilver, nongelatin processes is mainly divided into two categories: one producing dye images, and the other vesicular images. With the *dye image* process, ultraviolet radiation decomposes compounds known as diazonium salts. The remaining salts are converted to azo dyes with ammonia or heat. The ammonia may be applied either in solution form or as ammonia fumes. Since dye is formed in the areas protected from radiation, a positive image is formed from a positive original. No fixing is needed because the sensitivity is destroyed in the nonimage areas by the action of the exposing radiation.

A diazo material that produces black images on a white background is widely used for copying drawings, and the resulting prints are called white prints to distinguish them from blueprints. Diazo materials are available with paper, film, and foil bases and a wide variety of image colors. Limitations of the process are low sensitivity and the need for an ultraviolet-rich source of radiation for exposure. The high-contrast characteristics of the sensitized materials make them unsuited for copying continuous-tone originals. However, they are capable of high resolution, and they were at one time considered favorable for reproduction of microfilm images and optical sound recordings.

With the *vesicular* process, crystalline and noncrystalline diazonium salts are randomly dispersed in a thermoplastic film (see Figure 7–23). Upon exposure to ultraviolet-rich ra-

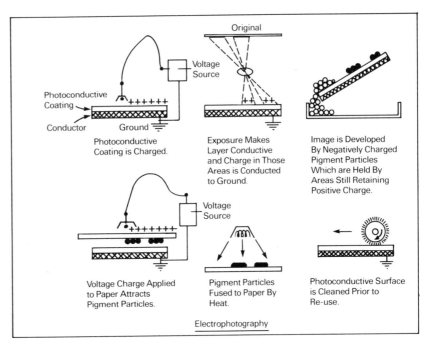

Figure 7–22 Electrophotography.

diation, the diazonium salt is decomposed and nitrogen gas is released into the thermoplastic film. Development consists of heating the film to change the light dispersal pattern mainly due to the tiny bubbles that are formed as the gas expands. Upon cooling, these larger bubbles remain and produce an image by scattering light (as contrasted to silver and dye images, which produce visible images by absorbing light). Fixation consists

With the diazo process, diazonium salts are decomposed by exposure to ultraviolet radiation and the remaining salts form a dye image when treated with ammonia.

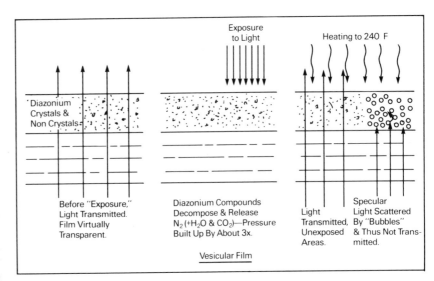

Figure 7–23 Vesicular film.

The image displayed on a television screen is produced by scanning a phosphor layer on the tube face with an electron beam, which causes the phosphors to emit light.

of a uniform exposure to radiation without heat which decomposes the residual diazonium compounds and the released nitrogen gas is allowed to diffuse from the film.

Although vesicular images appear low in contrast when viewed by diffuse illumination, the contrast is satisfactory when viewed by specular illumination. Since bubbles are formed where exposure occurs, a negative image is formed, but the processing procedure can be modified to produce a positive image. Typical applications are the production of positive images from microfilm negatives, black-and-white negatives from 35 mm color slides, and quick projectuals in audiovisual work. Before the displacement of black-and-white motion pictures with color, this process was a strong contender for making motion-picture release prints.

TELEVISION

Television is an electronic process that in its essential form does not produce a permanent image; but when tied to magnetic recording, the system does produce images that can be modified, edited, and recalled at any time. The *vidicon* tube contains an image-receiving area having a transparent electrically conductive coating upon which is deposited a photoconductive layer that is normally a good insulator (see Figure 7–24). When any area of the photoconductive layer is exposed to light, its resistance is decreased (it becomes conductive). When this is scanned with an electron beam, a signal is generated between the photoconductor and the transparent conductor in front of it, which depends on the scanned light pattern that was produced on

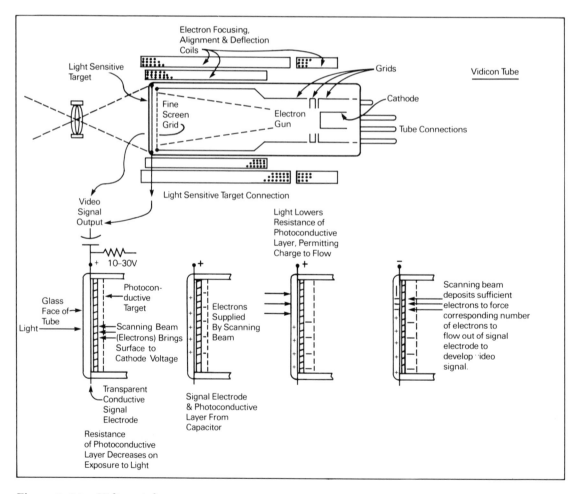

Figure 7–24   Vidicon tube.

it by the camera lens. This signal is then amplified and broadcast to the home receivers, where the signal is reconverted to an image on the TV screen. The signal can be recorded on a magnetic tape or disc to form a permanent image, or series of images. No silver or other materials such as colloids are involved in this type of process. Electronic imagery has become an important medium.

CHARGE-COUPLED DEVICES

There are several ways in which images formed by light can be electronically processed, including Charge-Injection Devices (CID), Charge-Priming Devices (CPD), and Charge-Coupled Devices (CCD). An example of the latter is the system announced by Sony in 1981, in which an image is placed on an array of small photosensitive detectors to be electronically processed and stored on a small magnetic disc for later electronic processing as a signal to a TV receiver for presentation of the picture. The same record can also be used to reproduce the picture, one such device making use of dye tissues that are heated to release colored dyes. The amount of heat in a given area is modulated by the signal from the magnetic disc to evaporate the dyes in amounts relating to the original image. The dyes then migrate to a plain paper sheet receiver. Three colored tissues plus a "black" tissue are used to produce a colored image with good shadows.

DYE TRANSFER

The gelatin-silver emulsion provides an intermediate matrix from which the image is produced by transferring dye to a mordanted gelatin coated film or paper support. (A mordant is a substance that binds a dye to a given material—in this case, gelatin.) With the Technicolor motion picture process, now defunct in the United States, a set of matrices (one each dyed with cyan, magenta, and yellow to represent the red, green, and blue records of the scene) transferred the dyed images to a gelatin coated "imbibition blank" film. Beginning in the 1930s the images were transferred to a processed positive film car-

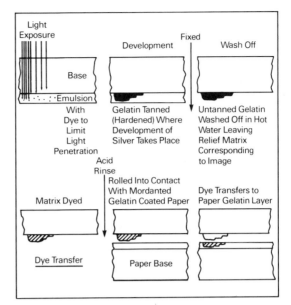

Figure 7–25 Dye transfer.

rying the silver sound track. For dye transfer prints on paper, see Figure 7–25.

DIFFUSION TRANSFER

With *diffusion transfer*, an exposed gelatin-silver negative emulsion on either a film or paper base is brought into contact with a receiver sheet, with a viscous activator solution containing a strong alkali and a silver halide solvent such as thiosulfate placed between them. The negative emulsion also contains a developing agent, and the receiver sheet, which is not light sensitive, contains nucleating particles such as silver sulfide, along with a developing agent. The activator causes development to take place in the negative, and the halide is reduced to silver to form a negative. At the same time, the undeveloped silver halide in the negative is dissolved by the thiosulfate, and diffuses to the receiving sheet. The silver sulfide particles serve as nuclei for development of the halide that diffuses to the receiving sheet, and it is reduced by the developing agent, now activated, to metallic silver to form a positive image. When the two sheets are separated, the remaining viscous activator—developing/solvent solution—stays with the negative, leaving the clean positive image on the receiving sheet.

With the diffusion transfer process, the silver halides that remain after development of a negative image are used to form a positive image.

No further fixing or washing is required. The process normally goes to completion, although varying the time and/or temperature can permit some variation in density or contrast.

This is essentially the type of process used in "instant" cameras for black-and-white photography, such as Polaroid. If the negative is on a paper base, it is discarded after the positive print has been formed; but if it is on a film base, the negative can be salvaged for normal printing by rinsing away the activator in a sodium sulfite solution. The Polavision® color motion picture process also utilized a variation of this type of process, but the negative is not removed from the positive since it is of relatively low density after processing. This is an additive screen process, in which the red, green, and blue filter elements are continuous fine lines, 4,500/inch.

The Polacolor® system of photography makes use of a multilayer film consisting of red-, green-, and blue-sensitive layers, interspersed with dye-developing layers containing developer molecules linked to cyan, magenta, and yellow dye molecules, as shown in Figure 7–26. In those areas that have been exposed to light, development takes place after the developer has been activated by the viscous material from the pod that has been broken to start the process. Where development occurs the developer/dye molecules are immobilized, but where no development occurs the developer/dye molecules migrate through the emulsion layers to the receiving layer of the adjacent paper, and are fixed in position to produce the positive color image. The Polacolor II system makes use of an opaque layer that protects the film during processing, and that permits the image to be visible after processing has taken place.

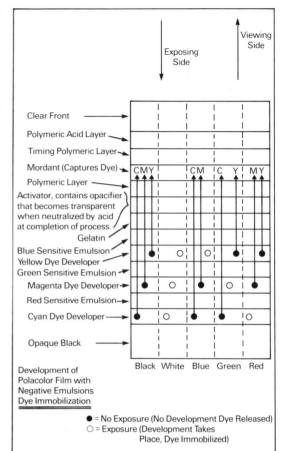

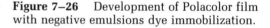

Figure 7–26   Development of Polacolor film with negative emulsions dye immobilization.

REVIEW QUESTIONS

1. The major disadvantage of cellulose nitrate as a support was that it . . . *(p. 188)*
   A. was expensive
   B. became brittle
   C. turned opalescent
   D. was combustible
   E. had a disagreeable odor

2. A disadvantage of polyester as a photographic emulsion support for motion-picture film is that it . . . *(p. 189)*
   A. is difficult to perforate for sprocket holes
   B. acts as a light pipe
   C. has a strong inherent color
   D. tears easily
   E. curls excessively

3. NC coatings on photographic films are applied to the . . . *(p. 190)*
   A. base side
   B. emulsion side
   C. edges

4. The name of non-RC photographic papers could be abbreviated . . . *(p. 190)*
   A. RF
   B. FC
   C. FB
   D. FF

5. Brighteners are used in . . . (p. 191)
   A. negative black-and-white films
   B. negative color films
   C. reversal color films
   D. black-and-white infrared films
   E. photographic papers
6. The number of groups of photographic paper thicknesses listed in the American National Standard on paper thicknesses is . . . (p. 191)
   A. 3
   B. 4
   C. 5
   D. 7
   E. 9
7. A statement that is not true for photographic gelatin is that gelatin . . . (p. 192)
   A. can be changed between liquid and solid states by changing the temperature
   B. has no effect on the sensitivity of the emulsion
   C. permits penetration by liquids
   D. serves as a receptor for silver ions in latent image formation
   E. is transparent
8. The more uniform the silver halide grains are in size in an emulsion, the . . . (p. 194)
   A. higher the contrast
   B. lower the contrast
   C. higher the speed
   D. lower the speed
9. The inherent spectral sensitivity of silver bromide is to . . . (p. 194)
   A. ultraviolet, blue, green, red, and infrared radiation
   B. ultraviolet, blue, green, and red radiation
   C. ultraviolet, blue, and green radiation
   D. ultraviolet and blue radiation
   E. blue radiation
10. If the gelatin is omitted during the precipitation stage of emulsion making . . . (p. 196)
    A. the chemicals will not react
    B. the silver halide formed will decompose rapidly
    C. the silver halide will not stay in suspension
    D. the silver halide formed will rapidly dissolve
11. Sensitizing dyes increase sensitivity of photographic emulsions to wavelengths of radiation . . . (p. 198)
    A. reflected by the dye
    B. absorbed by the dye
    C. transmitted by the dye
12. When exposed but undeveloped film is placed in a sodium thiosulfate solution, the latent image is . . . (p. 199)
    A. unaffected
    B. destroyed
    C. converted to a visible image
13. At the end of the primary process of latent image formation, the sensitivity center . . . (p. 199)
    A. has a positive electrical charge
    B. has a negative electrical charge
    C. is neutral
14. Movement during the secondary process of latent image formation consists of movement of . . . (p. 199)
    A. electrons
    B. sensitivity specks
    C. positive ions
    D. negative ions
    E. All of the above.
15. The minimum number of atoms of silver required to be formed during exposure for a silver halide grain to be developable is considered to be . . . (p. 199)
    A. one
    B. two
    C. three
    D. four
    E. five
16. A common characteristic of cyanotype, gum bichromate, and other nonsilver processes is that . . . (p. 203)
    A. none is capable of producing a neutral color image
    B. none of the images formed has as good archival permanence as silver images
    C. all are more expensive than silver processes
    D. all have lower film speeds than silver processes
    E. none is capable of producing an image of normal contrast
17. The first negative-positive photographic process is considered to be the . . . (p. 203)
    A. daguerreotype process
    B. cyanotype process
    C. calotype process
    D. gum bichromate process
    E. electrostatic process

18. An example of a photographic process that does not use silver is the . . . (p. 204)
    A. daguerreotype process
    B. cyanotype process
    C. wet collodion process
    D. printing-out-paper process

19. The distinctive characteristic of the wet collodion process at the time it was introduced was that it was the first . . . (p. 205)
    A. negative-positive process
    B. silver halide process
    C. process to use dyes
    D. process to use liquid development
    E. process to use a transparent base

20. A photographic process that is based on the hardening effect of exposure to light is the . . . (p. 205)
    A. cyanotype process
    B. thermography process
    C. electrostatic photography process
    D. gum bicromate process
    E. diazo process

21. In electrostatic photographic processes, light forms a latent image by . . . (p. 206)
    A. placing an electrical charge on a surface
    B. neutralizing an electrical charge on a surface
    C. hardening toner on a supporting surface
    D. creating a fluorescent image on the receiving surface

22. A distinguishing characteristic of the diazo dye process is that the images . . . (p. 207)
    A. have low contrast
    B. have high contrast
    C. have a magenta color
    D. are composed of silver dyes
    E. are fixed with ammonia gas

8 Black-and-White Photographic Development

John Sexton. *California Corn Lily*. Copyright © 1977 by John Sexton. All rights reserved.

Development amplifies the latent image by a factor of up to a billion times.

NEGATIVE DEVELOPMENT

When a silver halide emulsion on film or paper has been exposed to a light image, a *latent image* is formed in the emulsion; in most circumstances this image would not be visible because of the small amount of silver produced. The energy required to produce such a latent image is relatively small, permitting exposure times of 1/1,000 second or shorter under daylight illumination. The development process amplifies this image by a factor of up to 10^9, producing the final silver image. Assuming that optical density is approximately proportional to the amount of silver per unit area of the image, it is possible to calculate the density of a latent image. For example, the maximum density that can be obtained in a reflection print is about 2.0. This density divided by 10^9, the maximum development amplification factor, equals 0.000000002, the calculated density of the corresponding latent image. The smallest density difference that can be measured with conventional densitometers is 0.01.

The subsequent fixing step converts the unexposed, undeveloped silver halide remaining in the film to soluble silver complexes that are washed away to leave only the silver image. The photographic process is a negative-working one, in that dark areas in the scene record as light areas on the film, and light areas record as dark on the film. In the case of the original camera exposure, a negative image is created that serves as a light modulator to produce a positive image on another piece of sensitized material for viewing.

DIRECT-POSITIVE IMAGES

A positive image can be produced on some types of emulsions by overexposing them to such an extent that the reversal region of the characteristic curve is reached (see Figure 8–1). X-ray film, for example, could be used as a contact-printing medium for reproduction of an original negative X-ray image as a negative image simply by grossly overexposing it, and processing in the usual way. Emulsions have been produced that give positive images with normal development for copying and duplicating negatives.

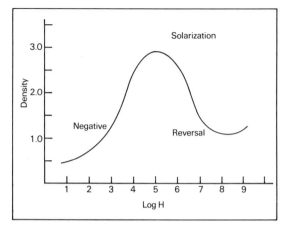

Figure 8–1   Solarization.

REVERSAL PROCESSING

The basic negative-working silver halide emulsion system can be used to produce positive images by means of reversal processing (see Figure 8–2). Some emulsions are more suited to this type of processing than are others. Reversal processing begins with a negative image, as in normal development. The silver image is then removed in a bleach, such as potassium dichromate-sulfuric acid, which reacts with the metallic silver to form soluble silver sulfate that is washed away, leaving the undeveloped silver halide in the emulsion. The remaining silver halide is then exposed to a strong light source, and developed again to produce a "reversed" or positive image. The regions that do not receive much

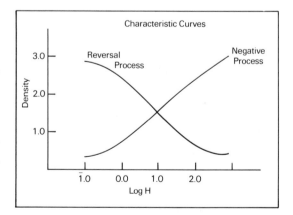

Figure 8–2   Densities increase as exposure increases in a negative process; while densities decrease with increased exposure in a reversal process.

original image exposure have very little density after the first development, but a great deal of halide is available to the second developer to produce a high density. Similarly, where there was high exposure in the beginning, there is little silver halide remaining for the second exposure and development, and low density is produced, as shown in Figure 8–3.

More recently, fogging developers have been used for the second or reversal development, which eliminates the need for second exposure.

MECHANISM OF DEVELOPMENT

Latent images can be divided into surface latent images and internal latent images. Most of the development takes place with the surface latent image, and the internal latent image normally contributes little to the total image. The latent image is formed when a certain number of quanta have produced a nucleus of silver atoms in or on the surface of a silver halide grain. Four atoms of silver are considered to be the minimum number that constitutes a latent image.

Development is the selective reduction to silver of additional silver halide molecules around the site of the silver latent image nuclei, with no appreciable reduction in those areas that have not been exposed to light. As development continues, the amount of me-

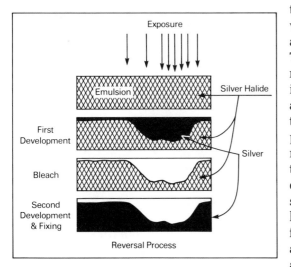

Figure 8–3   Reversal process.

tallic silver grows, and if carried far enough the entire silver halide grain is converted to metallic silver. There are basically two types of development: chemical development and physical development. In physical development, the silver is provided by a silver salt in the developing solution, whereas in chemical development the silver comes from the reduction of silver halide grains in the emulsion. In most instances, some physical development takes place along with chemical development as a result of the silver provided by the solvent action of sodium sulfite on silver halide grains, and the amount varies with the type of developer. Physical development can be accomplished after fixing the exposed but undeveloped image, in which case the silver nuclei not removed by fixing serve as the focal points for accumulation of silver from the developer, which contains a soluble silver compound. The acid in conventional fixing baths will destroy the latent image, however.

CONSTITUTION OF TYPICAL DEVELOPERS

The most important ingredient in a chemical developer is the developing agent or chemical reducer that converts the exposed silver halide to metallic silver. Most developing agents require a pH higher than 7 to function, and for this reason the developer also contains an alkali, sometimes referred to as the accelerator. In order to minimize oxidation of the developing agent by oxygen in the air, the solution also usually contains a preservative, most often a sulfite. A restrainer, usually a bromide, is also part of most developers. The restrainer has the effect of slowing the rate of development, but this effect is greater in the unexposed areas than in the exposed areas of the emulsion, thereby limiting spontaneous development, or chemical fog. The presence of a restrainer in the developer formula also tends to minimize variations due to the release of halide ions during development, which would themselves act as restrainers (see Figure 8–4). Bromides or other halides are also sometimes referred to as antifoggants, but this term is usually applied to a number of organic compounds that are used at much lower concentrations than bromide. A developer formula may also include other

Some physical development usually occurs with so-called chemical development.

Most developers contain a reducing agent, an alkali, a preservative, and a restrainer.

Developing agents provide electrons to reduce silver ions to silver atoms.

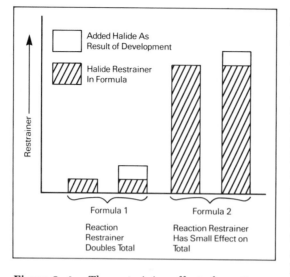

Figure 8–4 The restraining effect of reaction products of development is much greater when the developer formula contains a low amount of restrainer. A larger amount of restrainer in the developer formula makes the developer more tolerant of increases in restrainer due to the by-products of development.

compounds that accelerate development, provide more even development, prevent the formation of insoluble compounds, etc.

DEVELOPING AGENTS

Common developers today make use of organic developing agents, although in the past inorganic metallic agents such as ferrous sulfate have been used. Development takes place as a result of the transfer of an electron to the silver halide. With ferrous sulfate the iron of the developing agent goes to a higher valence, in the presence of an organic ion such as oxalate. (The loss of an electron to Ag^+ converts Fe^{++} to Fe^{+++}.)

$$Fe^{++} + Ag^+ \longrightarrow Ag^o + Fe^{+++}.$$

More recently, renewed interest has been shown in this and other metallic ions such as titanium and vanadium. The ferrous sulfate developers were most commonly acid with a pH range of 4 to 6, but they can operate in an alkaline solution.

HYDROQUINONE

The majority of silver developers utilize hydroquinone and Metol as their developing agents. Hydroquinone, discovered in 1881 by Abney, is used probably more than any other developing agent although usually in combination with another developing agent. It requires strongly alkaline solutions, since it has relatively low energy. Hydroquinone is slow to take effect but, once it has started, proceeds to develop rapidly and produces high contrast. It is quite sensitive to changes in temperature, and is practically inactive at temperatures below 15 C (59 F). In strongly alkaline solutions it is a good developer for high contrast and line copy work. Developers made with hydroquinone tend to produce stain and fog unless sufficient sulfite and bromide are included in the formula. Both hydroquinone and pyrogallol tan the gelatin of an emulsion adjacent to those areas where development takes place; and either of these developing agents, or a combination of both (sometimes with other agents), is used in tanning developers. Tanning developers are used to produce matrices that can be dyed, with the dye being transferred to a mordanted paper or film as in the dye transfer (or, previously, the Technicolor motion-picture film) processes.

METOL

Metol, introduced by Hauff about 10 years after the discovery of hydroquinone, is another popular developing agent. It is packaged, sometimes with a modification of the acid radical, under a variety of proprietary names other than Metol—such as Elon, Pictol, and Photol. Metol is unlike hydroquinone in its developing action. It is a soft-working developer, development is initiated almost from the start of the developing time, it retains much detail in the images produced, it is not very sensitive to restrainers such as bromide, it has good shelf life, and it is capable of developing relatively large quantities of photographic emulsion before becoming exhausted.

MQ DEVELOPERS

Developer formulas made with both Metol and hydroquinone (MQ) produce an effect superior to either one used alone or the sum of the two used separately. This superadditive effect seems to have the advantage of an earlier induction period, allowing the hydroquinone to come into play sooner than it otherwise would. The superadditivity, however, exists not only during the induction period, but also at any degree of development. One explanation is that Metol is the primary developing agent and its oxidation product is reduced back to Metol by the hydroquinone, as long as any hydroquinone exists in the developer solution. Several other developer combinations also produce an improved working characteristic, but Metol and hydroquinone are the best known. Phenidone and hydroquinone used together produce a similar superadditivity effect.

ACCELERATORS

In order for the developing agent to function properly, the pH of the solution must be maintained at some value, usually above 7 with organic developing agents. This is accomplished by adding an alkali compound, or a mixture of compounds designed to maintain the pH at the desired level over the life of the developer, or at least during the development time of a given film. Sodium hydroxide is a strong alkali that is used for some types of developers, but the amount required to produce the required pH for most developers would be small, and the hydrogen ions liberated during development would rapidly change the pH. If the required pH is above 12 or 13, however, the quantity would be sufficient to maintain the pH even after the hydrogen has neutralized a substantial proportion of the hydroxyl OH^- ions to H_2O.

Most developers require that the alkali be *buffered* so as to maintain the availability of OH^- ions over a period of use. The buffering compound is usually a salt of a weak acid, such as sodium phosphate, sodium metaborate, sodium sulfite, or sodium bicarbonate. Several factors, including other chemicals in the formula, affect the pH of the developer,

but the choice of alkali is the most significant one. The pH, then, is an indicator of the relative activity of the developer; the higher the pH, the higher the activity. Some types of developing agents require a strong activator to make them function, whereas others can function with relatively weak activators, and a few with none at all. The following typical accelerators are arranged in order of decreasing alkalinity:

Sodium hydroxide, pH $\longrightarrow$ 12 +
Sodium carbonate, pH $\longrightarrow$ 11.5
Sodium metaborate (Kodalk),
pH $\longrightarrow$ 10.8
Borax, pH $\longrightarrow$ 9.6
Sodium sulfite $\longrightarrow$ (weak alkali)

PRESERVATIVES

Being a reducing agent, the developer will also react with oxygen in the air and thus lose its capability to develop a silver image. A preservative is added to the solution to prevent aerial oxidation. Sodium sulfite is the most commonly used preservative for developers utilizing organic developing agents. In the case of Metol-hydroquinone developers, the sulfite also serves the purpose of removing the reaction products of regeneration of the Metol by hydroquinone, so that the process can go on without slowing down. Sulfite also acts as a weak silver halide solvent, forming complexes that provide for a degree of physical development—the effect being influenced by other ingredients in the formula such as bromide, certain developing agents, etc. A small amount of physical development contributes to "fine-grain" development. In color developers, where the formation of dyes requires further reactions with oxidation products, the amount of sulfite must be kept to a minimum to prevent removal of the intermediate reaction products.

RESTRAINERS

A simple developer formula (developing agent, activator, and preservative) may not differentiate adequately between the ex-

Metol and hydroquinone used together in a developer develop more silver than the sum of the silver produced by the two used separately.

The activity of most developers increases as the pH is increased by adding alkali.

Without a restrainer, most developers would develop an excessive number of the unexposed silver halide grains.

posed and unexposed grains of silver halide in the emulsion. That is, in addition to developing the image grain, it may also tend to develop the nonimage grains to produce silver or fog. (Restrainers sometimes cause some reduction in emulsion speed, especially if used in larger amounts.) A soluble halide, usually bromide, is added to the solution to act as a restrainer. It has the effect of controlling the reduction process so that there is a better differentiation between exposed and unexposed grains. In addition, since one of the reactions of development is the release of bromide or other halogen ions into the solution, these ions will act as a restrainer. As more and more film is processed in the developer, additional halide will be added to the solution, thus increasing the restraining effect. A quantity of bromide in the developer formula minimizes the variability produced by the addition of relatively much smaller quantities of halide resulting from development (see Figure 8–4).

In the processing of multilayer color films where at least three emulsions have to be quite precisely controlled relative to one another, the maintenance of the halogen ions in the developer may have to be closely monitored. The restraining action of various halides is considerably different—iodide, for example, requiring one-thousandth the quantity of bromide required to produce a given restraining action. On the other hand, the amount of chloride required is so high that it is not useful as a restraining agent.

ANTI-FOGGANTS

Since halides such as potassium bromide restrain the development of fog in nonimage areas, they are often referred to as anti-foggants. However, the term is usually reserved for another class of developer additives that minimize fog, usually with less effect on the overall speed of the emulsion. The term is applied to a group of organic compounds sometimes used in the preparation of emulsions, as well as developers, although the specific compounds are not necessarily interchangeable in the two applications. Some high-energy developers that require a high pH, for example, must of necessity incorporate an anti-foggant of this type to keep the fog at an acceptable level. Since the organic

anti-foggants usually require lower concentrations than does bromide, they may be used where high concentrations of bromide are out of the question because of its effect on the other characteristics of the emulsion. Typical anti-foggants used in developers include 6-nitrobenzimidazole and benzotriazole.

OTHER DEVELOPMENT ACCELERATORS

There are numerous compounds that increase the rate of development when added to developer formulas due to other than the effect of the pH increase produced by the developer accelerator. These compounds work in a variety of ways. Some wetting agents, for example, have a positive charge (cationic), and decrease the induction period by canceling out the negative charge on the silver grain. Some sensitizing dyes with positive charges can have a similar effect. Certain neutral salts such as potassium nitrate and sodium sulfate also have an accelerating effect, which is thought to be due to the improved diffusion of the developer ions through the gelatin of the emulsion. Another class of compounds that are weak silver solvents also act as development accelerators through the extra density provided by physical development (thiocyanates, thiosulfates, cyanides). Wetting agents are sometimes added also for the purpose of smoothing out the flow of developer over the emulsion surface to produce more uniform development.

Other substances, manufactured under proprietary names, are often added to developer formulas to increase shelf life, lower the freezing temperature of liquid developers, improve mixing and solution capabilities, etc. Thus a packaged formula may be considerably different from a mixed formula used for the same purpose, or sometimes even having the same name. These proprietary formulations often provide competitive advantages in handling, and they are not disclosed to the general public.

CONCENTRATION OF DEVELOPER

The components of a developer interact in a variety of ways. The development rate is in-

The concentration of developer solutions has a positive but not always predictable effect on the rate of development.

fluenced by the developing-agent concentration, the alkali and resulting pH, the restrainers, and the sulfite or preservative concentration. The interaction of all these components is not constant, and adds further to the complexity. For example, it is often difficult to predict the results of dilution of many developer formulas. Some of them are designed to minimize these interrelating effects, and carry recommendations for diluting to various strengths to meet specific photographic requirements. In general, these developers are such that the product of dilution and development time to provide a given degree of development is not always constant, and each dilution should be tested with the type of film being used. Many developer formulas are such that a concentrated stock solution is prepared, which is diluted to various strengths as required.

Other factors affecting the degree of development are time, temperature, and agitation. These are interrelated so that a change in one can, to a considerable degree, be compensated for by a change in the other. Thus, different developing times are recommended for tray development of film with constant agitation and tank development with intermittent agitation. Agitation should be selected to produce uniform development over the picture area, not to control the degree of development. Most photographic processes, however, are designed to function best at a prescribed combination of time, temperature, and agitation. This is particularly important for color processes, where the relationship of reproduction characteristics between the red, green, and blue records has to be precisely maintained.

DEVELOPMENT TIME

The chemical effects of development are not constant throughout the total time of the process. In the beginning there are no reaction products, and the developer has to penetrate the emulsion to gain access to the silver halide grains. As soon as the development process starts, reaction products begin to form and a restraining action comes into play, which is controlled to some extent by the removal of these products and their replacement with fresh developer by means of agi-

tation. In the beginning there is also a relatively low concentration of silver ions present as the result of the solvent action of the developer, and it is some time before any physical development begins to take place. Rapid-acting developers, therefore, have a minimum of physical development component, whereas developers requiring extended times tend to have a greater physical development component. Most fine-grain developers require relatively long developing times. The composition of the developer, the proportions of the developing agents, etc., have significant effects on the development characteristics with time. Most developer formulas in common use have been optimized for performance throughout the range of times that produce the desired contrast indexes for films.

DEVELOPER TEMPERATURE

Development time is also a function of developer temperature. But again, there is no hard and fast rule for quantifying the relationship that would apply to all developer formulas. In practical situations with well-designed developers, and within the range encountered in normal practice, the rate of development approximately doubles for every 10 C (18 F) increase in temperature. The *temperature coefficient* is the ratio of increase in development time that occurs when the temperature is decreased by 10 C (18F). It can be determined by observing the times for first appearance of the image in the developer at the two different temperatures and applying the following formula:

$$\text{Temperature coefficient} = \frac{T_1}{T_2}$$

When the developer contains two developing agents, the coefficient is useful over only a limited range. Some typical coefficients are:

Metol	1.3
Pyrogallol	1.9
Hydroquinone	2.5
Metol-hydroquinone	1.9

The temperature coefficient is the development time factor that compensates for a decrease in the temperature of 10 C.

If the time to achieve a constant gamma at two temperatures is plotted on a graph in which the time scale is a ratio scale, a chart is produced that will show the time of development for any temperature required to produce the gamma, as shown in Figure 8–5.

Developers with hydroquinone as a developing agent have a relatively high response to changes in temperature, whereas those with Metol or Phenidone have a lower response. With the combination of Metol and hydroquinone this is evened out to some extent, but at lower temperatures there is less developing effect of the hydroquinone than Metol, and at higher temperatures the reverse is true. The concentration of bromide in the developer affects its response to temperature variations, as does pH. A higher pH value or a lower bromide concentration makes the developer less responsive to temperture variation (it has a lower temperature coefficient).

DEVELOPER AGITATION

Agitation affects both the degree and the uniformity of development. When development starts, the solution penetrates into the emulsion to react with the silver halide grains. As this proceeds, the developer is used up, and has to be replaced with fresh developer. The presence of the gelatin retards the diffusion, but in time the partially exhausted developer

> **Agitation removes used developer from the surface of the emulsion and replaces it with fresh solution.**

diffuses to the surface of the emulsion. If development is stagnant, this exhausted developer accumulates at the surface and slows down development. Agitation of the developer removes the exhausted developer and replaces it with fresh solution that diffuses into the emulsion to continue the development process. Agitation rate should be geared to obtain maximum uniformity of development. With high-acutance developers, where the image enhancement is achieved by the accumulation of exhausted developer and of fresh developer, at the edges of high densities and low densities, respectively, little or no agitation is recommended. With short developing times agitation has a greater effect on degree of development and overall density than it does with long developing times.

To ensure uniformity of development, the manner of agitation is important. There should be a nonlinear flow of the developer over the film, otherwise large areas of density differing from the surrounding area will allow exhausted (or relatively fresh) developer to be swept across adjacent areas of differing exposure. A motion-picture film, for example, that is carried through the process in a linear fashion will show "drag," areas of light density trailing dark areas on the film. The effect would be more prominent with short developing times. Likewise, intermittent agitation in trays can produce standing waves that will cause patterns on the developed film. Constant gas-burst agitation will tend to produce linear patterns, but if the bursts are intermittent, the combination of the bubbles' sweeping action followed by a period of stagnation will produce a more uniform result on the developed film. With tray processing constant agitation is recommended, but the tray or the materials in it should be moved in a random way (the tray is rocked, first one corner down, then the second) so as to minimize standing waves or directional effects.

The type of processing apparatus used has a strong influence on the degree of agitation (see Figure 8–6). Deep tanks in which the film is manipulated by hand can produce minimum agitation, as well as can stagnant tray development. Shallow tanks, spiral reels, and trays offer an intermediate degree of agitation and require well-defined agitation techniques to minimize nonuniformity problems. Processing drums, on which the film

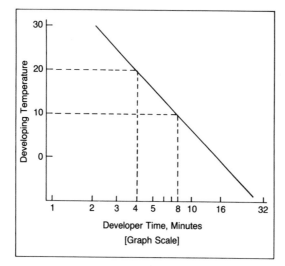

Figure 8–5   Time-temperature graph.

or paper is held in close proximity to the rotating drum surface with the developer carried between the drum surface and the emulsion surface, produces close to maximum agitation—the developer is exchanged at a very high rate. The roller transport processor produces a similar high degree of agitation. Tubes that carry the sensitized material inside, with the developer and other solutions flowing over the emulsion as the result of rotating or tilting the tube, also produce a high degree of agitation. Processing machines for long lengths like motion picture films, with spiral configurations also provide good agitation, but these can lead to unevennesses due to the linear passage of the film through the solution. Here extra agitation of the solution by means of circulating pumps and jets can be made to sweep the film to minimize these problems.

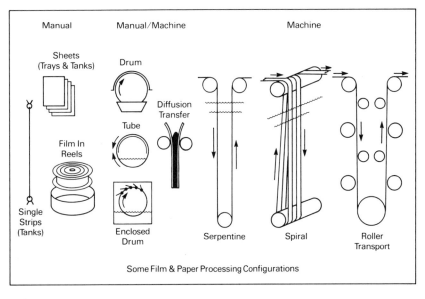

Figure 8–6   Some film and paper processing configurations.

REPLENISHMENT

When processing small volumes of photographic materials, it is best to use the developer once, then discard it. In this way uniform fresh developer is always used, and the resulting degree of development is more uniform from one time to the next.

When larger quantities of materials are being processed, it is not economical to discard the relatively large volumes of solutions required, and so replenishment can maintain the constant activity of the developer (or other solutions used in the photographic process). In determining the makeup of the replenisher formula, and the rate of addition, several factors have to be considered. These include: (1) the loss of developer resulting from carry-out by the film or paper as it is removed from the solution; (2) loss in developer activity due to accumulation of reaction products of development in the solution, mainly soluble bromide, and to some extent other halides; (3) loss of activity due to exhaustion of developing agent and sulfite consumed in the developing process; (4) increase in developer concentration due to evaporation; and sometimes (5) carry-in of water or other chemicals if a pre-bath or pre-rinse is used prior to development. (Other steps in a process will also have to take into account the carry-over of the previous solution or rinse.)

The replenishment formula is calculated to accommodate these changes due to processing, and to maintain the activity of the developer at a constant level. Such a formula is established by measuring the changes in the developer's chemical composition as measured quantities of film or paper are processed; calculating the amount of each chemical required to maintain the developer at its original strength, and adjusting this formula by actual performance tests. Since there are differences in the variables, it is often necessary to adjust the replenishment rate and/or formulation, especially if the process is to be maintained over a long period of time.

Some moderate volume replenishment systems are designed to be replenished over a period that would require the addition of an amount of replenisher equal to the volume of the original developer solution. Then the developer is discarded, and a new developer, along with an equal volume of replenisher, is prepared for use with additional films. This generally avoids or minimizes the problems that can arise when attempting to maintain the process over a long period without re-mixing. With some systems, only a replenisher chemistry set is sold, and a starter solution is provided that adjusts the replenisher to make it similar to a "seasoned" developer by adding bromide and reducing pH with an acid, and without processing any film to "condition" the developer. This becomes

Developer replenishers are designed to maintain a constant level of activity in developers used to process large amounts of film or paper.

the properly compounded developer, and the replenisher itself is used for replenishment as intended.

Replenisher formulas generally supply additional sulfite and developing agent that are consumed in the process, provide additional alkali to make up for loss, but reduce the amount of or omit bromide to compensate for that added to the developer as the result of the development reaction. In general, long-running processes can be maintained by suitable control of the chemistry through sensitometric monitoring and analysis of the constituents of the developer in use. In many instances, the carryover of solution from the tank is sufficient to accommodate the replenisher solution introduced into the tank; but in some instances control is achieved by having the formulas adjusted so that some additional developer has to be withdrawn to permit the introduction of a greater amount of replenisher. This makes it possible to adjust the withdrawal rate as another means of controlling the developer's chemical composition. In some cases it may be necessary to remove more than is carried over by the film to compensate for excessive buildup of halides beyond that required in the original formula.

Large-scale operations with continuous processing require constant chemical and sensitometric monitoring, as well as careful adjustments in the replenisher formula and rate in order to maintain the result within narrow limits. Since the rate of replenishment is dependent upon the type of film being processed, and the type of images on the film (high density vs. low density, etc.) these factors must be considered in maintaining uniformity, particularly in the case of color processes.

PAPER PROCESSING

The developers used for papers generally follow the same principles as those for films, with additional problems due to the physical differences (which govern carry-over, among other things). Alkalinity and other aspects of the developer may have an effect on the sizing, curl, and/or brittleness of the paper. The content of the developer—restrainers and reaction products—can have a significant ef-

Process control procedures used to maintain consistent quality involve sensitometric tests and chemical analysis of the processing solutions.

fect on the tone of the image produced, and this can become more apparent if the prints are subsequently chemically toned.

ANALYSIS OF DEVELOPERS

The best control of processing operations involves the use of both sensitometric tests and chemical analysis of the processing solutions, techniques of process control (see Chapter 3). However, the analysis of most photographic chemistries involves rather complicated procedures and requires a well-equipped laboratory operated by a trained chemist. In addition, some methods are subject to error due to the presence of other chemicals in the solution. It is best to continually make reference to a properly mixed operating solution that has not been used, to make sure the calibrations of the analyses are accurate. Simple procedures include the measurement and control of pH, measurement of specific gravity, and notation or measurement of the color of the solutions. These can be plotted and serve as an indication of change that may call for further study or more advanced analysis. And, of course, since all chemical processes are a function of time, temperature, and, in the case of photography, agitation, careful monitoring of these aspects is probably the most important control that can be exercised.

The importance of good darkroom habits cannot be overstressed. Most failures of photographic processes can be attributed to accidents involving contamination of the processing solutions, even in trace amounts.

SPECIFIC BLACK-AND-WHITE PROCESSES

Processing of black-and-white films, papers, and plates normally involves development, followed by a rinse in an acid stop bath, fixing in an acid thiosulfate solution, washing to remove the soluble silver complexes and processing chemicals, and then drying. The wide choice of developer formulas depends on the photographic objectives, economic considerations (time, overhead), quality advantages (real or imagined), safety, ecological considerations, etc.

Table 8–1　Some developer formulas compared

	DK-50	DK-50R	D-76	D-76R	D-76d	A-17	D-25	DK-20	DK-20R	D-72	D-85
Water (liters)	.750	.750	.750	.750	.750	.750	.750	.750	.750	.750	.500
Metol	2.5 g	5.0 g	2.0 g	3.0 g	2.0 g	1.5 g	7.5 g	5.0 g	7.5 g	3.1 g	
Sodium sulfite	30.0 g	30.0 g	100.0 g	100.0 g	100.0 g	80.0 g	100.0 g	100.0 g	100.0 g	45.0 g	30.0 g
Hydroquinone	2.5 g	10.0 g	5.0 g	7.5 g	5.0 g	3.0 g				12.0 g	22.5 g
Sodium metaborate[a]	10.0 g	40.0 g						2.0 g	20.0 g		
Borax			2.0 g	20.0 g	8.0 g	3.0 g					
Sodium carbonate[b]										80.0 g	
Boric acid					8.0 g						7.5 g
Sodium bisulfite							15.0 g				2.2 g
Paraformaldehyde											7.5 g
Sodium thiocyanate								1.0 g	5.0 g		
Potassium bromide	0.5 g					0.5 g		0.5 g	1.0 g	2.0 g	1.6 g
Water to make	1 liter	1 liter	1 liter	1 liter	1 liter	1 liter	1 liter	1 liter	1 liter	1 liter	1 liter

[a]Kodalk
[b]Monohydrated

Paper processing formulas are designed to produce pleasing image colors and densities, both before and after chemical toning; to have good tray or machine tank life, and freedom from unwanted stain or fog in the whites of the print. Color formulas have to be capable of forming appropriate dyes of good density. In some applications, film speed is the primary criterion, and the formula may be chosen to maximize this characteristic, even at the sacrifice of some aspects of image quality. In most cases there has to be a "trade-off" between desired attributes and those that are required to meet some condition. Table 8–1 compares some typical developer formulas.

REVIEW QUESTIONS

1. The development process amplifies the latent image silver by as much as . . . (p. 214)
   A. one hundred times
   B. one thousand times
   C. one million times
   D. one billion times
   E. one trillion times
2. The basic steps in reversal processing, in the correct order, are . . . (p. 214)
   A. develop, fix, bleach, re-expose, redevelop, fix
   B. develop, bleach, re-expose, redevelop, fix
   C. re-expose, develop, bleach, redevelop, fix
   D. develop, re-expose, bleach, redevelop, fix

3. In chemical development, the silver that forms the final image is provided by . . . (p. 215)
   A. a pre-development toning bath
   B. the developer
   C. the latent image
   D. the silver halide grains
   E. the fixer
4. Developing agents are classified as being . . . (p. 216)
   A. oxidizers
   B. reducers
   C. ambivalent
5. "Superadditivity" refers to the effect produced by using . . . (p. 217)
   A. Metol and Phenidone in the same developer
   B. Metol and hydroquinone in the same developer
   C. double the normal amount of Metol in a developer
   D. Metol and sodium sulfite in the same developer
6. The chemical most commonly used as a preservative in developers is . . . (p. 217)
   A. potassium bromide
   B. sodium sulfate
   C. sodium sulfite
   D. sodium sulfide
   E. sodium thiosulfate
7. Benzotriazole, which is sometimes used in developers, is classified as . . . (p. 218)
   A. a developing agent
   B. a preservative
   C. an alkali
   D. an anti-foggant
   E. an accelerator

Selecting a film developer commonly involves a trade-off, such as fine grain versus short developing time.

8. In comparison with the original developer, developer replenishers typically contain . . . (p. 222)

A. less preservative
B. less restrainer
C. less alkali

9 Post-Developmental and Archival Considerations

William Henry Jackson. Courtesy of Smithsonian Institution National Anthropological Archives, Bureau of American Ethnology Collection.

STOP BATHS

A stop bath consists of a mild acid, with a pH of 3 to 5, that neutralizes the alkali of the developer and stops development. An acid fixer also provides stopping action; but after a quantity of developer has been carried over to the fixer, the acid is neutralized, the aluminum of the hardener will precipitate out as a sludge, and the stopping action is no longer precise. A water rinse before either the stop bath or the acid fixer will remove some of the developer, and prolong the life of either, but then the stopping action is drawn out and imprecise. A stop bath may also contain some hardening agent, and thus becomes an acid hardening stop bath. The most commonly used stop baths consist of 1.4%–4.5% solutions of acetic acid, although a solution of sodium or potassium bisulfite is sometimes used. Some contain other additions.

INDICATOR STOP BATHS

An indicator dye used in an acid stop bath will change color when the acid has been neutralized by developer carry-over.

The two-bath fixing procedure enhances the archival quality of photographic prints.

Indicator stop baths contain a dye that changes color after a substantial increase in pH as the acid is neutralized, and the stop bath can then be renewed before trouble occurs. It is important to select a dye that changes to a color that can be distinguished under the red, orange or amber color of a safelamp. An alkaline solution of bromocresol purple dye can be used to test the effectiveness of an acetic acid stop bath. When a few drops of such a solution made up as KODAK Stop Bath Test Solution SBT-1 is added to a sample of about 30 ml of the stop bath, it will have a yellow color if satisfactory, and a purple color if the acid has been neutralized. Some photographic chemical manufacturers sell a proprietary indicator stop bath that has the indicator dye incorporated in the bath. The dye is not strong enough to have any effect on the appearance of prints or films processed through the bath.

DICHROIC FOG

A fixer whose acid strength is at or near exhaustion will no longer effectively stop the developing action. As a result, some development will proceed at the same time as some silver halide is being dissolved by fixation; and dichroic stain, which is mostly silver in the colloidal or near-colloidal state, will be produced. *Dichroic* means that it has two different colors, one by reflected light and the other by transmitted light.

TIMING LAYERS

Another system of stopping development is used in instant photography systems. One of the many layers in the film consists of a "timing" layer that dissolves through a membrane in a definite time and neutralizes the activator component of the system to stop development.

FIXING BATHS

A fixing bath converts the relatively insoluble silver halide in the film or paper into a soluble complex of silver that can be washed out of the emulsion layer, baryta layer, and paper fibers, to make the negative or print stable. Otherwise, the undeveloped silver halide would not only reduce the contrast of a negative, but it would darken in time and destroy the image. Sodium, ammonium, or (sometimes) potassium thiosulfate are the compounds most commonly used in fixing baths; although sodium thiocyanate, and sodium cyanide have been used. (The latter is a hazardous compound and therefore is not recommended for use in photography.)

When a fixing bath is used to fix films or prints, silver ions accumulate in the solution, which exert a force in the opposite direction and thus slow down the rate of fixation. If this is carried far enough, a point is reached where the soluble complex is not completely formed in the time allotted to fixation. If the fixing time is increased, the insoluble complex is adsorbed to the gelatin of the emulsion, the baryta coating if there is one, and the fibers of the paper base. The insoluble complexes are thus not removed by washing, and the silver can later be reduced by environmental factors to produce unwanted color and/or density. The presence of this reduced silver can affect the metallic silver image and cause it to become sulfided, thus changing color and losing density. Therefore, to ensure stable photographic images, it is important that the fixing solution not be worked beyond its capacity. The use of a two-solution fixing technique, whereby the prints are first bathed for a time in a fixing solution, then transferred to a second, fresher fixing solution, assures that the silver halide is converted to a soluble complex. After some time, the first fixer is discarded (or consigned to silver recovery), the second fixer becomes the first fixer, and a fresh fixing bath is added.

Most fixing baths make use of the sodium salt of thiosulfate and are formulated to be acid in nature, in order to neutralize any remaining alkali from the developer, and because most hardeners require an acid condition. Potassium alum is commonly used as a hardening agent. Boric acid serves as a buffer to maintain the acidity of the bath as it is used. Sulfite reacts with any free sulfur that is formed in the process, converting it to thiosulfate. A typical acid hardening fixing bath is the following:

KODAK Fixing Bath F-5

Water, about 52 C (125 F)	600 ml
Sodium thiosulfate (pentahydrated)	240.0 g
Sodium sulfite (anhydrous)	15.0 g
Acetic acid (28%)	48.0 ml
Boric acid crystals	7.5 g
Potassium alum, fine granular (dodecahydrated)	15.0 g
Cold water to make	1.0 liter

Films should be fixed for not less than double the clearing time, usually 5 to 10 minutes. The recommended fixing time for prints is also 5 to 10 minutes. Excessive fixing times will promote adsorption or retention of the fixer complexes in gelatin, or more particularly on the fibers of paper base, and should be avoided. In addition, excessive fixing can bleach silver images, especially those on paper prints. Where the photographic application requires a nonhardening fixer, the formula contains only sodium thiosulfate, sodium sulfite, and sodium bisulfite to produce an acid condition.

If a fixing bath is made with ammonium thiosulfate, or with ammonium chloride added to sodium thiosulfate, the fixing action is much more rapid. The following formula is more rapid-working than the F-5 formula,

and is capable of fixing about 50% more film, or paper. Overfixation should be avoided, especially of prints and fine-grained films, to prevent bleaching.

KODAK Rapid Fixing Bath F-7

Water, about 52 C (125 F)	600 ml
Sodium thiosulfate (pentahydrated)	360 g
Ammonium chloride	50.0 g
Sodium sulfite (anhydrous)	15.0 g
Acetic acid (28%)	48.0 ml
Boric acid crystals	7.5 g
Potassium alum, fine granular (dodecahydrated)	15.0 g
Cold water to make	1.0 liter

Caution: With rapid fixing baths, do not prolong the fixing time for fine-grained film or plate emulsions or for *any* paper prints; with prolonged fixing, the image may have a tendency to bleach, especially at temperatures higher than 20 C (68 F). This caution applies particularly to warm-toned papers.

REJUVENATION AND MAINTENANCE OF FIXING BATHS

Fixing baths cannot be replenished, since the problem is not exhaustion of the chemicals, but the accumulation of silver ions in the solution. More thiosulfate could be added, but this would not overcome the excess silver. Electrolytic silver recovery, however, can remove silver from the solution, and the other components of the fixer (hardener, sulfite, etc.) can be made up by replenishment to prolong the useful life of the bath from which silver has been removed.

WASHING

After fixing, photographic materials have to be washed to remove any of the fixing chemicals remaining as well as the silver compounds that have been formed during the fixing reaction. The problem is not a simple one. The chemical reactions leading to a soluble silver complex pass through steps in which insoluble complexes are formed, and these have to be fully converted to the soluble compounds. In addition, these complexes

Excessive fixing can bleach silver images.

Removal of the by-products of fixation from fiber-base prints is made more difficult by overworking the fixing bath or overfixing the prints.

A 2% solution of sodium sulfite is an effective washing aid for fiber-base prints.

have a pronounced tendency to adhere to or be adsorbed by the fibers and baryta coatings of ordinary papers (see Figure 9–1). Films and polyethylene- or resin-coated (RC) papers do not generally present this problem except to the small extent that it may occur in gelatin coatings. For these reasons the fixing step should be for the minimum time that will ensure that the reaction goes to completion to prevent any appreciable adsorption of the chemicals.

As the fixing process continues, and more silver complexes accumulate in the fixing bath, the tendency increases for the reaction to be forced in the reverse direction. In general, the fixing bath will perform well when the total amount of silver (in the form of ions from residual silver thiosulphates) in the bath is below 2.0 grams/liter. Films and resin-coated papers are readily washed, as they do not have fibers or baryta coatings exposed to the solutions for adsorption to take place. Fiber-base papers may not be completely washed even after 60 minutes of washing under ideal conditions, whereas films are usually washed in 20 to 30 minutes under similar conditions and resin-coated papers in 4 minutes, provided the wash-water flow rate is sufficient to give the required number of changes. The specific gravity of a typical fixer is in the vicinity of 1.18. However, with a reasonable flow of water the fixer does not have a chance to settle to the bottom of the wash tank.

HYPO ELIMINATORS AND WASHING AIDS

Hypo removal by washing can be facilitated by the use of *washing aids,* and residual hypo and thionates can be eliminated with *hypo eliminators.* The hypo-clearing agents consist of salts that act on the relatively small amount of thiosulfate or complexes remaining after a short washing, replacing them with a radical that is more easily removed by further washing. A 2% sodium sulfite solution has been found to be one of the best for this purpose, but other salts such as sodium sulfate can be used.

The sulfite replaces the thiosulfate and thionates, and is itself readily washed out. The following would be a typical procedure in practice: The fixed materials are rinsed for a short time in water to remove excess fixer solution; then they are soaked for about 2 minutes for films and 3 minutes for papers in a 2% solution of sodium sulfite, followed by a longer wash—10 minutes for single-weight papers, 20 minutes for double-weight papers, and 50 minutes for film.

Hypo eliminators work on a different principle, that of oxidizing the residual thiosulfate to sulfate, which is more readily removed by washing. This should be carried out only after most of the hypo or thionates have been removed by washing, possibly with the aid of a hypo-clearing agent.

PREPARATION FOR DRYING

After films or papers are washed for the time necessary to remove residual thiosulfate or thiosulfate-silver complexes, they should be carefully dried. This aspect of the photographic process is often neglected to the extent that it subtracts substantially from the quality and acceptability of the final product. Fiber-base papers absorb considerable water during washing, and this has to be removed in the drying process. RC (resin-coated) papers, on the other hand, absorb very little water, and this is restricted to the thin emulsion and NC (non-curl) coatings. These materials dry very rapidly—in a matter of a few minutes.

After washing and prior to drying there are several important techniques to note. The

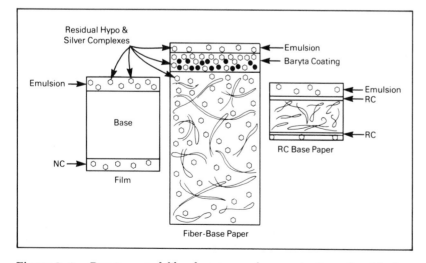

Figure 9–1   Baryta-coated fiber-base paper favors retention of residual hypo and silver complexes; whereas film and RC-base paper do not have the fiber areas or baryta for retention of these compounds.

wash water sometimes contains sediment that has been allowed to pass into the water lines from water-treatment plants. At certain times this may be more of a problem than at others. For example, changes in water demand cause disturbances in settled particles. When municipal filter beds are flushed, fine sand or other sediment may be noticed in the tap water. If this sediment is allowed to remain on a processed negative, it may become attached to the emulsion or NC surface and produce small white spots on the final print. For this reason it may be required to rinse the negatives in a separate bath of filtered water. If this is a persistent problem, the water line leading to a laboratory should be fitted with a filter. Excess water droplets and sediment in small amounts can be removed with a clean moist sponge or chamois by carefully wiping the surfaces. This has the disadvantage of sometimes putting more contamination on the negative than it removes.

Films washed in clean water can be hung up to dry without wiping. However, if the excess surface water is not removed from the negative, and if it is in the form of droplets, drying marks or deformities in the image will form, ruining the negatives. In large laboratories where the film is processed in a serpentine fashion in continuous lengths, excess moisture is blown off by an air jet or squeegee. Water spots on sheet films and individual rolls can be avoided by using a wetting-agent bath before drying, which spreads the surface layer of water so that no droplets are formed. A commercial wetting agent, Photo-Flo, is one sold for this purpose. This bath, if made with filtered or otherwise clean water, can also serve as a rinse to remove some sediment. Gently rubbing the film surfaces as they are removed from the wash water can dislodge particles that may tend to cling to them.

DRYING FILMS

It is best to dry films at room temperature; but this procedure may require an inconveniently long time in some situations so that heat is often applied, along with air circulation, to hasten the drying process. At the start of drying, relatively warm air can be used, as evaporation of the moisture in the film has a cooling effect. As drying progresses, high temperatures should be avoided. If the temperature is too high, drying will become excessive, and the film will become brittle and have a tendency to excess curl. Other damage to the film may occur. The air should be filtered to prevent the accumulation of dust on the film surfaces. A large volume of relatively pure air can carry a substantial number of particles that could be attracted to the tacky film. This attraction can be accentuated by the electrostatic charge built up in the film by the air flowing over its surfaces.

DRYING PRINTS

Similar precautions apply to the removal of excess water from the surfaces of prints that are to be dried. Print-drying machines usually have squeegee rollers at their entrances that remove excess surface water. RC or "plastic"-coated papers carry relatively little water and can be dried quite rapidly, within a few minutes without heat, after the excess water has been removed. Fiber-base papers, on the other hand, can carry a relatively great load of water, which has to be removed. Air drying the prints at room temperature on clean cheesecloth or racks is probably the least destructive to the image and the physical quality of the prints. However, if the relative humidity is high, this kind of drying process is apt to be excessively long. For this reason, most photographic installations make use of heat dryers of one form or another (see Figure 9–2).

PRINT-DRYING PROBLEMS

Control of heat in the drying process is necessary to prevent excessive drying and undesirable changes in the nature of the prints. Physical problems include curl, brittleness and buckling—nonuniform dimensional change. In addition, ordinary fiber-base papers expand when wet and shrink on drying to a final dimension that is usually smaller than that existing at the time of exposure. These dimensional changes are not uniform for the machine and cross-machine directions (referring to the original paper-making-machine web). In ordinary photography such

Drying marks from drops of water on film while drying are difficult, if not impossible, to remove.

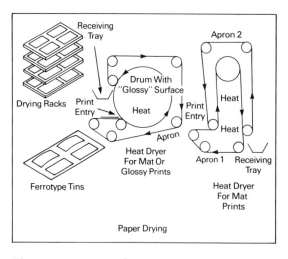

Figure 9–2   Paper drying.

changes are insignificant, but they can be a factor where the print's dimensional characteristics are important, as in photogrammetry. However, if the prints are dried by ferrotyping or on a dryer that utilizes a similar principle, they retain their expanded dimension when dry, and this dimensional change is considerably greater than the result of drying without restraint. This change also is of no consequence, unless a collection of prints is made with the grain of the paper in two directions. Then the differential between the expansion in one direction and that in the other will make images that appear to be considerably different in size—up to 3/16 inch on an 8×10-inch print. (In ferrotyping the wet print is squeegeed onto a smooth metal or laquered metal surface for drying. The gelatin on the print's surface is molded by the glossy surface of the ferrotype plate, producing a high gloss. The ferrotyping surfaces are usually treated with a polish that prevents the gelatin from sticking to the plate when dry.)

COLOR CHANGES

A further potential problem in print drying is the color change that sometimes occurs in the image. Some prints dried on heat dryers will change from a good black image tone, particularly in the maximum densities, to a bluish tone of less density. This effect is called *plumming*. Sepia-toned prints often change

color on heat drying, and this is more noticeable with selenium-toned prints than with sulfide-toned prints. In either case, if the print is not entirely dry at the time it is dry-mounted with heat, the difference in color within a given print can be as much as to make it totally unacceptable.

Prints dried without heat also change tone and apparent density, compared to their visual appearance when wet. The effect depends to some extent on the type of emulsion, but mostly on the nature of the print surface and appears to be greatest on mat-surface prints. The photographer or darkroom printer has to take this into account when judging the exposure and development of prints as observed under the darkroom illumination. The amount of adjustment is governed by experience with the particular type of paper being used. One guide is to observe the just-perceptible difference between a nonspecular white highlight in the print and the non-image area (margins).

The amount of moisture removed by drying photographic materials has to be considered in the design and layout of photographic finishing establishments. If the drying process saturates the air, the removal of water is impeded. This can lead to attempts to correct the problem with increases in drying temperature, which can have a destructive effect on the final product.

PRESERVATION OF PHOTOGRAPHS

The chief factor in the preservation of silver photographic images is the total removal of silver complexes and hypo from the print by washing, before drying. After that it is important that the print not be exposed to any contaminant that would react with the silver image. The choice of adhesives for mounting, if the print is mounted at all, is important. Some kinds of glues and pastes, especially those that might contain sulfur, should never be used. Hygroscopic materials (substances that absorb water from the air) should not be used. Dry-mounting tissues especially designed and marketed for mounting photographs are generally acceptable. But even this type of mounting may eventually deter from the value of the photograph if it becomes a

Some photographic prints appear darker when dry than when wet.

collector's item. The best recommendation seems to be that of using "archival" materials (mounting and matting boards, and cloth tape) to suspend the print under a mat, then in a frame under glass for display. If the mat has to be replaced after some handling, it is a simple matter to remove the print and remat it.

Sometimes the photographic print is only an intermediate image for use in preparation of plates for printing in catalogs, magazines, or newspapers, and once these plates have been prepared, the photograph is no longer needed. In the motion-picture industry, the release prints need only last as long as the physical capability of the film to sustain screenings—about 500 to 1,000 projections—after which it is destroyed. However, if the images are expected to have some value in later years, other precautions have to be taken. The photographer will need to make sure the print is made to have archival permanence, and the motion-picture companies will have to have their prints made on film with dyes that will hold up under long-term storage, or have separations made on black-and-white film that can be stored and then brought out for regeneration of the color films at a later date.

Storage containers, mat boards, and mounting boards are potential sources of agents that are deleterious to photographs. For this reason, if photographs are to be stored for any time, they should be packed in properly designed containers made of inert materials. Old film and paper boxes should not be used, neither should wooden and cardboard cases. Metal cases coated with paints that do not give off any fumes, or cases made of inert or "acid-free" board, should be used. Mount and mat boards should be inert and acid free; that is, buffered to a slightly higher than neutral pH for black-and-white photographs. However, there may be some question in using this type of material for color photographs, since most of these dyes may survive best under slightly acid or nearly neutral conditions.

It is important that the storage area's atmosphere be maintained with moderate or lower temperatures, and low relative humidity (below 50%). The cyclic effect of day and night conditions—wet then dry, winter then summer—contributes more to rapid deteri-

oration than when these conditions are maintained at a desirable low humidity and temperature with small cyclic variations.

ARCHIVAL CONSIDERATIONS

Since resin-coated papers have not existed long enough to establish their long-term keeping characteristics, prints intended for archival storage or for permanence are usually made on fiber-base papers, which have a longer history for forecasting the permanence of their images. However, since these papers will retain some small amounts of residual silver complexes even with prolonged washing, additional consideration has to be given to their fixing, washing, and after-treatment.

There are several ways to process for permanence, proposed by photographic materials manufacturers and by the American National Standards Institute. The recommendation of the Eastman Kodak Company is to follow the procedure given in the following table after the prints are developed:

Even properly processed prints are susceptible to being damaged if exposed to certain contaminants.

One procedure for increasing print permanence is to treat the processed print with a hypo eliminator and then with a gold protective solution.

Steps After Development

Step	Time	Comments
Rinse	30 sec.	Fresh acid stop bath.
First fix	3–5 min.	Use two fixing baths with
Second fix	3–5 min.	frequent agitation.
Rinse	30 sec.	Running water.
Hypo clear	2–3 min.	KODAK Hypo Clearing Agent.
Hypo eliminator	6 min.	HE-1 formula with occasional agitation.
Wash	10 min.	Running water flowing rapidly enough to replace water at least every 5 minutes.
Gold protective solution	10 min.	GP-1 formula with occasional agitation.*
Wash	10 min.	Same as first wash.
Dry	As required	In clean, dust-free space.

*A selenium toner or a sulfide toner can also be used at this stage. Toners are considered to be more effective than the gold protection.

KODAK Hypo Eliminator HE-1

Water	500 ml
3% hydrogen peroxide	125 ml
10% ammonia solution[1]	100 ml
Water to make	1 liter

[1]Prepared by adding 1 part 28% ammonia (concentrated) to 9 parts water.

Prepare the eliminator solution immediately before use, and do not store in a stoppered bottle, since the gas that develops may break the bottle.

KODAK Gold Protective Solution GP-1

Water	750 ml
1% gold chloride solution	10 ml
Sodium thiocyanate solution[2]	15.2 ml
Water to make	1 liter

Add the gold chloride stock solution to the volume of water indicated. Dissolve the sodium thiocyanate separately in 125 milliliters of water. Then add the thiocyanate solution slowly to the gold chloride solution, stirring rapidly.

The American National Standard Method for Evaluating the Processing of Black-and-White Photographic Papers with Respect to the Stability of the Resultant Image, PH4.32-1980, gives a somewhat different treatment method. The hypo eliminator precedes a 1% sodium sulfite clearing bath, and the eliminator formula contains four times as much hydrogen peroxide, and one gram of potassium bromide/liter to stabilize the tone of the print, which is sometimes modified by the HE-1 formula.

> Some toners increase the archival qualities of photographic prints, others decrease it.

ALTERNATIVE METHODS

Various other methods of archival processing are suggested. One of these is the Ilford Ilfobrom Archival Chemistry and Processing Sequence (from the Ilfobrom Galerie brochure), which places emphasis on a short fixing time. The rate of fixing is substantially faster than the rate of build-up of thiosulfate complexes in the fibers of the paper, and the time is not sufficient to permit thiosulfate complexes to attach themselves to the fibers of the base. The recommended Ilfobrom Archival Chemistry is as follows:

Step	Material	Time
1. Development	Ilfobrom Developer (1 + 9)*	2 min.
2. Stop bath	Ilfobrom Stop Bath (1 + 9)	5–10 sec.
3. Fixation	Ilfobrom Fix (1 + 4)	30 sec.
4. First wash	Good supply of fresh running water	5 min.
5. Wash aid	Ilfobrom Archival Wash Aid (1 + 4)	10 min.
6. Final wash	Good supply of fresh running water	5 min.

*One part of concentrate added to 9 parts of water. All processing times are at 68 F (20 C). This sequence totals less than 23 minutes.

TONING

Toning is the chemical modification of the silver image to give it a different color. The term *tone* is sometimes applied to the lightness of an image or area, and to the color of the silver image prior to chemical modification, which may be affected by modifications of development (see Figure 9–3).[3] The most common toning technique is to form a silver sulfide or silver selenide image by one of several methods. Less common is to convert the silver image to a metallic salt that is other than silver, or that is in combination with the silver. Some combination toning processes involve sulfide or selenide and a metallic toner.

The sulfide- or selenium-toned images are quite stable—in fact, usually more stable than the silver image alone—since the sulfide compound is one of the most common results of deterioration when residual thiosulfate is present or when sulfur compounds in the air have attacked the silver image. These compounds are resistant to further oxidation. The non-noble metallic compounds (iron, copper, uranium, etc.—those whose salts are more soluble than those of the noble metals such as silver, gold, platinum, etc.) form images less stable than the original silver images. This is particularly true of the iron compounds, which are easily attacked by impurities in the air and by alkalis in the water during washing. Gold- and platinum-toned images are, along with selenium and sulfur, the most stable of photographic images.

SEPIA TONES

The sepia tones produced by selenium and sulfur are less popular than they were some years ago. This may be partially due to the extra processing steps involved, the difficulty of maintaining and reproducing a given tone, the public association of sepia tones with antiquity; and to less emphasis on permanence along with the improved knowledge of proper processing of silver images for

---

[2] 1 ml of thiocyanate solution is equal to 0.66 grams of powdered sodium thiocyanate.

[3] Current, "Some Factors Affecting Sepia Tone." *PSA Journal 1950 Annual*, pp. 684–87.

greater permanence. There was a time when photographers emphasized sepia-toned images in their marketing efforts, and sometimes adopted a proprietary tone of their own for sales promotion.

REDUCTION AND INTENSIFICATION

Modification of a negative to improve its printing characteristics may involve reduction or intensification in one form or another. This is less likely to be required in the present day than it was in the past, before modern film, exposure meters, and more precise processing methods were used. These procedures are useful at times to salvage a negative that has been incorrectly exposed or processed, where reshooting is not possible. Some reducers and intensifiers may be used in modified form to make corrections or improvements in prints, but formulas and techniques have to preclude staining. Reduction consists of lowering the densities and contrast of the negative. Intensification involves increasing the negative's densities and contrast. Some toners lighten or darken print images.

Reducers can be classified according to the effect they have on the negative densities. *Proportional* reducers are those that lower densities by a given percentage of the original densities, thereby removing more density from the dense areas and lowering contrast (see Figure 9–4). *Sub-proportional* reducers are those designed to remove equal amounts of silver from all densities, producing a less-dense negative with no change in contrast. With *super-proportional* reducers, the per-

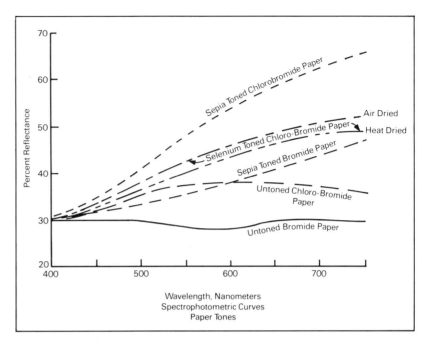

Figure 9–3 Spectrophotometric curves for paper tones.

centage of reduction is greater in the higher densities than in the lower densities, resulting in considerably less contrast with little change in shadow densities.

Intensifiers generally produce *proportional* intensification of the negative's densities; that is, they add density in proportion to the amount of silver already present, thereby increasing contrast (see Figure 9–5). Intensifiers that add proportionately more density to highlights than to shadows are called *super-proportional*, and are used where extreme high contrast is desired. Those that increase densities in the shadow region (toe of the curve) by a larger percentage (but not by a larger amount) than in the highlight region are called *sub-proportional* intensifiers.

Proportional type reducers and intensifiers produce effects on negatives similar to those that would have been obtained with shorter and longer developing times.

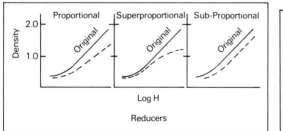

Figure 9–4 Reducers.

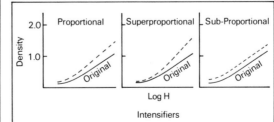

Figure 9–5 Intensifiers.

Other procedures have been used successfully to retrieve information from negatives that have been badly underexposed. One of these is to soak the developed negative in a fluorescing dye, such as Rhodomine B, that is absorbed more strongly in areas where silver is present even in minute quantities. The dyed negative is subjected to ultraviolet radiation to produce a fluorescent image, which is then rephotographed.

PRINT MODIFICATION

Reduction and intensification techniques can be applied to paper prints or film transparencies as well as to negatives, but it is usually not worthwhile to resort to these treatments for overall changes in density or contrast (although Ansel Adams used selenium toner for blacker blacks and increased contrast). Instead, the methods lend themselves to local treatment to enhance the tonal makeup. The formulas for reducers and intensifiers for this purpose are usually different, and the concentrations are lower. It is important that no residual traces of the chemical treatment show in the final print or regenerate later.

NONCHEMICAL TREATMENT

A nonchemical method of modifying a negative is to make a positive transparency by printing, and from this make a new negative with exposure and development in both steps adjusted to produce the desired negative characteristics. Contrast and other printing characteristics can also be altered by producing simple masks that are bound in register with the negative for printing.

REVIEW QUESTIONS

1. Failure to use a stop bath for quantity processing of prints may result in . . . (p. 226)
   A. fogged prints
   B. contrasty prints
   C. discolored prints
   D. sludge in the developer
   E. sludge in the fixer
2. The pH values of stop baths are usually in the range of . . . (p. 226)
   A. 2 to 4
   B. 3 to 5
   C. 4 to 6
   D. 5 to 6
3. In the two-solution fixing procedure, the fresher fixing bath is in the . . . (p. 227)
   A. first position
   B. second position
4. The recommended method of extending the life of a fixing bath is to . . . (p. 227)
   A. heat the bath to 100 F
   B. add replenisher
   C. remove the silver
5. Hypo eliminators work on the principle of . . . (p. 228)
   A. causing thiosulfate ions to precipitate
   B. neutralizing the static electrical charge on the thiosulfate ions
   C. converting thiosulfates to sulfates
   D. producing a molecular oscillation
6. With respect to image stability, toned images . . . (p. 232)
   A. are more stable than silver images
   B. are less stable than silver images
   C. vary considerably in stability
7. If a negative of a normal-contrast subject that was correctly exposed but overdeveloped is to be reduced, it should be reduced in a . . . (p. 233)
   A. proportional reducer
   B. sub-proportional reducer
   C. superproportional reducer

10 Tone Reproduction

Minor White. *Ritual Branch*. 1958. Courtesy of the Minor White Archives, Princeton University. Copyright © 1982 by Trustees of Princeton University.

PURPOSES OF TONE REPRODUCTION

Except where the photographer is creating a photograph that bears no resemblance to an original subject, it is important to study the relationship between the tones in the original scene and the corresponding tones in the reproduction. Thus, by studying the factors influencing the reproduction of tone, the photographer can learn how to control the process to give the desired results.

Chapter 4 showed how the tone-recording properties of films and papers can be described through the use of the characteristic curve. If the conditions of measurement used to obtain the data simulate the conditions of use, the characteristic curve provides an excellent description of the tone-rendering properties of the film or paper emulsion. However, the result of the photographic process is a reproduction that will be observed by a person, which therefore involves the characteristics of the human visual system. In this respect, perceptual phenomena, including brightness adaptation, appear to play a significant role. Thus the determination of the objective tone-reproduction properties required in a photograph of excellent quality is related to the perceptual conditions at work when the image is viewed. The final image that is viewed normally includes the cumulative tone-reproduction characteristics of the camera film plus the camera optics and the printing material plus the printing optics or, in the case of slides and motion-picture films, the projector optical system.

OBJECTIVE AND SUBJECTIVE TONE REPRODUCTION

Luminance is light per unit area, an objective concept. Lightness is a subjective perception that correlates only roughly with luminance.

When determining the tone-reproduction properties of the photographic system, there are objective and subjective aspects to be considered. In the objective phase, measurements of light reflected from the subject are compared to measurements of light reflected or transmitted by the photographic reproduction. These reflected-light readings from the subject are called luminances and can be measured accurately with a photoelectric meter having a relative spectral response equivalent to that of the "average" human eye. The logarithms of these luminances are then plotted against the corresponding reflection or transmission densities in the photographic reproduction, and the resulting plot is identified as the objective tone-reproduction curve.

The perception of the lightness of a subject or image area corresponds roughly to its luminance; but unlike luminance, lightness is not directly measurable. Because the perception of lightness involves physiological as well as psychological factors, an area of constant luminance can appear to change in lightness for various reasons, including whether the viewer had previously been adapted to a high or a low light level. Psychological scaling procedures have been devised to determine the effect of viewing conditions on the perception of lightness and contrast. It has been established, for example, that a photograph appears lighter when viewed against a dark background than when viewed against a light background; and that the contrast of a photograph appears lower when viewed with a dark surround than when viewed with a light surround.

A graph showing the relationship between the perceived lightnesses of various areas of a scene or photograph and the measured luminances of the corresponding areas would be called a subjective tone-reproduction curve. It is because the perception of a photograph's tonal qualities can vary with the viewing conditions that the American National Standards Institute and other organizations have prepared standards for the viewing conditions in which photographs are produced, judged, and displayed. The influence of viewing conditions should not be underestimated when considering the specifications for optimum tone reproduction.

OBJECTIVE TONE-REPRODUCTION/PREFERRED PHOTOGRAPHS

Most photographers realize that tone-reproduction characteristics required for satisfactory photographs depend to a great extent upon the lighting conditions under which the photographs are viewed. Unfortunately, reflection-type prints are commonly viewed under illumination levels ranging from less than 100 lux to more than 5,000 lux. (The Photographic Society of America currently recommends an illuminance of 800 lux for judging and displaying reflection prints [PSA Uniform Practice No. 1, 1981].) Large transparencies are usually viewed with back illumination, with a recommended illuminance of approximately 1400 candelas per square foot.

However, since these transparencies are viewed under the same general room illumination as reflection-type prints, the eye is in a similar state of adaptation. Slides and motion-picture films projected onto a screen and observed in a darkened room are viewed with the eye in a considerably different state of adaptation. Since there are three widely varying viewing conditions, different objective tone-reproduction curves are required if the photographs are to meet the subjective standards of excellence in each case.

In order to determine the objective tone-reproduction curve that gives the best subjective tone reproduction for black-and-white reflection prints, an experiment was performed under the conditions typically encountered by photographers. Various outdoor scenes were photographed at different exposure levels. The exposed films were processed to different contrasts, and from the resulting negatives black-and-white prints were produced. The prints varied in contrast, density, and tone-reproduction properties. Numerous observers were asked to view these prints and judge them for their subjective quality. The objective tone-reproduction curves for all of the first-choice prints were averaged, and the resulting curve is illustrated in Figure 10–1, where the reflection densities of the print are plotted against the log luminances of the scene.

The 45° reference line is arbitrarily located so that its lowest point corresponds to a diffuse white object in the scene and the minimum density of the photographic print material. If the luminances in the scene were reproduced exactly in the print, the resulting tone-reproduction curve would have a slope of 1.00, matching that of the reference line. In fact, the average curve for first-choice prints is located 0.2 to 0.3 density units below the 45° line except in the highlight region, where the curve cannot go below the minimum den-

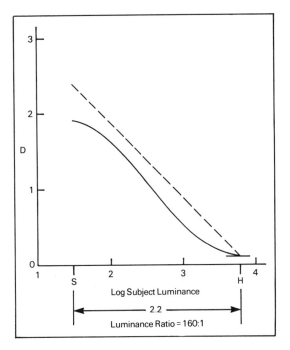

Figure 10–1 Objective tone-reproduction curve for a preferred reflection print of an average outdoor scene.

sity of the paper. The curve shows low slope in both the highlight and shadow regions, but has a gradient of 1.1 to 1.2 in the midtone area. This indicates that the print's highlights and shadow regions are compressed compared to the original scene, while the midtones have been slightly expanded in contrast. The experiments that lead to this tone-reproduction curve involved scenes of low and high contrast, as well as interior and exterior subject matter. The results in all cases were remarkably similar to that shown in Figure 10–1. Whenever the departure from the desired slopes was greater than 0.05, the observers invariably judged the prints as being unacceptable in contrast, indicating very narrow tolerance levels. Consequently, it appears that the desired objective tone-reproduction properties in a black-and-white reflection print are essentially independent of the characteristics of the original subject.

Similar studies have been performed with large transparencies viewed under room-light conditions on a back illuminator. The objective tone-reproduction curve that gave the best objective tone reproduction for this condition is illustrated in Figure 10–2, where the diffuse transmission density of the image is

The tone-reproduction curve for facsimile reproduction would be a 45° straight line.

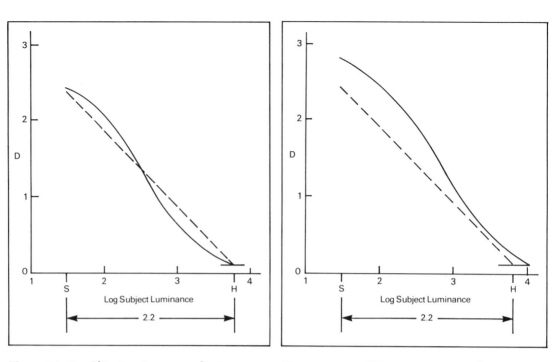

Figure 10–2   Objective tone-reproduction curve for a transparency of preferred quality, viewed on a bright illuminator under average room light.

Figure 10–3   Objective tone-reproduction curve for motion pictures (and slides) of preferred quality, projected on a screen in a darkened room.

plotted against the log luminance of the scene. The desired objective tone-reproduction curve is very nearly the same as the 45° reference line but with a midtone slope still greater than 1.0.

The differences in curve shape between the transparency and the reflection print are due primarily to the increased density range of which the transparency film is capable, since the eye is in a similar state of adaptation. As the brightness of the illuminator used to view the transparency increases, the desired tone-reproduction curve moves slightly above and to the right of the one illustrated in Figure 10–2.

The preferred objective tone-reproduction curve for a transparency or a motion-picture projected onto a screen and viewed in a darkened room is illustrated in Figure 10–3, where the effective screen viewing densities (measurements made with a spot photometer of the projected image on the screen) are plotted against the log luminances of the scene. This curve shows a slope considerably greater than that of the previous two and is principally the result of the dark-adaptation, lateral-adaptation, and visual-flare characteristics of

the eye. Psychophysical experiments of the human eye's response indicate that the perceived brightness (lightness) contrast is lower when the eye is dark adapted than when it is light adapted.

Since it is desirable to obtain a subjective lightness contrast in the projected image on the screen similar to that which would occur in a reflection print, the objective tone-reproduction curve necessary for dark-adapted conditions must have a greater slope. If a higher-intensity lamp is used in the projector, the tone-reproduction curve shifts slightly to the right and above that which is shown in Figure 10–3. Therefore, the preferred density level and optimum exposure for a transparency film are related to the amount of light supplied by the projector and screen used when viewing the images. This means that the effective film speed for a reversal film is in part related to the conditions under which the images will be viewed.

The three conditions illustrated represent those most commonly encountered in photographic reproductions and pertain only to pictorial representations. In each case, the (objective) curve represents optimum subjec-

A projected image in a darkened room that had the same luminances as a good-quality reflection print would appear to be unacceptably flat.

tive tone reproduction, and thus an aim for the system. It is the task of the photographer to select the appropriate materials and processes to achieve these aims.

OBJECTIVE TONE-REPRODUCTION ANALYSIS

The flow diagram in Figure 10–4 isolates some of the more important factors to be considered in a study of the photographic reproduction of tones. Since the subject represents input in terms of log luminance values and the print represents output in terms of densities, it is the stages in between that must concern us. It should be noted that a study of tone-reproduction characteristics excludes many important properties of the photographic system. For example, no attention is paid to the reproduction of small detail nor to the reproduction of various colors in the scene. Thus, when considering the various stages of the process, only those factors that influence the large-scale tonal relationship should be considered.

To simplify this task it is necessary to consider only three principal stages in the process, as illustrated in Figure 10–5. Each of these stages will be represented by a graph illustrating its influence on the tonal relationships. Each graph has a set of input values and a set of output values. The output from the first stage becomes the input to the second, and so on through the system, thus forming a chain of input/output relationships. In this fashion, the major stages of the process can be studied for their effects upon the end result. The important phases of the photographic process can be synthesized by means of this tone-reproduction study. For such an approach to work, data must be obtained about these phases. Methods for obtaining data about the subject, film, and photographic papers were discussed in Chapter 4 and so will not be repeated here. However, the problem of optical flare was only briefly mentioned in Chapter 2, and so a discussion of it follows.

OPTICAL FLARE

When a camera is aimed at a subject, the subject luminances and therefore the physical

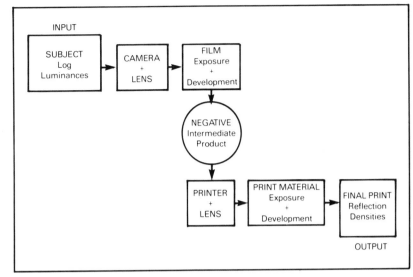

Figure 10–4 Flow diagram of the photographic process.

values of light reaching the lens represent the initial input data for the process. If the luminance of the lightest diffuse highlight is divided by the luminance of the darkest shadow with detail, the resulting value is termed the *luminance ratio*, and the log of that ratio is termed the *log luminance range*. In an important experiment conducted by Jones and Condit of the Eastman Kodak Research Laboratory, the luminance ratios of more than 100 outdoor scenes were measured. The frequency distribution of the luminance ratios is shown in Figure 10–6. The smallest ratio encountered was 27:1 and the

Since densities, which are used to measure image tones, are logarithms, it is appropriate to express the measurements of subject tones as log luminances for tone-reproduction graphs.

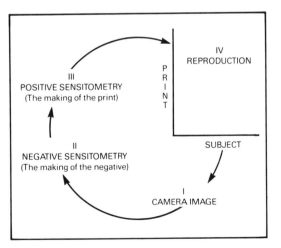

Figure 10–5 Simplified tone-reproduction system.

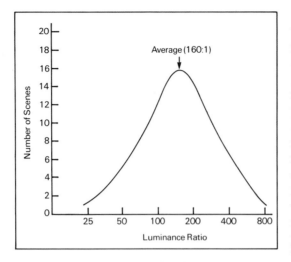

Figure 10–6   Frequency distribution of luminance ratios of outdoor scenes.

The luminance ratio of average outdoor and indoor scenes is considered to be approximately 160 to 1.

largest was 760:1, with an average ratio of 160:1. The frequency distribution in Figure 10–6 indicates that luminance ratios of less than 100:1 and greater than 300:1 are encountered much less often.

Much less information exists about the luminance ratios of interior scenes, but some experiments indicate that the average luminance ratio of an interior scene is not greatly different from 160:1. The average scene ratio of 160:1 has special significance since most manufacturers optimize exposure and development conditions for their products based on this value. A set of luminance readings from a typical outdoor scene is shown in Table 10–1. Nine different areas of the scene were measured with a spot meter to obtain the data. The luminance ratio of the scene is 160:1.

When the camera is pointed at such a scene, the lens receives the light from each area and focuses it on the film plane at the back of the camera. Thus, in this first stage of tone reproduction, the subject luminances are transformed to image illuminances at the film plane. In an ideal situation there would be a direct proportion between the subject luminances and the image illuminances; the tonal relationships would be maintained at the back of the camera. However, at any point in the image plane of a camera, the illuminance is the result of two different sources: (a) the illumination focused by the lens and projected to the film plane, which constitutes the image-forming light; and (b) light that is the result of single and multiple reflections from the lens surfaces, diaphragm, shutter blades, and additional interior surfaces of the camera, providing an approximately uniform illuminance over the whole image area.

This second source of light is referred to as flare light, or simply flare, and provides non-image-forming light, since it is not directly focused by the lens. Figure 10–7 illustrates how flare occurs in a simple camera. Since the light is reflected off any interior surface, flare light is present on any projected image, whether it is formed by the lens or a pinhole.

Since flare is non-image-forming light and occurs in the image plane as a uniform veil, it increases the illuminance of every point on the camera image and thus results in a loss of image contrast. The effect is similar to viewing a projected transparency on a screen

Table 10–1   Luminance values of an outdoor scene

Area No.	Description of Area	Luminance Candela/Square Foot	Footlambert
1	White cloud	1,114	3,500
2	Clear sky	637	2,000
3	Grass in sunlight	350	1,100
4	Side of house in sunlight	200	630
5	Front of house	115	360
6	Car in open shade	67	210
7	Tree trunk	38	120
8	Grass in shade	22	70
9	Base of tree in heavy shade	7	22
	Luminance ratio of subject = 160:1		

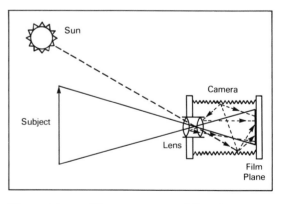

Figure 10–7   The occurrence of flare in a camera. Solid lines in camera represent image-forming light, while dashed lines in camera represent non-image-forming (flare) light.

in a darkened room and then viewing the same image with the room lights on. The loss of image contrast on the screen is due to the additional illuminance produced by the room lights on the screen surface.

To illustrate the effect in the camera, assume that the camera is being pointed at an average subject with a 160:1 luminance ratio. If there were no flare-light in the camera, the range of image illuminances at the back of the camera would also be 160:1. However, if in addition to that image-forming light there was one unit of flare light uniformly distributed over the film plane (i.e., the corners are receiving one unit, the edges are receiving one unit, the center is receiving one unit), the ratio would now be 161:2, which reduces to approximately 80:1. Thus the 160:1 luminance ratio in front of the camera has been reduced to an illuminance ratio of 80:1 at the back of the camera.

It is perhaps obvious that the reduction in contrast is the result of different percentage increases at the opposite ends of the scale. For example, the additional unit of flare light in the shadows provides a doubling of the light in that region, while the additional unit of light in the highlights represents only a very small percentage of increase in that region. The result is that the ratio of illuminances at the back of the camera is always less than the ratio of luminances in the subject. Flare can be expressed as a flare factor, which is derived by dividing the luminance ratio in front of the camera by the illuminance ratio at the back of the camera:

$$\text{Flare factor} = \frac{\text{Subject luminance ratio}}{\text{Image illuminance ratio}}$$

The flare factor will be 1.0 when there is no flare light, which would occur only in a contact-printing situation. The results of many experiments indicate that most modern lenses and cameras have an average flare factor of nearly 2.5 under typical conditions. Under some conditions the flare factor may be as high as 4 or 8, or as low as 1.5.

Some of the more important factors that influence the amount of flare are:

1. *Subject characteristics.* Subjects with high luminance ratios or subjects with large areas of high luminance tend to produce large amounts of flare. Snow scenes, beach scenes, and high-key scenes (white on white) are all examples of subject matter that would give large amounts of flare.
2. *Lighting conditions.* Subjects that are back lit (with the light source in the field of view) give greater amounts of flare than subjects that are front lit.
3. *Lens design.* By designing a lens to contain the smallest number of elements possible and coating the elements with antireflection materials, its flare characteristics can be greatly minimized. However, lens design and coatings cannot decrease the flare factor to 1.00 (zero flare), since considerable stray light still reaches the image plane by reflection from the lens mount and other inner areas of the camera, and even the surface of the film itself.
4. *Camera design.* If the camera's interior is black and the surfaces are mat, stray light reflections will be minimized. Any light leaks in the camera body will also act as flare and further reduce the image contrast.
5. *Dust and dirt.* By keeping the lens surfaces clean and minimizing dust particles in the camera body, the number of surface areas on which light reflections may occur can be minimized and the flare reduced.

Of all the factors listed, the two with the greatest influence on flare are the subject characteristics and the lighting conditions. A number of studies suggest that approximately 80% of the flare encountered in typical photographic situations is governed by these two factors. Consequently, even when high-quality lenses and cameras are used and kept in good condition, the photographer cannot avoid the loss of image contrast that results from flare. Thus it is important for the photographer to obtain an estimate of the amount of flare, in the form of a flare factor, that is affecting his or her system.

The direct method for measuring flare involves the use of a gray scale and a small spot meter. The camera is focused on the gray scale and the small spot meter is used to measure the corresponding image areas at the back of the camera. The ratio of luminances on the original gray scale can then be compared to the ratio of illuminances at the back of the camera, and the flare factor can be computed. Although this method is direct, it is a diffi-

> **Camera flare reduces contrast mostly in the shadow areas. Enlarger flare reduces contrast mostly in the highlight areas.**

> **A typical camera flare factor of 2.5 reduces a scene luminance ratio of 160:1 to 64:1.**

cult procedure to follow, and in small-format cameras it is a practical impossibility.

An alternative method is to work backwards using a photographic film as the measuring device. First, a negative of the gray scale is made with the camera. A second piece of film is exposed in a sensitometer, which is a flare-free instrument since it involves a contact print. Both pieces of film are processed together to ensure identical development and the resulting characteristic curves are drawn. The highlight densities of the curves are matched, and any difference in the rest of the curves can be attributed to flare. The actual amount of flare can be found by measuring the horizontal displacements between the two curves, at the bottom of the top curve. The antilog of the difference in log exposures is the flare factor.

Such a set of curves is illustrated in Figure 10–8, which shows that the effect of flare on the characteristic curve is to increase the length of the toe, add curvature to the lower part of the straight line, and reduce the shadow contrast (slope) in the negative. The greater the amount of flare in the system, the greater the differences will be between these two curves.

The purpose for obtaining the flare factor is to estimate the range of illuminances that will occur at the back of the camera, and thus the range of tones to which the film will be exposed. For example, if a scene with a lu-

minance ratio of 200:1 is photographed with a lens-camera combination with a flare factor of 4, the film will actually receive a ratio of illuminances of 200 divided by 4, which equals 50:1. In this fashion the exposure ratio (or, more usefully, the log exposure range) for the film, and ultimately the development time necessary to produce a normal-contrast negative, can be determined.

In order to study more thoroughly the influence of flare in the tone-reproduction cycle, consider the graph in Figure 10–9. The data for such a graph are derived from the flare-measurement method first described. A small spot photometer was used to meter the illuminances provided by the gray scale at the back of a camera, and the relationships between subject luminance and image illuminance were plotted. Since logarithms are commonly used to describe the input and output properties of photographic materials, the scales are log luminance and log illuminance on the graph. If there were no flare in the system, the relationship between subject log luminance and image log illuminance would be a 45° straight line, indicating a direct proportion.

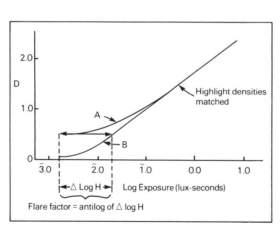

Figure 10–8 The effects of flare on the characteristic curve of a black-and-white negative material. Curve A was generated by photographing a reflection gray scale with a camera, while curve B came from a sensitometer (contact printed to a step tablet).

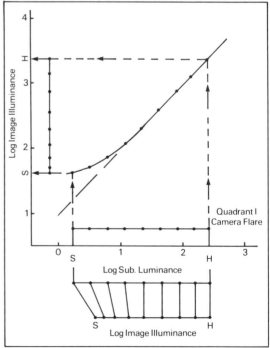

Figure 10–9 The effect of optical flare (or other stray light) on the image illuminances at the back of the camera.

However, as discussed earlier, the low-illuminance shadow regions at the back of the camera suffer a greater percentage increase than do the highlight illuminances. Since the highlights are relatively unaffected, the relationship is at a slope of 1.0 (45°) in that area. As the image illuminances decrease (the shadows become darker), the percentage increase becomes greater and thus reduces the slope of the curve considerably, resulting in a loss of contrast in the shadow area. This indicates that flare reduces shadow contrast in addition to reducing the overall ratio of tones at the film plane.

In Figure 10–9, the nine dots on the log luminance axis represent the nine luminances of a typical outdoor scene listed in Table 10–1. By extending straight lines directly up from these nine points until they intersect the flare curve, and then extending them to the left until they intersect the log illuminance axis, the nine dots on that axis can be generated.

A diagram at the bottom of Figure 10–9 compares the log subject luminances in front of the camera to the log image illuminances at the back of the camera for a normal amount of camera flare. The log image illuminance range is shorter, indicating an overall loss of contrast, and the relationship between the darker tones has been compressed, indicatng a loss primarily of shadow contrast.

The graph in Figure 10–9 is important because it represents the first stage of the objective tone-reproduction process and illustrates the effect on tonal relationships occurring at that stage. Ultimately, the log image illuminances will be fixed in position when the camera shutter is tripped, converting them to log exposures, which become the input to the film. In this example, the two dashed lines on the graph indicate the location of the darkest shadow with detail and the lightest diffuse highlight. These two tones will be followed through the four-quadrant objective tone-reproduction cycle.

THE MAKING OF THE NEGATIVE

The next stage of the process involves the making of the negative and is represented by the characteristic curve of the film-development combination used. Since the output of stage I (the optical flare quadrant) becomes the input of stage II (the making of the negative), it is necessary to rotate the characteristic curve of the film 90° counter-clockwise so that the log illuminance axis of quadrant I matches the log exposure axis for quadrant II, as illustrated in Figure 10–10. Here the broken line indicating the shadow detail area (S) has been located at the log exposure for the film that results in a minimum useful density of 0.10 above film base plus fog. This

Flare in the human eye reduces contrast in the shadow areas, the same as flare in a camera does.

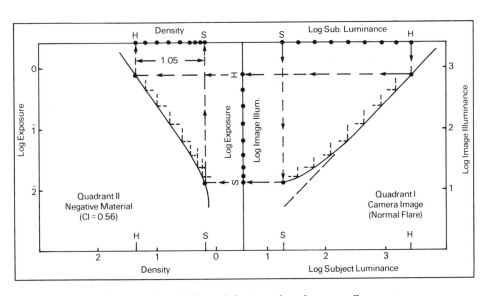

Figure 10–10   The combined effect of the transfer of camera flare onto the negative's characteristic curve.

is reasonable as this density level is considered to be the minimum density for maintaining minimum shadow detail.

Since this is the ISO/ASA speed point for pictorial films, the shadow detail of the scene should be exposed in a way that produces this density. If less camera exposure is given, the dashed shadow detail line from quadrant I would intersect the negative curve at a much lower density (quadrant II would be shifted upward). If more camera exposure were given, the dashed shadow line would fall at a higher density on the characteristic curve (quadrant II would be shifted downward).

The shape of the characteristic curve shown in quadrant II indicates that it was developed to a contrast index of 0.56, resulting in a highlight density of 1.15 above base plus fog, as indicated by the dashed diffuse highlight line (H). The film was developed to this contrast index because it was desired to produce a negative with a density range of approximately 1.05, since it is known that such a negative will easily print on a grade 2 paper in a diffusion enlarger. If a condenser enlarger were to be used, the necessary range would be less, as discussed in Chapter 4. The nine dots on the density axis of the negative are generated by continuing the lines from quadrant I into quadrant II until they strike the characteristic curve, and then reflecting them upward until they intersect the density scale.

The relationship between the log exposures on the film and the resulting densities can be illustrated in a fashion similar to that of quadrant I, as shown in Figure 10–11. This diagram indicates that compression of tone is occurring at all levels in the image, with the greatest amount of compression occurring again with the shadow areas. This is typ-

A negative density range of approximately 1.05 is recommended for printing on grade 2 paper with a diffusion enlarger, but a density range of 0.80 is recommended with a condenser enlarger.

ically the case for pictorial negative films, since they are invariably processed to contrast indexes less than 1.0. Thus, significant compression of tones occurs at the negative making stage, with the darker shadow tones suffering the greatest amount of compression.

At this stage of the process the photographer has many image controls available. Among the more important are:

1. *Film type.* To a great extent, the properties of the emulsion determine the shape of the characteristic curve. For example, lithographic (or lith) emulsions generally will produce curves with slopes greater than 1.0 under most development conditions, while pictorial films will produce slopes less than 1.0 under most development conditions. Therefore, if the input log exposure range is very short, the photographer would do well to select a lith-type emulsion. If, however, the log exposure range is long (1.3 or greater), a pictorial film would be a better choice. Specialized emulsions have been designed to handle log exposure ranges of excessive lengths (1,000,000:1 and greater).

2. *Development.* As discussed in Chapter 4, the length of development time provides a useful control for obtaining a variety of curve shapes (slopes) in the film. For pictorial applications, the concept of contrast index provides the most useful guide for estimating the required slope. The recommended development times found in most manufacturers' literature are those which will produce a contrast index of approximately 0.56, since this is the slope that will convert an average outdoor scene into a normal-contrast negative for diffusion enlargers. Flatter scenes will provide a shorter input range to the film and require a steeper slope to maintain a constant density range (contrast) in the negative. Contrasty subjects will provide the film with a longer range of log exposures and consequently require a lower slope to maintain the density range at the normal level. Thus, development provides the photographer with a powerful tool for contrast control in tone reproduction.

3. *Exposure.* The task here is to select the

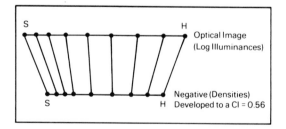

Figure 10–11 The relationship between the optical image at the film plane and the densities in the negative.

appropriate f-number and shutter speed to place the shadow-detail area of the subject at an exposure that will produce the minimum useful density for shadow detail in the negative. If less exposure is given, less shadow detail will result. If more exposure is given, greater shadow detail will result. Thus the f-number and shutter speed selected will have the greatest influence upon reproduction of the shadow tones. It is important to note that underexposure by more than one-half stop introduces a compression of shadow tones that cannot be adequately compensated for in later stages of the process. On the other hand, with typical pictorial subjects and films, overexposure by as much as four stops can be compensated for in the later stages but with the result of increased print exposure and grainier prints.

THE MAKING OF THE POSITIVE

The third major stage of this system is the making of the positive. For the example that follows, the reproduction will be in the form of a reflection print. When printing a negative, the negative's densities control the exposures that the paper will receive. The thinner shadow areas of the negative will allow more light to strike the paper than will the denser highlight areas. Consequently, the output of the negative in the form of transmission densities can be related to the input of the photographic paper in the form of log exposures; this explains the positioning of quadrant III in Figure 10–12. Quadrant III represents the characteristic curve of a normal grade of black-and-white photographic paper.

The optimum contrast index for film development depends on scene contrast, type of enlarger, and paper contrast grade.

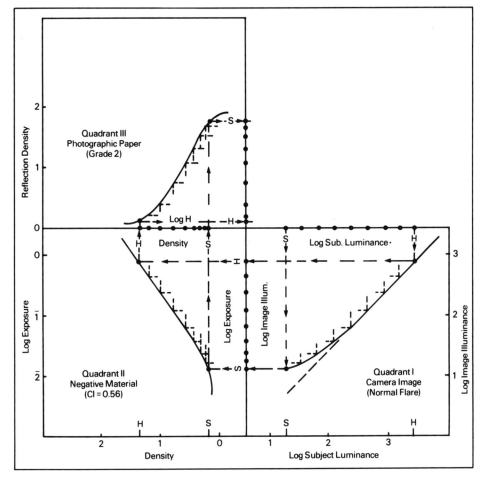

Figure 10–12 The combined effect of the transfer of camera flare through the negative characteristics and onto the print characteristics.

The curve shape selected for this quadrant is based upon the relationship between the density range of the negative and the useful log exposure range of the paper. In Chapter 4 it was shown that pictorial negatives generally are best printed on a paper where the negative density range is equivalent to the paper's useful log exposure range. In this example, the negative has a density range of 1.05 and therefore requires a paper curve having a useful log exposure range of 1.05, as does the curve in quadrant III of Figure 10–12.

The f-number and printing time of the negative determine the location of the characteristic curve in quadrant III relative to the left and right position. The best reproduction is generally obtained when the useful shadow density of the negative produces a density in the print equal to 90% of the maximum density of which the print material is capable; this is the basis for the location of the paper curve in quadrant III.

If the relationship between the negative and the paper is correct, the diffuse highlights of the negative should generate a density in the print that is approximately 0.04 greater than the base density of the print. The broken shadow-detail line is extended upward from quadrant II, where it is reflected from the negative curve until it intersects with the print curve at its density of 90% of the maximum density.

Likewise, the dashed diffuse-highlight line is extended upward until it strikes the paper curve at the density equivalent to 0.04 above the base plus fog of the paper. Both lines are then reflected to the right, which is the way in which all of the dots on the print density line were generated. Figure 10–13 illustrates the relationship of the nine input tones to the paper (from the negative densities) and the nine output tones from the paper in the form

A print density of 0.04 above base density is considered appropriate for diffuse subject highlights, reserving paper white for images of light sources and specular reflections.

The curve in quadrant IV of a tone-reproduction diagram represents the relationship between the print tones and the subject tones.

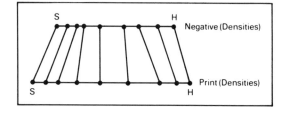

Figure 10–13 The relationship between the negative densities and print densities.

of reflection densities. The midtones of the negative are expanded in the print, while the highlights of the negative are compressed.

GENERATION OF THE TONE-REPRODUCTION CURVE

In the last phase of the system the tones generated in the print are compared to the tones that existed in the original scene. As illustrated in Figure 10–14, this is achieved by extending the print tone lines from the photographic paper quadrant to the right into the fourth quadrant titled Reproduction, and noting where they intersect the corresponding lines projected upward from the appropriate subject log luminances.

For example, in Figure 10–14 the line representing the diffuse highlight on the photographic paper curve has been extended into the fourth quadrant until it intersects the line extended upward from the diffuse highlight of the subject. The intersection of these highlight lines in quadrant IV determines the highlight point on the objective tone-reproduction curve. Likewise, the line representing the detailed shadow tone in the print is extended into the fourth quadrant until it intersects the line extended upward from the same tone in the subject, generating the shadow detail point of the tone-reproduction curve.

This procedure is repeated for each of the intermediate points to obtain the complete objective tone-reproduction curve in Figure 10–14 in the fourth quadrant. The shape of this curve can provide insight into the nature of the tone reproduction occurring in the photograph.

Alert readers will have noticed that a flare curve has not been included for the printing stage of the tone-reproduction process. All optical systems have flare, which, as we have seen with camera flare, can have a significant effect on image contrast. In this discussion of tone reproduction it is assumed that contact prints are being made so that flare is not a factor at the printing stage. If prints are to be made with an enlarger, two additional factors must be considered—flare and the Callier effect. Enlarger flare reduces contrast and

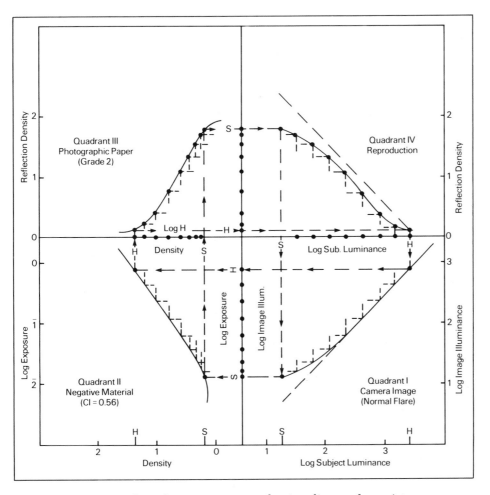

Figure 10–14   Complete objective tone-reproduction diagram for a pictorial system.

the Callier effect increases contrast, but since the net effect of these two factors is different for diffusion and condenser enlargers, different data must be used for the two types of enlargers. The easiest way to incorporate this information into the four-quadrant tone reproduction system is to make the D-log H paper curves by projection-printing a step tablet with the type of enlarger that is to be used for future printing, being careful to mask the negative carrier down to the edges of the step tablet.[1]

In this idealized case, since correct exposure and development of the negative were achieved and the negative was correctly printed, the resulting tone-reproduction curve closely resembles that of the desired tone-reproduction curve for black-and-white reflections prints, illustrated in Figure 10–1. The 45° dashed line representing facsimile reproduction has been included in quadrant IV for comparison purposes only. Examination of the curve in quadrant IV reveals that the shadow region of the curve has a slope less then 1.0 (45°), indicating that the subject tones in the print have been compressed relative to those in the scene. This is also the case with the highlight region of the curve. In the midtone region the slope is slightly greater than 1.0 (45°), indicating a slight expansion of midtone contrast in the print compared to the original scene.

Thus, in preferred tone reproduction for black-and-white reflection prints, there is often compression of shadow and highlight detail with an accompanying slight increase

Color films scatter very little light and therefore have a Callier Q factor of approximately 1.0.

[1]Stroebel, "Print Contrast as Affected by Flare and Scattered Light." *PSA Journal*, 48:3 (March 1982), pp. 39–42.

in midtone contrast. This relationship between tones in the photographic print and the tones in the original scene can also be illustrated as shown in Figure 10–15. In this representation, the nine different subject tones are equally spaced, indicating equal tonal differences. However, in the bar representing the print tones, the highlight and shadow tones have been significantly compressed while the midtones have been slightly expanded. Again, this is typical of the negative-positive process when a reflection print is the final product.

At this point it is useful to review the properties of the four-quadrant tone-reproduction diagram illustrated in Figure 10–14. This system represents a graphical model of the actual photographic process with respect to the reproduction of tone. The major limitations and strengths of the photographic system can be determined by such a diagram. For example, quadrant I represents the camera image as primarily affected by optical flare. Optical flare is an inescapable part of any photographic system that incorporates the projected image. Thus the photographer must live with the compression of tone (loss of detail and contrast) that results from this problem. Little control can be exerted on the process at this stage except to use a lens shade and keep the lens clean.

Quadrant II represents the results of exposing and processing the film, and thus the production of the negative. It is at this point that the photographer can exert the greatest amount of control through the selection of the film type and the corresponding exposure and development conditions.

Quadrant III represents the making of the print and, consequently, is the last step in controlling print tones. Since the choice of

The four-quadrant tone-reproduction diagram reveals that for normal tone reproduction image contrast is reduced in quadrants I and II.

paper grade is primarily dictated by the nature of the negative, and since development variation does not have any significant effect upon paper contrast, the photographer has only limited control of the outcome at this stage of the process. Additionally, if a reflection print is being made, the photographer must accept the fact that the print's tonal range will be less than that of the original subject.

The use of the four-quadrant tone-reproduction diagram also allows the photographer to see the expansion and compression of tones resulting at each stage of the process. For example, there is an inevitable loss of shadow contrast in the camera image caused by optical flare. Shadow contrast is further decreased when the shadow exposures are placed in the toe of the film's characteristic curve, which is typically the case. A third reduction in shadow contrast occurs when the negative is printed and the shadow tones of the negative are placed on the shoulder of the paper's characteristic curve. This explains the lowered contrast in the shadows of the final black-and-white reflection print.

The subject's midtones are relatively unaffected by the camera's flare characteristics and are typically exposed onto the straight-line section of the film's characteristic curve. Although the midtone slope of the negative curve is usually less than 1.0 (0.5 to 0.6), the midtones of the negative are printed onto the mid-section of the paper's characteristic curve where the slopes are considerably greater than 1.0 (1.5–3.0). The result is a midtone contrast in the print that is only slightly greater than that of the original subject.

The highlights are also unaffected by optical flare at the camera stage and, when the film is exposed, are placed on the upper straight-line section of the characteristic curve. Thus the only distortion introduced into the highlights when the negative is made is associated with the lowered slope in the negative's characteristic curve. However, when the negative is printed, the highlight tones are placed in the toe section of the characteristic curve for the photographic paper and, as a result, suffer their greatest compression. This is the cause of the lowered slope and lessening of highlight detail in the tone-reproduction curve in quadrant IV.

It should be evident that all four quadrants

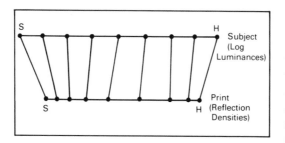

Figure 10–15 The relationship between the subject tones and the print tones.

can be described in terms of the slopes in the various areas (shadows, midtones, highlights) in each quadrant. Slope can be considered to be the rate of output for a given input, and the cumulative effect can be predicted by multiplying the slope of the curve in quadrant I by the slope of quadrant II, by the slope in quadrant III, to predict the slope that would result in quadrant IV. If the concept of the average gradient $(\overline{G})$ is substituted for slope, the relationship may be expressed by the following formula:

$$\overline{G_I} \times \overline{G_{II}} \times \overline{G_{III}} = \overline{G_{IV}}$$

The average gradient in each quadrant represents the average slope between the diffuse highlight point and the shadow detail point. For example, in Figure 10–14 the average gradient for quadrant I is 0.82, for quadrant II it is 0.56, and for quadrant III it is approximately 1.78. Substituting these values in the above formula gives the following result:

$$0.82 \times 0.56 \times 1.78 = 0.82$$

The calculations indicate that the average gradient for quadrant IV will be approximately 0.82, which is the case when the average gradient in quadrant IV is measured. A similar study may be done individually for the shadows, midtones, and highlights to assess the effect of each stage of the process on these areas. Often, this relationship between gradients can be used to work backwards in the system to predict the necessary gradient in any quadrant. For example, if it is desired to obtain an average gradient of 0.82 in quadrant IV (the reproduction quadrant), and it is known that a photographic paper with an average gradient of 2.0 will be used when making the print, and it is further known that the average gradient in the flare quadrant will be 0.82, the values can be substituted in the formula and the average gradient necessary in quadrant II can be predicted as follows:

$$0.82 \times \overline{G_{II}} \times 2.0 = 0.82$$
$$\overline{G_{II}} = 0.50$$

In this fashion the photographer can predict the necessary contrast index for the production of an excellent print under these conditions. By knowing the relationship be-

tween contrast index and development time, the photographer can actually determine the proper length of development time for any set of conditions. The concept of relating the various stages of the photographic process through the gradient at each stage often serves as the basis for some very useful nomographs that assist the photographer in predicting such things as the correct development time for the negative and the proper printing conditions of the resulting negative.

It should be evident that these nomographs represent abstractions or simplifications of the objective tone-reproduction diagrams discussed in this chapter and, as such, are more useful to the photographer. However, it is important to have an understanding of the input-output relationships at the various stages of the process and the rates affecting those stages; that understanding is most directly obtained through an understanding of the four-quadrant tone-reproduction diagram.

TONE REPRODUCTION AND THE ZONE SYSTEM

The concepts of tone reproduction and the accompanying tone-reproduction diagrams are intended to provide the photographer with a basis for understanding the nature of the photographic process. Experience has shown that photographers who have a firm understanding of the materials and how to control them are those most likely to consistently produce high-quality images. However, it is recognized that the tone-reproduction diagrams presented in this chapter do not provide a convenient way for photographers to exert control over the process at the time images are being made.

During the past several decades, there have been countless procedures proposed for the control of image quality through the manipulation of the various stages of the photographic process. For example, the early Weston exposure meter dial had U and O positions marked to represent shadow and highlight positions for a normal scene. A third position labeled N represented the average midtone position, and therefore the normal exposure. The U position was located four stops below the normal arrow, the O position three stops above the normal arrow. The dif-

The average gradient of tone-reproduction curve is the slope of a straight line connecting the highlight and shadow points on the curve.

The origins of the Zone System can be traced to the work done by engineers at the Weston meter company in 1939.

"Without visualization the Zone System is just a 5-finger exercise." —*Ansel Adams*

ference between the U and O locations corresponded to a seven-stop range or a luminance ratio of 128:1 (or 160:1 rounded to the nearest whole stop).

Perhaps the most comprehensive system—certainly the best known—is the Zone System, which was proposed by Ansel Adams and further refined by Minor White and others. In its most elementary state, the Zone System provides the photographer with a vocabulary for describing the various stages of the photographic process from the original scene through the completion of the final print. In its more advanced form the Zone System will lead the photographer to proper exposure, development, and printing conditions in order to reproduce the scene in a given fashion. Thus the basic premise of the Zone System is that the photographer must visualize the final print from the appearance of the original subject before taking the photograph. Through knowledge of the capabilities and limitations of the photographic system, the photographer can then manipulate its various components to achieve this visualized result.

In tone-reproduction studies of pictorial systems, the subject properties are expressed in terms of luminances (candelas per square foot) and log luminances. In the Zone System, the subject is described in terms of subject values, which relate to different subject luminances that are labeled by Roman numerals for easy identification. The values are related to each other by factors of two (one camera stop); Value II reflects twice as much light as Value I, Value III reflects twice as much light as Value II and four times as much light as Value I, and so on.

Table 10–2 contains the 10 subject values commonly used in the Zone System and their definitions. Using these definitions, the contrast of the original subject can be described in terms of the number of values it contains. For example, the scene-luminance measurements by Jones and Condit that led to the average luminance ratio of 160:1 were made on the darkest and lightest areas of the scene. These areas most nearly correspond to Values I and VIII in the Zone System. This means that the typical outdoor scene contains eight values: I, II, III, IV, V, VI, VII, and VIII. Furthermore, since each value is related to the next by a factor of two (one camera stop), such a scene contains a seven-value range (VIII − I = VII); and the ratio of the extreme values can be found by multiplying 2 times itself seven times (2^7). The resulting ratio of 128:1 compares closely with the results of the Jones and Condit study.

Thus, in Zone System terminology, the typical outdoor scene contains eight values and is said to cover a seven-stop range. If a scene contains fewer than eight values, it is a flatter-than-normal scene, while a scene with more than eight values would be described as a contrasty scene. It is important to note that many references describe the average outdoor scene as containing five (or some other number) values ranging typically between Value III and Value VIII. The discrepancy is due to at least two factors. The first is associated with defining exactly what is meant by the terms *detailed shadow* and

Table 10–2   The relationship between various parts of a scene and the corresponding subject value in the Zone System

Subject Value	Description
0	Absolute darkest part of scene. Example: when photographing the side of a cliff containing a cave.
I	A very dark portion of the scene where the surface of an object can be detected. Examples: dark cloth in shade; surface is visible with slight texture.
II	A dark area of the scene showing strong surface texture; perceptible detail. Examples: surface of building in heavy shade; shadow of tree cast on grass in sunlight.
III	Darkest part of scene where tonal separation can be seen; texture and shape of objects are evident. Example: tree trunk in shade where texture of bark and roundness of trunk with shading are evident.
IV	The darker midtones of a scene; objects in open shade. Examples: landscape detail in open shade; dark foliage; Caucasian skin in shadow.
V	The middle gray tone of the scene; 18% reflectance. Examples: dark skin; gray stone (slate); clear north sky.
VI	The lighter midtones of a scene. Examples: average Caucasian skin in direct light; shadows on snow in sunlit scene.
VII	A very light area of the scene containing texture and detail. Examples: light skin in direct light; snow-covered objects with sidelighting.
VIII	The diffuse highlights (nonspecular) of the scene; white surfaces showing texture. Examples: white shirt in direct light; snow in open shade; highlights on Caucasian skin.
IX	The specular (glare) highlights of the scene; surface is visible but no texture. Examples: glare reflections from water, glass, and polished metal; fresh snow in direct sunlight.
X	The absolute brightest part of the scene. Example: a light source (sun, tungsten bulb, flash lamp, etc.) included in the field of view.

textured highlight. Obviously, the opinions of photographers will differ as a result of esthetic judgments. The second reason (and perhaps a more fundamental one) is related to the definition of a value. Some references define a value as an interval—a difference between two tones—while others state that it is a specific tone or subject luminance. This problem will cause a one-value discrepancy, since the number of intervals between tones will always be one less than the number of tones. In all discussions of the Zone System in this text, a value is defined as a specific tone and the difference between tones as a value range. The difference between tones is also expressed as a stop range, log luminance range, and luminance ratio.

The value range of a subject can be converted to a log luminance range by simply multiplying the number of camera stops (factors of two) it contains by 0.30, the log value of one stop. Thus the average outdoor scene containing eight values covers a seven-stop range, which will give a log luminance range of 2.1 (7 × 0.3 = 2.1) and a luminance ratio of 128:1. A flat scene containing only five values covers a four-stop range and gives a log subject luminance range of 1.2 (7 × 0.3 = 1.2) or a ratio of 16:1. A contrasty scene containing nine values will cover an eight-stop range and give a log range of 2.4 (8 × 0.3 = 2.4) or a ratio of 256:1. Therefore, in its initial application the Zone System provides the photographer with a method of describing and quantifying the tonal values of the original scene. In the more technical tone-reproduction system, this correlates with the determination of the subject luminance ratio and ultimately the log subject luminance range as the principal description of the subject contrast.

In its second phase, the Zone System provides a way to determine the proper exposure and film development to obtain a normal-contrast negative. However, as with all systems of image control, it is necessary to test the film extensively to determine its characteristics. In the Zone System this invariably involves determining the camera exposure necessary to obtain the details desired in a particular shadow value, and the development necessary to correctly reproduce a given highlight value. Typically, the photographer would arrive at a personal

working film speed and would determine the relationship between development time and contrast in the negative.

In this context, the subject values are translated into exposure zones, which are described by a similar set of Roman numerals. Although subject values can be measured in candelas per square foot (c/ft^2), they are typically determined through visual (subjective) evaluation of the scene. Exposure zones (or, more simply, zones) are used to describe the input exposure scale of the film and are more technical. Thus a given subject value may be placed in any exposure zone, with the other values falling on the exposure scale where they may. Typically, an area of important shadow detail is placed in a lower exposure zone (Zone II or III), with the higher values (highlights) falling in higher zones, depending upon their lightness.

In Zone System terminology, the development time necessary to make a typical outdoor scene (eight values) result in a negative of normal contrast (density range equal to 1.10 for printing with a diffusion enlarger) is identified as "N" development. The development time necessary to make a nine-value subject result in a normal-contrast negative (density range of 1.10) is defined as "N minus one zone," or simply, "N minus 1." Similarly, a subject containing only six values would require N plus 2 development to yield a normal-contrast negative.

The concept of zone development is related to the use of a normal contrast index for film development. "N minus" development is associated with lower-than-normal contrast-index values, while "N plus" development is associated with higher-than-normal contrast indexes. In the Zone System, "N plus" development is often referred to as expansion, while "N minus" development is referred to as compaction (see Table 10–3).

F-numbers are positions on a scale while f-stops are the intervals between positions.

The Zone System Personal Speed Index is based on a negative density of 0.10 above base-plus-fog density for Zone I.

Table 10–3

Development Index Zone System	Contrast Index	
	Diffusion	Condenser
N + 2	.72	.55
N + 1	.63	.47
N	.56	.42
N − 1	.53	.39
N − 2	.50	.37

In the late 1800s, the photographer Dr. P. H. Emerson put this question to Hurter and Driffield, the fathers of photographic sensitometry: "Suppose I want to photograph three houses, a white one, a gray one, and a black one. What is it you say I have to do to secure a truthful rendering of tone; what is it you say I can alter by development, and what is it I cannot alter?"

Therefore, the function of the Zone System at this stage is to assist the photographer in placing the important shadow value at some minimum useful position (Exposure Zone II or III) in the toe of the characteristic curve, and to provide the proper development that will result in a normal-contrast negative, regardless of the original scene contrast. This phase of the Zone System is often summarized by the following statement: "Expose the film for the shadows and develop the negative for the highlights." If the photographer has provided correct exposure and development, the density values produced in the negative will just fill the reproduction capacity (useful log exposure range) of a normal grade of photographic paper.

The third phase of the Zone System involves the printing of the negative and, therefore, the production of print values. At this stage the concept of zones can, again, be used to describe the appearance of the tones of

gray in the print; however, the term *print value* is used instead of *zone* to differentiate it from the exposure scale of the negative. Since reflection prints are limited in the range of tones that can be produced, and since there is distortion present in the photographic process, the values of the print have different definitions. These are listed in Table 10–4. The goal is to produce a print in which the values of the original subject are reproduced as the desired values in the finished print. This phase of the Zone System is analogous to the third quadrant of the tone-reproduction diagram, in which the characteristic curve of the photographic paper is inserted and the output expressed in terms of reflection densities.

At this point the Zone System comes full cycle; the photographer views the final print and determines the closeness with which it matches the visualized print. The results of this evaluation provide the basis for future refinements and improvements of the system. At this stage photographers rely upon their subjective opinion regarding the quality of the final image. However, the Zone System provides photographers with a vocabulary and a set of measurement concepts that can provide a clear understanding of the factors influencing the final result. In the objective tone-reproduction system, this relates closely to the fourth quadrant, which contains the tone-reproduction curve resulting from the combination of materials and processes used to make the photograph. In the tone-reproduction system the quality of the final print can be assessed in terms of how closely it matches the aim curve for that particular system. In the Zone System, the quality of the reproduction is assessed relative to what the photographer visualized in the subject.

The similarities between the tone-reproduction system and the Zone System are not the result of chance. The Zone System actually represents a practical simplification of the concepts of tone reproduction. It is not surprising then, that many photographers have trouble understanding and implementing the Zone System without understanding the tone-reproduction system. The relationship between the two systems can be illustrated in a variety of ways. First, consider Figure 10–16, which represents an adaptation of Zone System terminology to the con-

Table 10–4 The relationship between the various reflection densities of the photographic paper and print values

Value	Description
0	Maximum density of paper; principally determined by surface characteristics of print; absolutely no tonal separation.
I	In theory, a just-noticeable difference from the maximum density. Sensitometric tests indicate that this value is located at 90% of the maximum density. At print densities higher than this, texture is suppressed.
II	Darkest area of the print where strong surface texture is still maintained. Approximately equal to a density that is 85% of the maximum density.
III	Darkest area of the print where detail can be seen; in the shoulder of the curve where there is sufficient slope for tonal separation.
IV	The darker midtones of the print. The upper midsection of the curve is used, giving obvious tonal separation.
V	The middle gray tone of the print; 18% reflectance; reflection density of 0.70. Occurs in the middle of the curve, producing maximum tonal separation.
VI	The lighter midtones of the print. The lower midsection of the curve is used where the slope is still quite high, yielding good tonal separation.
VII	A very light area of the print containing texture and detail. The toe section of the curve is employed where the slopes and densities are low, causing tonal compression.
VIII	The diffuse (nonspecular) highlights of the print. The just-noticeable density difference above the paper's base-plus-fog density. A reflection density of approximately 0.04 above base plus fog.
IX	The specular highlights of the print. This is the base-plus-fog density of the paper and is the absolute whitest tone obtainable in the print.
X	Base plus fog density of the paper; maximum white (same tone as print zone IX).

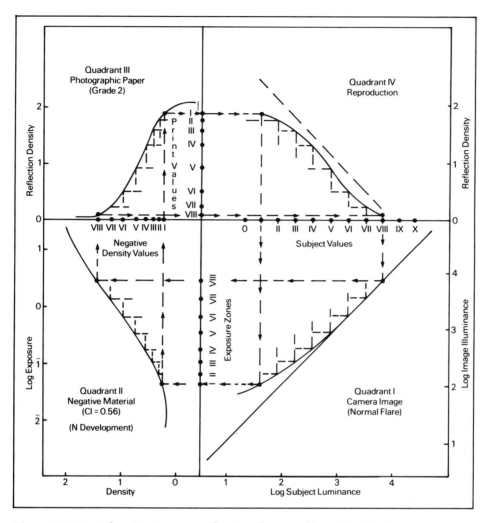

Figure 10–16 Objective tone-reproduction diagram illustrating the Zone System concept.

ventional four-quadrant tone-reproduction diagram. This figure reveals the distortions encountered by the original subject zones in their travels through the photographic process. Notice, for example, that the shadow values are compressed first because of the camera's flare characteristics, second because they are placed in the toe section of the film curve, and third because they are placed in the shoulder section of the paper curve.

As a result, the print's shadow values are considerably compressed compared to the shadow values in the subject. Although the highlight values in the subject are unaffected by camera flare, they are somewhat compressed by the minimum slope in the upper portion of the film curve, and greatly com-pressed as a result of being exposed onto the toe section of the paper curve. Thus the print's highlight values are greatly compressed compared to the highlight values of the original subject.

The midtone subject values are relatively unaffected by camera flare; they are somewhat reduced in contrast because of the medium slope (less than 1.0) in the mid-section of the film curve; and they are greatly expanded in contrast because they are placed in the steep mid-section of the paper curve. As a result of these conditions, the midtone values in the print are slightly expanded compared to the midtone values in the subject. Thus there is not a simple one-to-one relationship between the values of the subject and the values in the final print.

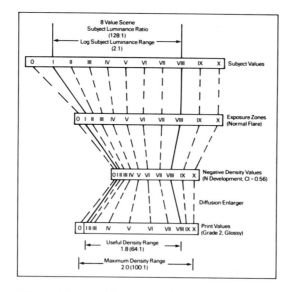

Figure 10–17   Distortion of zones that occurs at the major stages of the photographic process. (Adapted from Eastman Kodak Publication F-5.)

Zone System. In no way can such diagrams be substituted for the actual testing of the photographic system and for the creative use of the results. The four-quadrant tone-reproduction system, however, can be correlated with the Zone System provided the exposure meter and shutter are known to be accurate (or the errors are known and compensated for); and the flare, film, and paper curves are appropriate for the materials and equipment.

The effects of various stages of the photographic process on the subject values can also be illustrated through the use of a bar diagram as shown in Figure 10–17. The horizontal bar designated Subject Values has been divided into 10 equal values, with the appropriate Roman numerals. Value I and Value VIII have been marked to indicate that they cover the range nearly equivalent to that of a typical outdoor subject. The Exposure Zones bar illustrates the effect of camera flare on the optical image at the back of the camera. Here only the lower (shadow) zones are being compressed.

A third bar represents the density values in the negative after it has been given "N" development (contrast index equal to 0.56). Notice that the shadow zones have been further compressed and now the midtone and highlight zones are compressed. The fourth bar illustrates the relationship of the print values for a glossy paper, assuming the negative was printed in a diffusion enlarger. The highlight values are considerably compressed compared to the corresponding zones in the original subject. The midtone values show a slight expansion, indicating they have a slightly higher contrast than the corresponding subject zones. Figures 10–16 and 10–17 are presented with the belief that an understanding of tone-reproduction principles will facilitate an understanding of the

REVIEW QUESTIONS

1. One of two identical photographic prints is viewed in front of a white background and the other is viewed in front of a black background, with identical illumination on the two prints. The print in front of the white background will tend to be judged as being . . . (p. 236)
   A. lighter
   B. darker

2. The density of the center portion of the "preferred" tone-reproduction curve is . . . (p. 237)
   A. the same as the density of the corresponding subject area
   B. less than the density of the corresponding subject area
   C. more than the density of the corresponding subject area

3. If a scene luminance ratio is 150:1 and the image illuminance ratio at the film plane of a camera is 50:1, the flare factor is . . . (p. 241)
   A. 1
   B. 2
   C. 3
   D. 50
   E. 100

4. For normal exposure of the film, the shadow should be located on the curve in quadrant II where the . . . (p. 243)
   A. log exposure is 0.04 above the base-plus-fog log exposure
   B. log exposure is 0.1 above the base-plus-fog log exposure
   C. density is 0.04 above the base-plus-fog density
   D. density is 0.1 above the base-plus-fog density
   E. density is 90% of Dmax

"Caucasian skin in sunlight is rendered with an amazing illusion of reality as a Print Value VI."
—*Minor White*

5. When selecting a grade of paper in quadrant III, the selection should be made so that the . . . (p. 246)
   A. density range of the paper matches the density range of the negative
   B. log exposure range of the paper matches the log exposure range of the negative
   C. density range of the paper matches the log exposure range of the negative
   D. log exposure range of the paper matches the density range of the negative

6. The curve in quadrant IV represents the . . . (p. 246)
   A. print
   B. relationship between the print and the negative
   C. relationship between the print and the subject
   D. relationship between the negative and the subject

7. In the Zone System, diffuse highlights are represented by a subject value of . . . (p. 250)
   A. VI
   B. VII
   C. VIII
   D. IX
   E. X

8. In the Zone System, "N minus 1" development is recommended for . . . (p. 251)
   A. a six-value subject
   B. a seven-value subject
   C. an eight-value subject
   D. a nine-value subject
   E. a ten-value subject

11 Micro-Image Evaluation

Michael J. McNamara. *Silver Halide T-Grains.*

IMAGE CHARACTERISTICS

The small-scale image characteristics of graininess, sharpness, and detail (collectively called definition) strongly influence picture quality. One can produce a print that has good tone-reproduction qualities but suffers in overall quality because of excessive graininess, lack of sharpness, or insufficient detail. For this reason it is important to understand small-scale image characteristics and how they relate to the choice of photographic materials, equipment, and processes. The micro-image attributes of sharpness and detail apply to the optical image formed by a lens as well as the photographic image recorded on light-sensitive film or paper. Graininess, however, is a characteristic imposed on the image by the recording material.

GRAININESS/ GRANULARITY

Graininess is a subjective property. Granularity is an objective property.

Photographic emulsions consist of a dispersion of light-sensitive silver halides in a transparent medium such as gelatin. Figure 11–1 represents a small area of unexposed photographic film magnified 2,500 times. Figure 11–1A shows the emulsion in its pris-

tine state, before exposure and development. The crystals vary in shape and size, and crystals with similar shapes can be oriented in any direction within the thickness of the emulsion. The spacing between the crystals varies, with some crystals touching and others overlapping. Although the crystal size typically varies widely within a given emulsion, the average size is generally larger in fast films than in slow films. It should be noted, however, that the size of the silver halide crystals is not the sole determinant of film speed.

The exposed silver halide crystals are transformed during development into grains of silver that are more or less opaque, depending upon their size and structure (see Figure 11–1B). The developed silver grains seldom conform exactly to the shapes of the silver halide crystals; the size and shape of each silver grain depends upon the combination of exposure, developer type, and degree of development in addition to the size and shape of the original silver halide crystal. As the silver grains increase in size, the spaces between the grains through which light can pass freely become smaller. The overlap in depth of the individual silver grains results in a rather haphazard arrangement of silver grain clusters. This nonhomogeneity of the

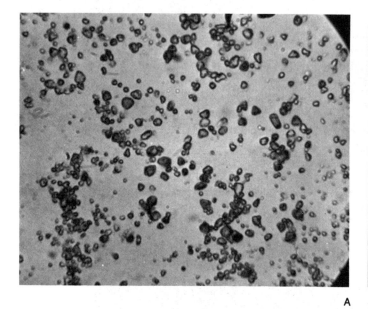

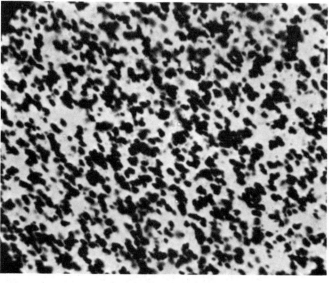

A   B

Figure 11–1   Silver halide crystals. (A) Grains of silver halide are randomly distributed in the emulsion when it is made. (B) Silver is developed at the sites occupied by the exposed silver halide.

negative image usually can be detected in prints made at high magnifications as an irregular pattern of tonal variation superimposed on the picture image. Figure 11–2A compares enlarged prints and transparencies made from different 35 mm films. Notice the loss of fine detail as graininess increases.

Graininess is an important variable since it sets an esthetic limit on the degree of enlargement, as well as a limit on the amount of information or detail that can be recorded in a given area of the film. Graininess can be thought of as noise or unwanted output that competes with the desired signal (the image). All recording mediums have some type of noise that limits the faithfulness of the desired signals, including television, photo-

mechanical reproduction, and video and audio tapes and discs. Graininess, which can be esthetically pleasing, adversely affects sharpness and resolution. Even though graininess, sharpness, and resolution are defined as three distinctly different image characteristics, in practice they are not entirely separate and independent.

Since graininess is generally considered to be an undesirable characteristic, why not make all photographic film fine-grained? Fine-grained film represents a trade-off in which some film speed has been sacrificed. There tends to be a high correlation between graininess and film speed, as shown in Table 11–1. It should be noted, however, that significant advances have been made in emul-

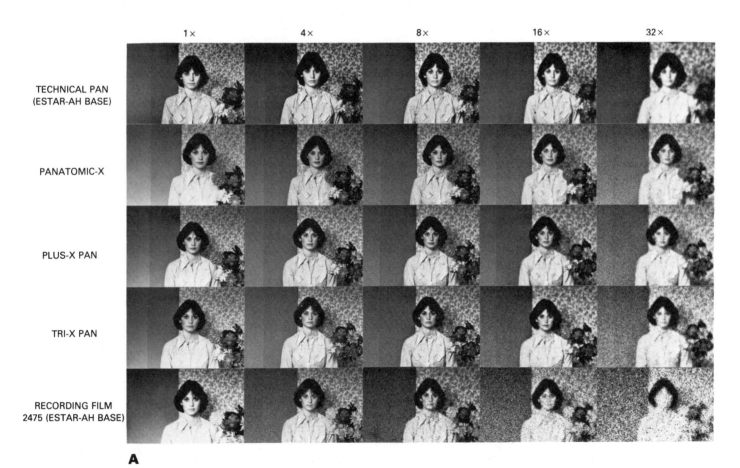

A

Figure 11–2   The effect of film graininess on detail. (© Eastman Kodak Company.) (A) Comparison prints from five different 35 mm black-and-white negatives at various magnifications. (A 32× magnification represents a 30 × 45-inch print.) (B) Comparison of four different 35 mm color transparency films at various magnifications (1×, 4×, 8×, 16×, and 32×). (C) Comparison prints from four different 35 mm color negatives at various magnifications (1×, 4×, 8×, 16×, and 32×). *(Continued on next page.)*

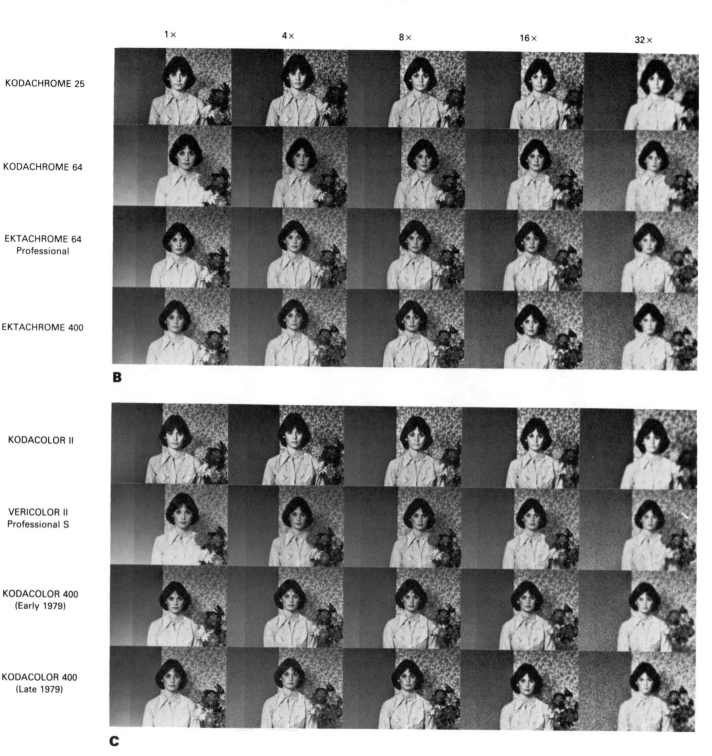

Figure 11–2   *Continued.*

sion technology so that the fast films of today
are much less grainy than they were some
years ago. Since the measurement and specification of graininess have not been standardized internationally, caution needs to be
exercised when comparing data from different manufacturers.

MEASURING GRAININESS

It is relatively simple to look at an enlarged
print and determine whether or not graininess can be detected. The extent of the graininess, however, is more difficult to assess. It
is tempting to make greatly enlarged prints

Table 11–1 Comparison of speed and graininess for several panchromatic films intended for pictorial use. (Manufacturer's graininess classifications must be compared with caution since the methods of measurement and specification are not standardized.)

Film Name	ISO (ASA) Speed	Graininess
Panatomic-X	32	Extremely Fine
Plus-X Pan	125	Extremely Fine
Super-XX	200	Fine
Tri-X Pan	400	Fine
Royal Pan	400	Fine
Royal-X Pan	1250	Medium

from two negatives and compare them side-by-side for graininess. Although such a method is commonly used in studies of graininess described in photographic magazines and is appropriate for some purposes, it does not actually measure graininess. It is a subjective qualitative method that enables one to say print A looks grainier than print B, but not to indicate the magnitude of the difference. Even when making such simple comparisons it is difficult to make the prints so that there are no other differences to confuse the issue, such as subject, density, contrast, and sharpness.

Various methods have been used in an effort to obtain a reliable and valid measure of graininess. The most widely used method of measuring graininess directly is a procedure called *blending magnification*. Under a set of rigidly specified conditions, a negative that has been uniformly exposed to produce a uniform density is projected onto a screen at increasing levels of magnification, beginning at a level where graininess is not visible. A viewer, seated in a specified position with controlled room-light conditions, is asked to determine when graininess first appears. A numerical value is then obtained by taking the reciprocal of the blending magnification and multiplying it by 1,000. For example, if the blending magnification were 8, then 1/8 × 1,000 = 125. The number 125 represents the graininess of that negative under the conditions specified, and as the conditions are repeatable, other negatives could be so measured.

A practical alternative to the blending magnification method is to keep the magnification constant but to vary the viewing distance. One can move toward a print until it begins to look grainy. The minimum distance at which graininess is not evident is the *blending distance,* and the number is a measure of graininess. Such blending distance measurements, although practical, are subject to variability.

When judging whether a print is excessively grainy one must keep in mind its use. One does not normally view a 16 × 20-inch print at a reading distance of 10 inches, nor an 8 × 10-inch print at a distance of 10 feet. How the final print will be viewed or reproduced is important. Photographs reproduced in newspapers using a 60 lines/inch screen will look grainier than those reproduced in a quality magazine using a 150 lines/inch screen. Motion pictures present an interesting situation, since each frame is in the projector gate for only 1/24 second. When one views a motion picture at a large magnification, as from the front row in a theater, graininess is commonly experienced as a boiling or crawling effect, especially in areas of uniform tone. This is a result of the frame-to-frame change of grain orientation.

It is essential that the density level be specified when making graininess measurements. As shown in Figure 11–3, graininess is highly dependent upon the density level. With negatives, maximum graininess occurs at a density between 0.3 and 0.6, depending to some extent upon the luminance level of the test field. A density of 0.3 corresponds to a transmittance of one half, which represents equal clear and opaque areas. One would expect little graininess at very low and very high density levels, but for different reasons. At

When comparing prints or negatives for graininess, the prints or negatives should match in density, contrast, and sharpness.

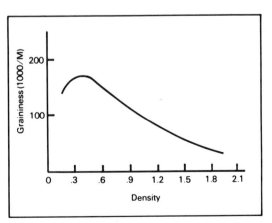

Figure 11–3 Negative graininess is highest at a density of 0.3 to 0.6 and then decreases rapidly as density increases. (The illuminance on the sample is constant.)

very low densities there is little grain structure. As the density of the negative increases, the amount of light transmitted decreases. Since the ability to see detail and tonal differences decreases at low light levels, the perception of graininess decreases rapidly as the density increases.

At a very high density, graininess is not perceptible even though the silver in the negative has a very grainy structure. This can be demonstrated by selecting a negative having a uniform density of about 1.5 and increasing the intensity of the viewing light source 16 times. The appearance of graininess will jump considerably, for in effect the density of 1.5 behaves as though it were a density of only 0.3 with respect to viewing luminance. In fact, if the light level is adjusted so that the transmitted light remains constant regardless of the density level, the result will be the curve shown in Figure 11–4, which indicates that the graininess increases gradually with increasing density.

There is another method of demonstrating that graininess increases with negative density when compensation is made for the lower transmittance; it involves making a series of photographs of a gray card with one-stop variations of exposure, from four stops below normal to four stops over. Developing the film normally will produce negatives having nine different density levels ranging from about 0.2 to about 1.5. Each negative is printed at a fixed high magnification with the necessary adjustment in the printing exposure time to produce the same medium gray density on

Graininess is most easily detected in uniform midtone areas.

Condenser enlargers increase graininess.

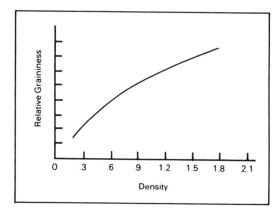

Figure 11–4 Graininess increases as density increases if the light reaching the eye is adjusted so that it is the same regardless of the density of the *negative*.

all of the prints. The prints should be viewed side-by-side under uniform illumination and at the same distance to observe differences in graininess.

Considerable variation in the density level at which graininess is most apparent has been found among different types of images such as film negatives, prints, projected black-and-white transparencies, projected color transparencies, black-and-white television, and color television. The density level is higher for color images than for black-and-white, for transparencies as well as television, as shown in Figure 11–5. This is partly due to the scattering of light by silver particles, known as the Callier effect. The Callier coefficient, a measure of this effect, is the ratio of the density with specular illumination to the density with diffuse illumination. Since color dyes scatter the light very little, the Callier coefficient is approximately 1.0 with color images. In black-and-white images, the scattering of light and the Callier coefficient increase with density and the size of the silver grains, producing values higher than 1.0.

Recent measurement of graininess for black-and-white and color *prints* reveals that the black-and-white prints exhibit peak graininess at a density of about 0.65 and color prints at an average density of about 0.93, with some prints peaking at a low density of 0.80 and others at a high density of 1.05. Further, it was found that varying the level of illumination on prints had little effect on the density at which peak graininess was observed. "An increase in illumination level by a factor of 6.5 × increases the critical density by about 0.07. The shift in critical density that would occur between viewing a print in sunlight and in indoor home lighting would be less than 0.2 density."[1] (The illumination level does of course have a pronounced effect on the perceived tone reproduction of a print.)

GRAIN SIZE AND IMAGE COLOR

Although we think of the silver image as being black, as the term *black-and-white* implies, the color can vary considerably between im-

[1] Zwick, D., "Critical Densities for Graininess of Reflection Prints," *Journal of Applied Photographic Engineering* 8:2 (1982), p. 73

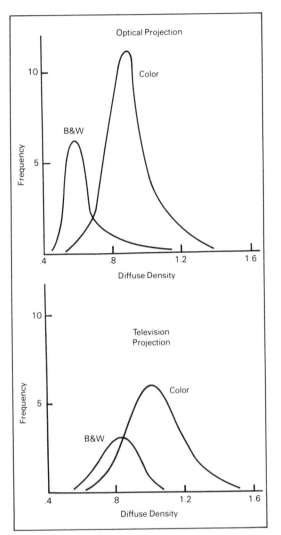

Figure 11–5 The density at which graininess is most objectionable is different for black-and-white and for color films, and for different systems of projection. (The difference between black-and-white and color is due mostly to the Callier Q factor; silver images scatter more light.) (Redrawn from Zwick, D., "Film Graininess and Density—A Critical Relationship," Photographic Science and Engineering 16:5 (1972), p. 345.)

ages depending upon the size and structure of the silver particles. Large silver particles absorb light almost uniformly throughout the visible spectrum, producing neutral color images that appear black where the density is high. Smaller silver grains absorb relatively more light in the blue region than in the green and red regions, causing the images to appear more brownish. By making the silver particles sufficiently small, it is possible to obtain a saturated yellow color, as exem-plified by the yellow filter layer between the top two emulsion layers in some color films.

When the smallest dimension of small par-ticles approximates the size of the wave-length of light, they scatter light selectively by wavelength, an effect known as Rayleigh scattering. Since the scattering varies in-versely with wavelength to the fourth power, blue light is scattered most, and the trans-mitted light appears reddish. Fine-grained negatives that have a brownish color can have different printing characteristics than would be predicted by visual inspection, because of the lower Callier coefficient when printing with a condenser enlarger, and because of the different response of some printing materials to light of different colors — for example, variable-contrast papers.

GRANULARITY

Granularity is the objective measure of den-sity nonuniformity that corresponds to the subjective concept of graininess. Granularity determination begins with density measure-ments made with a microdensitometer—a densitometer that has a very small aperture—across an area of a negative that has been uniformly exposed and developed. The data are automatically recorded and mathemati-cally transformed into statistical parameters that can be used to describe the fluctuations of density due to the distribution of the silver grains in the negative, and therefore to de-termine the granularity.

The process can be described graphically as illustrated in Figure 11–6. The fluctua-tions are greater at a density of 1.0 than they are at a density of 0.3. If all of the many points indicating the variation from an assumed av-erage density are collapsed as shown on the left of the illustrations, a near-normal distri-bution is generated. Such a curve allows the specification of granularity in statistical terms that may be generalized. For any mean or average density (D), the variability of density around that mean is specified by the standard deviation σ (D) (read sigma D). Because σ (D) is the *root mean square* (rms) of the devia-tions, such granularity measurements are called *rms granularity*. (RMS is the standard deviation of micro-density variations at a particular average density level.)

Controlled graininess can be used to create a textured or impres-sionistic effect.

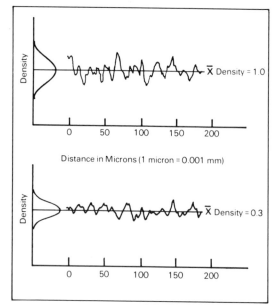

Distance in Microns (1 micron = 0.001 mm)

The best way to minimize graininess is to use a fine-grain film.

Figure 11–6   Microdensitometer traces of a negative material at a high density, and at a low density. The fluctuation of density over micro-areas on the negative describes the granularity of the negative. The higher density of 1.0 has greater fluctuations and therefore higher granularity.

Table 11–2   Some typical Kodak films and plates and their rms granularities. The data represent a very specific set of conditions of testing. Any comparison of rms granularity for other films must be made under the same set of conditions.

Film or Plate	RMS Granularity
High-resolution plate	0.0025
Kodachrome 64 (daylight)	0.010
Plus-X Aerecon Type 8401	0.034
Tri-X Aerecon Type 8403	0.048

MINIMIZING GRAININESS

Much has been written and claimed about developers for reducing graininess. The same could be said for such claims as was said years ago by Mark Twain when he read his obituary in a newspaper: "Claims . . . are highly exaggerated." Two major problems with claims of reduced graininess are the questionable authenticity of the testing procedures, and the failure to completely report difficulties and losses associated with such reduction. Validity of graininess comparison tests requires negatives that have been exposed and developed to produce the same density and contrast, and printed to produce the same tone reproduction. This is a difficult task, and processing the film for the manufacturer's recommended time and temperature seldom produces negatives that meet these requirements. The fact of the matter is that little can be done to improve graininess through development unless one is willing to sacrifice one or more other attributes such as film speed, image sharpness, and tone-reproduction quality. Depending upon the requirements for specific photographic situations, such trade-offs may be warranted, but there is no substitute for beginning with a fine-grained film.

Fine-grain development always results in a loss of film speed. In situations where minimum graininess and maximum film speed are both important, the photographer is presented with a choice of using a high-speed film and a fine-grain developer or a slower film that has finer inherent grain with normal development. The control over graininess by the choice of developer tends to be small compared to the control by the choice of film. Films that are especially designed for micro-

Granularity values for Kodak negative films, and reversal and direct duplicating films are made at a density of 1.00. Microdensitometer traces are made with a circular aperture having a diameter of 48 micrometers (0.048mm).

Since there is a good correlation between granularity measurements and graininess, rms granularity numbers are used to establish the graininess classifications found in some data sheets and data books: microfine, extremely fine, very fine, fine, medium, moderately coarse, coarse, and very coarse grain. In addition to providing a good objective correlate for graininess, rms granularity is analogous to the way noise is specified in electronic systems. This provides an important linkage for photographic-electronic communication channels and systems. Some typical values of rms granularity for several Kodak films and plates are shown in Table 11–2. They were made with a microdensitometer having a scanning aperture of 48 micrometers, corresponding to a magnification of 12, on a sample having a diffuse density of 1.0.

filming, where high contrast is appropriate and slow speed is tolerable, have a low level of graininess that is truly remarkable. Such films have even been used for pictorial photography by developing them in low-contrast developers in situations where the lack of graininess is a more important consideration than the slow speed.

GRAININESS OF COLOR FILMS

Color films exhibit graininess patterns having both similarities and differences, compared with the graininess patterns in black-and-white films. With dye-coupling development, dye is formed in the immediate area where a silver image is being developed so that the dye image closely resembles the grain pattern of the silver image, and the dye pattern remains after the silver image is removed by bleaching. The image structure is more complex with color film due to the three emulsion layers, each containing a different color dye image. Graininess and granularity are measured and specified in the same manner with color films as with black-and-white. The perception of graininess with color films results from the luminosity (brightness) contribution of the three dye images rather than from the hue and saturation attributes, although the appearance of graininess is not equal for the three colors. The relative contribution of each of the dye layers to granularity is as follows:

Green record	(Magenta dye layer)	60%
Red record	(Cyan dye layer)	30%
Blue record	(Yellow dye layer)	10%

These percentages will vary depending upon the dyes used, but the magenta layer will always be the major contributor to the perception of graininess, mainly because our eyes are most sensitive to green light, which the magenta dye layer controls. Figure 11–2B compares four different 35 mm color-transparency films at various magnifications while Figure 11–2C compares prints made from four different 35 mm color negatives at various magnifications (1×, 4×, 8×, 16×, and 32×).

SHARPNESS/ACUTANCE

In art the terms *hard-edged* and *soft-edged* are sometimes used to describe the quality of an edge in an image. In photography the term *sharpness* describes the corresponding abruptness of an edge. Sharpness is a subjective concept, and images are often judged to be either sharp or unsharp, but sharpness can be measured. If one photograph is judged to be sharper than another, the sharpness is being measured on a two-point ordinal scale. If a large number of photographs are arranged in order of increasing sharpness, the sharpness is being measured on a continuum on an ordinal scale.

The sharpness of a photographic image in a print is influenced and limited by the sharpness characteristics of various components in the process, including the camera optics, the film, the enlarger optics, and to a lesser extent the printing material. Sharpness and contrast are closely related, so any factor that increases image contrast tends to make the image appear sharper. Similarly, anything that decreases image sharpness, such as imprecise focusing of the enlarger, tends to make the image appear less contrasty.

Although it is not possible to obtain a sharp print from an unsharp negative simply by focusing the image with an enlarger as some wags suggest, there are various procedures available for increasing the sharpness of photographic images. Some film developers, for example, produce higher local contrast at the boundaries between areas of different density, which causes the image to appear sharper. The anomaly on the denser side of the boundary is called a *border effect*, the anomaly on the thinner side is called a *fringe effect*, and the two together are referred to as *edge effects* (see pages 126–27). There are also electronic and photographic techniques for either making an image edge more abrupt or increasing the local contrast at the edge.

Developing factors that alter edge effects, and therefore image sharpness, include developer strength, developing time, and agitation. Diluting the developer and developing for a longer time with little or no agitation enhances edge contrast and sharpness. Unfortunately, this procedure can also cause uneven development with some film-developer

Textured paper surfaces obscure graininess.

Sharpness and contrast go hand-in-hand.

Glossy print surfaces enhance contrast and sharpness.

combinations even though it is used routinely with high-contrast photolithographic materials. One developer specifically formulated to enhance image sharpness with normal-contrast films is the Beutler High Acutance Developer, named after the man who pioneered developers of this type. The formula is:

Sodium sulfite (Na_2SO_3)	5.0g
Metol	0.5g
Sodium carbonate (Na_2CO_3)	5.0g
Water	1 liter

ACUTANCE

Acutance is a physical measurement of the quality of an edge that correlates well with the subjective judgment of sharpness. It is measured in a laboratory by placing an opaque knife edge in contact with the photographic material being tested and exposing the material with a beam of parallel (collimated) light. This produces, after development, a distinct edge between the dense and clear areas, but because of light scatter in the emulsion the edge has a measurable width. The edge, which is less than 1 mm (1,000 microns) wide, is then traced with a microdensitometer. The result is a change in density with distance. A typical density-distance curve is shown in Figure 11–7. The rate at which density changes as a function of the microdistance of the edge is a graphical description of acutance (sharpness).

Since the rate of change of density can be expressed in terms of slope or gradient, the average gradient of the curve between two specified points becomes a numerical expression of acutance, as shown in Figure 11–8. Cutoff points A and B establish the part of the curve over which the geometric average gradient will be determined. The gradient $\Delta D/\Delta x$ is found for each of the intervals between A and B. The gradients are then squared, and the squared gradients are added together and divided by the number of intervals to determine the average square gradient. The formula for these calculations is:

$$\overline{G^2} = \frac{\Sigma(\Delta D/\Delta x)^2}{n}$$

Acutance is calculated by dividing the average square gradient by the density scale (DS) between points A and B on the vertical scale: Acutance = $\overline{G^2} \div$ DS.

RESOLVING POWER

Resolving power is the ability of the components of an image-formation process (camera lens, film, projector lens, enlarging lens, printing material, etc.) individually or in

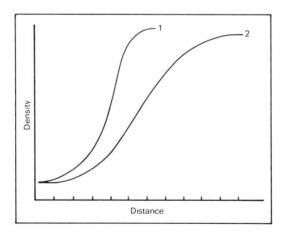

Figure 11–7   The rate at which the density changes across the very edge of an image is described by the slope of the curve. The film represented by curve 1 has a higher slope or gradient and therefore has higher acutance and sharpness.

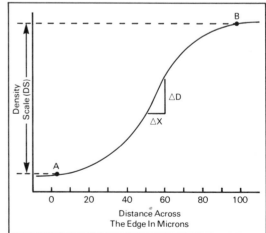

Figure 11–8   To obtain a numerical value for acutance, the average gradient of the curve is calculated and then divided by the density scale (DS).

combination to reproduce closely spaced lines or other elements as separable. Resolving power is the correlate of the image quality referred to as detail, so that a lens having high resolving power is capable of reproducing fine detail. Targets used to measure resolving power typically have alternating light and dark elements either in the form of parallel bars as in the United States Air Force (USAF) and American National Standards Institute (ANSI) Resolution Targets, or in the form of angular letters and numbers as in the Rochester Institute of Technology (RIT) Alphanumeric Resolution Test Objects (see Figure 11–9). Since subject contrast influences resolving power, such test targets are commonly supplied in high-contrast (black-and-white) and lower-contrast forms.

To test the resolving power of a lens, a target is placed on the lens axis at the specified distance (for example, 21 times the focal length) to produce an image at a specified scale of reproduction (for example, 1/20). The aerial image is examined with a microscope at a power about equal to the expected resolving power, and the finest set of lines or elements that can be seen separately is selected. A conversion table translates this information into lines-per-millimeter resolving power, usually in terms of dark-light pairs of lines. Alternative methods of expressing resolving power are in lines per unit distance in object space rather than in the image, and as angular resolving power.

Variations can be made in the above procedure to obtain additional information. A row of targets can be used so that the images fall on a diagonal line at the film plane to determine the resolving power for various off-axis angles as well as on-axis (see Figure 11–

Alphanumeric resolution targets produce less variability among observers.

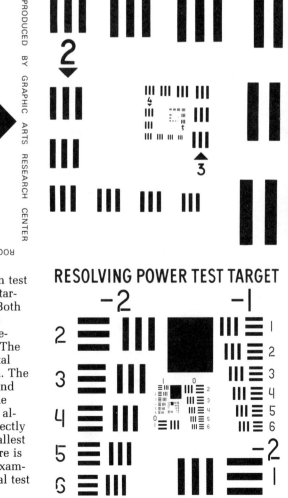

Figure 11–9 Examples of three resolution test targets. Although the USAF and the ANSI targets look different, they are quite similar. Both take the shape of a spiral made up of three black bars that have a square format and decrease in size (increase in line frequency). The USAF target has both vertical and horizontal bars, which allows a check for astigmatism. The ANSI target would have to be rotated 90° and photographed twice to accomplish the same thing. The RIT target is unique in that it is alphanumeric. Since the evaluator must correctly identify the letters and numbers in the smallest set of characters judged to be resolved, there is less variability when different evaluators examine the same images than with conventional test targets.

10). The test can be repeated with the lens set at different f-numbers to determine the setting that provides the best compromise between reducing lens aberrations and introducing excessive diffraction. The effect of filters on the optical image can be determined by measuring the resolving power with and without the filter, also with and without compensation for focus shift with the filter.

The relationship between the resolving power of the components of a system and the resolving power of the entire system is commonly expressed with the formula $1/R = 1/r_1 + 1/r_2 + 1/r_3 \ldots$ where R is the resolving power of the system and each r_n is the resolving power of a component. This formula reveals that the resolving power of the system cannot be higher than the lowest resolving power component. In fact, the resolving power of the system is always lower than the lowest component. If, for example, a lens with a resolving power of 200 lines/mm is used with a film having a resolving power of 50 lines/mm, the resolving power of the combination is 40 lines/mm. Thus it would be a mistake to believe that the quality of the camera lens is unimportant for photographs made for reproduction in newspapers and on television, where the maximum resolving power is relatively low. The only question is whether differences in resolving power of two lenses

Resolving power and film speed tend to vary inversely.

would produce a noticeable difference in the reproduction.

Since it is not possible to measure the resolving power of photographic film directly, it is common practice to use the film with a high-quality lens of known resolving power, determine the resolving power of the system, and then calculate the film's resolving power. The resolving power of the lens-film combination is referred to as photographic resolving power to distinguish it from film resolving power and lens resolving power.

RESOLVING-POWER VARIABLES

There are a number of factors that can enter into the testing procedure to produce variable resolving-power values, including the following:

Test target contrast. The higher the contrast of the test target, the higher the measured resolving power will be. Black-and-white reflection targets have a luminance ratio contrast of approximately 100:1. The ratio for low-contrast targets may be as low as 2:1 or even less. The transmittance ratio for transparency test targets is generally about 1,000:1. High- and low-contrast test targets are illustrated in Figure 11–11, and comparison data for the two types are presented in Table 11–3.

Focus. Focusing inaccuracies can be caused by human error or mechanical deficiencies. With lenses having residual curvature of field, the position of optimum focus will not be the same for on-axis and off-axis images. Other common problems are a difference between the position of the focusing screen and the position of the film plane, and film buckle. Bracketing the focusing position is sometimes advisable.

Camera movement. Any movement of the camera during exposure will have an adverse effect on resolving power, including the jarring caused by mirror movement in single-lens reflex cameras.

Exposure level. There is an optimum exposure level, with a given film, to obtain maximum resolving power. This level does not necessarily correspond to the published film speed.

Development. The degree of development

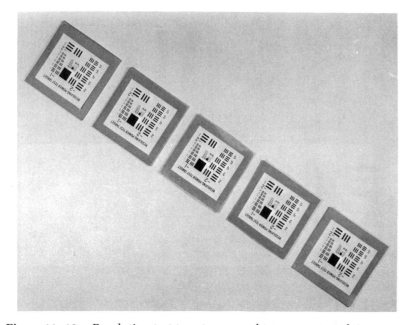

Figure 11–10   Resolution test targets arranged to measure resolving power on and off axis.

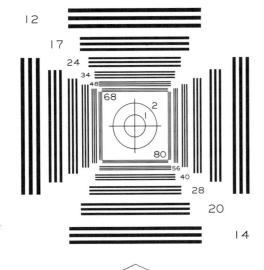

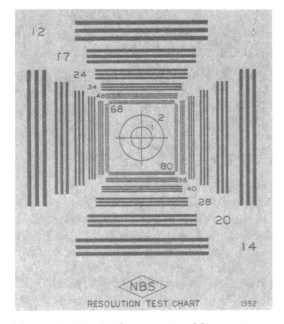

Figure 11–11 High-contrast and low-contrast resolving-power test targets. The target on the top has a contrast ratio of 100:1, while the one on the bottom is 2:1.

affects negative density, contrast, and granularity, all of which can influence resolving power.

Light source. Exposures made with white light and with narrow-wavelength bands of light, ultraviolet radiation, or infrared radiation can produce different resolving-power values for a variety of reasons including chromatic aberration, diffraction, and scattering in the emulsion.

Table 11–3 Resolution, word, and numerical data for several Kodak films. In general, as the speed of the film increases, resolution decreases. Resolution measured with a high-contrast target is two or more times that measured with a low-contrast target. (Special films such as spectroscopic plates have resolution values up to 2,000 lines/mm but speed values of about 1.0 or less.)

Kodak Film	ISO (ASA) Speed	Low Contrast	High Contrast	Resolution Word Category for H.C.
Panatomic X	32	80	200	Very high
Plus-X	125	50	125	High
Super-XX	200	40	100	High
Tri-X Pan	400	50	100	High
Royal Pan	400	40	80	Medium
Royal X Pan	1250	40	100	High

Human judgment. Different viewers may disagree on which is the smallest set of target lines that can be resolved in a given situation. With experience, a person can become quite consistent in making repeated interpretations of the same images. For this reason, resolving power is more useful for comparison tests made by one person than in comparison tests of resolving-power values to published values. Table 11–4 shows the variations in resolving power when six different laboratories tested the same film using high-contrast, medium-contrast, and low-contrast targets. The values represent an average of three tests for each condition. The upper table shows the results before the numbers were rounded off to fit the established ANSI categories. Much of the variation can be attributed to the difficulty viewers have in judging which set of lines is just resolvable.

"The resolving power of a photographic material is not measurable apart from the other components of the photographic process. What we invariably estimate is the res-

The resolving power of a film is higher when measured with a high-contrast test target.

Resolving power changes with exposure, and is highest with the correct exposure.

Table 11–4 Resolution values from six outstanding photographic organizations testing the same film under the same specified conditions. The upper values represent raw data while the lower values are rounded off to fit categories specified by ANSI.

Test Target Contrast	A	B	C	D	E	F
High Contrast	130	100	140	112	125	126
Medium Contrast	105	89	100	100	93	126
Low Contrast	53	35	45	36	43	40
High Contrast	125	100	125	100	125	125
Medium Contrast	100	80	100	100	100	125
Low Contrast	50	32	50	40	40	40

olution of a *system*, including the target, the illumination method, the optical system, the photographic material and its treatment, and the readout method."[2] The same can be said in testing resolution for any system—the resolution of paper copies made on machines such as the Ektaprint, Xerox, and IBM machines; the resolution of video systems; the resolution of printing plates used in graphic arts printing; the resolution of the human eye; and so on.

RESOLVING-POWER CATEGORIES

Since resolving-power numbers mean little to those who are inexperienced in working with them, words are sometimes used to describe resolving-power categories. Eastman Kodak Company has set up the following relationship between words and numbers for films:

Ultra high	630 lines/mm and higher
Extremely high	250, 320, 400, 500 lines/mm
Very high	160, 200 lines/mm
High	100, 125 lines/mm
Medium	63, 80 lines/mm
Low	50 lines/mm and below

RESOLVING POWER AND GRAININESS

The effect of graininess on resolution can be seen in Figure 11–12. The sequence of prints are of a person and a resolution target photographed simultaneously at three increasing distances between subject and camera. All three photographs were made on the same 35 mm film and processed at one time. To obtain the same image size, different amounts of enlargement were required. The first print can be considered normal. Notice that there is detail and texture in the woman's coat and that many of the pine needles are easily dis-

tinguishable. Similarly, fine detail is maintained in the resolution target as seen in the small distinguishable lines in the center array. In the second and third prints the fine detail in both pictures has been increasingly obscured by the increase in graininess. The fine detail in the woman's coat, her facial features, the pine needles, and the small bars in the resolution target are no longer distinguishable. The coarser areas of the photograph, however, are still distinguishable as seen by the large array of three-bars and the solid black square in the resolution target, and the vertical black area of the woman's coat. Graininess, which is the major contributor to visual noise in a photographic system, takes over and the signal (information) is diminished or completely lost.

RESOLVING POWER AND ACUTANCE

Although image detail and sharpness are commonly perceived as similar, and the corresponding measurements of resolving power and acutance often correlate well, they can vary in opposite directions. That is, one film or lens can have higher resolving power but lower acutance than another film or lens. There was some dissatisfaction with the heavy reliance on resolving power used as a measure of image quality in the past because occasionally a photograph that tested higher in resolving power than another photograph was judged by viewers to be inferior in small image quality. A small loss of fine detail, which corresponds to a reduction of resolving power, may be less objectionable esthetically than a loss in sharpness or contrast (see Figure 11–13).

When trying to obtain a reliable and valid measure of image quality, it is tempting to search for a single number that will describe that characteristic. As early as 1958, two research scientists, George C. Higgins and Fred H. Perrin, cautioned: "No single number can be attached to any system to describe completely its capability of reproducing detail clearly."[3]

Resolving power is not a dependable indicator of the appearance of relative sharpness of images.

Land resolution from an orbiting satellite is approximately 20 meters per line pair from a height of about 400 miles.

---

[2] Todd, H., and Zakia, R. *Photographic Sensitometry.* 2nd ed. Dobbs Ferry, NY: Morgan & Morgan, 1974, p. 273.

[3] Higgins and Perrin, *Photographic Science and Engineering* (August 1958), p. 66.

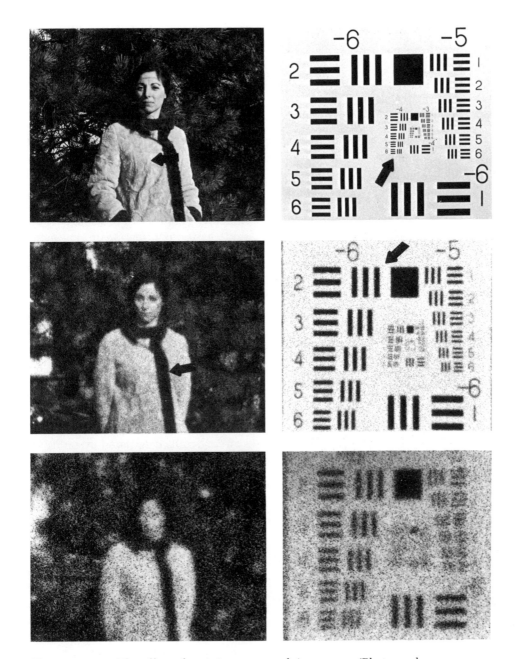

Graininess tends to lower film resolving power.

Figure 11–12   The effect of graininess on resolving power. (Photographs by Carl Franz.) 4× magnification. 16× magnification. 32× magnification.

MODULATION TRANSFER FUNCTION (MTF)

The modulation transfer function is a graphical representation of image quality that eliminates the need for decision making by the observer. The MTF system differs from resolving-power measurement in two other important ways. First, the test object generates a sine wave pattern rather than the square wave pattern generated by a resolving-power target; second, the percent response at each frequency is obtained. Resolving-power values are threshold values, which give only the maximum number of lines resolved; i.e., the highest frequency. Figure 11–14 shows a sine wave test target, a photographic image of the target, and a microdensitometer trace of the

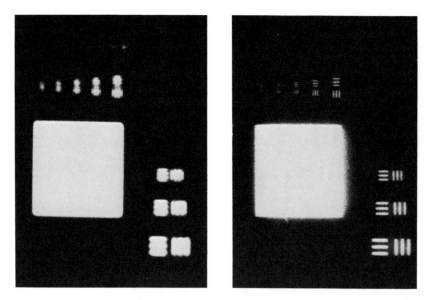

Figure 11–13   Enlargements of small images of a test object made on the same film. The left image was made on the optical axis of the lens where acutance was high but resolving power was low. The right image was made 15° to the right of the optical axis where the acutance was low but resolving power was high.

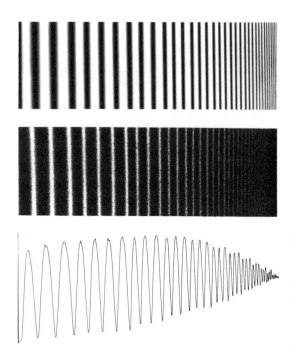

Figure 11–14   To measure modulation transfer functions a sinusoidal test target is photographed and then a microdensitometer trace made of the photographic image. (© Eastman Kodak Company.) (Top) Sine wave test target. (Middle) Photographic image of test target. (Bottom) Microdensitometer trace of the photographic image of the test target.

image. Note that the test target increases in frequency from left to right and that the "bars" do not have a hard edge but rather a soft gradient as one would expect from a sine wave.

The modulation transfer function is a graph that represents the image contrast relative to the object contrast on the vertical axis over a range of spatial frequencies on the horizontal axis, where high frequency in the test target corresponds to small detail in an object. If it were possible to produce a facsimile image, the contrast of the image would be the same as the contrast of the test target at all frequencies, and the MTF would be a straight horizontal line at a level of 1.0. In practice, the lines always slope downward to the right, since image contrast decreases as the spatial frequency increases. Eventually the lines reach the baseline, representing zero contrast, when the image-forming system is no longer able to detect the luminance variations in the test target. As with resolving power, an MTF can be determined for each component in an image-forming system or for combinations of components. The MTF for a system can be calculated by multiplying the modulation factors of the components at each spatial frequency.

The advantage of modulation transfer functions is that they provide information about image quality over a range of frequencies rather than just at the limiting frequency as does resolving power; but there are also limitations and disadvantages. For example, photographers cannot prepare their own modulation transfer functions, and the curves are more difficult to interpret than a single resolving-power number. MTFs are most useful for comparison purposes. Figure 11–15 shows published curves for two pictorial-type films having speeds of 32 and 1250. Although both films have the same response up to about 5 cycles/mm, the response falls off more rapidly for the faster film as the frequency increases, indicating that the slower film will hold fine detail better. It should be noted also that modulation transfer functions for lenses do not provide the desired information about off-axis aberrations such as coma and lateral chromatic aberration. When this information is needed, it is necessary to go through an additional procedure to produce an optical transfer function.

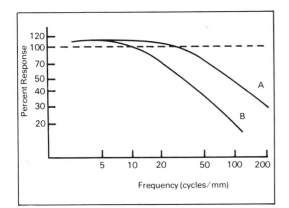

Figure 11–15   MTF curves for two pictorial films with quite different film speeds. Film A is a slow film (ISO 32) and B is a fast film (ISO 1250). One could expect intermediate speeds to fall between A and B. (Output values greater than 100% are the result of adjacency effects in the film.)

The following is an example of how a system MTF can be determined by multiplying the modulation factors of the components. If the modulation factors at 50 cycles/mm are .90 for the camera lens, .50 for the film, .80 for the enlarger lens, and .65 for the printing paper, the product of these numbers, and therefore the modulation factor for the system, is .23.

PHOTOGRAPHIC DEFINITION

Photographic scientists and engineers stress the fact that no single number satisfactorily describes the ability of a photographic system to reproduce the small-scale attributes of the subject. The term *definition*, however, is used to describe the composite image quality as determined by the specific characteristics of sharpness, detail, and graininess (or the objective correlates acutance, resolution, and granularity). *Definition* is applied to optical images as well as images recorded on photographic materials, although the term *photographic definition* is preferred for the latter. Since the measured resolving power of a photographic lens, material, or system did not always correlate well with perceived image definition, the modulation transfer function was introduced as a more comprehensive objective measure of image quality. Because it is difficult to translate modulation transfer function curves into small-scale image quality, photographers have still felt the need for a simple but meaningful measure of definition. One film manufacturer has attempted to satisfy this need by rating its films on the basis of the degree of enlargement allowed, which takes into account the combined characteristics of graininess, sharpness, and detail.

It should be noted that photographic definition is not determined entirely by the quality of the photographic lens and the photographic materials, but also by factors such as focusing accuracy, camera and enlarger steadiness, subject and image contrast, exposure level, and use of filters. It is not realistic to demand the same level of definition for all types and uses of photographs, including exhibition photographs, motion-picture films, portraits, catalog photographs, and photographs to be reproduced in magazines, newspapers, and on television. Many publications refused to accept black-and-white or color photographs made with small-format cameras long after it had been demonstrated that such cameras were capable of making photographs with much better definition than the photomechanical processes were capable of reproducing.

PIXELS

In electronic photography a light-sensitive, electrically charged microchip serves as the equivalent of film in a camera. Silicon is generally used as a light-sensitive surface, and the chip is designed as a mosaic containing hundreds of thousands of discrete minute areas that act as photoreceptors. Each area is capable of registering picture information and is called a *pixel* (picture element). (A microchip is also analogous to the retina of the eye, which contains a mosaic of minute light-sensitive receptors called rods and cones, or simply photoreceptors.) Figure 11–16 illustrates a light-sensitive silicon chip. For reproduction purposes only 400 receptors or pixels are shown, whereas in reality they contain as many as half a million pixels within an area of one-half-inch square. As light falls on the chip, each pixel generates electrical signals that are proportional to the illuminance. The signals from each pixel are transmitted

Definition is a broad term that includes sharpness, detail, and graininess.

Pixel means picture element.

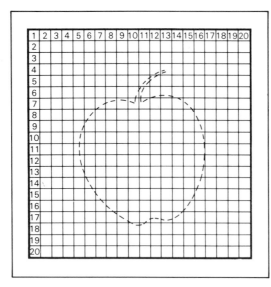

Figure 11–16   A representation of a silicon charge-coupled device (CCD) with 400 squares or pixels. The camera's lens focuses the image of an object or scene on the many light-sensitive discrete surfaces (pixels) of a charge-coupled device. The CCD converts the image into an array of electrical charges that are proportional to the intensity of the light falling on each pixel (light-sensitive square).

to a magnetic recording disc. The disc magnetic image can then be converted to a positive or negative visible image on a television screen, or it can be recorded on paper to form "hard copy."

To specify and compare the image-quality capabilities of a light-sensitive chip such as a charge-coupled device (CCD), a charge-injection device (CID), or a charge-priming device (CPD), the term *pixel* has come into use. The greater the number of separate or discrete pixels possible per unit area of a chip, the higher the potential image quality. Put differently, the smaller the pixels, the more pixels per unit area, and the better the image quality. One way to compare the image-quality potential of a CCD, CID, or CPD system to a photographic system or other system of picture recording, such as television and graphic-arts printing, is in terms of point spread functions, described earlier. The smaller the point spread function of a pixel, the more information that can be packed per unit area.

With this in mind, imagine the 400-pixel chip in Figure 11–16 as having 1,000 pixels within each of the 400 small squares shown.

Further imagine the overall square being reduced to an area of one-half square inch. This would provide a microchip having 400,000 pixels. Each pixel would be about 0.0008 inch in diameter (about half the diameter of a human hair). In 1983 a one-half-inch-square chip (12.7 × 12.7 mm)—about the size of a fingernail—had a total of about 280,000 pixels. By comparison, Kodacolor film of about the same size used in the 1982 disk camera had a total of about 3 million pixels, each pixel being about 0.0003 inch in diameter. With photographic emulsions the number of pixels possible per unit area is determined by the granularity and acutance of the emulsion. Projections for light-sensitive electrically charged chips estimate 1 to 10 million pixels by 1988. As a comparison, the human eye contains about 127 million pixels consisting of 120 million rods and 7 million cones all contained in an area about 2 inches square (see Figure 11–17). The cones are much smaller than the rods, the smallest cone receptor being about 0.00008 inch (2 micrometers) in diameter.

Common to all systems of picture recording is the breaking up of a larger continuous optical image into small discrete elements. In photography the discrete picture elements are played back optically, whereas with silicon microchips (CCD, CID, CPD devices) the information is played back electronically. Since the pixels are discrete photoreceptors, they provide a digital system of recording

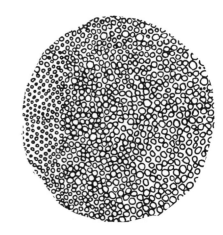

Figure 11–17   The 120 million rods and 7 million cones in the retina of the human eye can be thought of as a total of 127 million pixels.

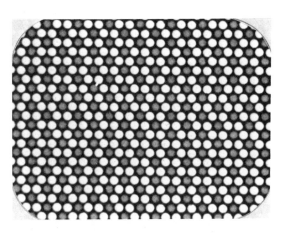

Figure 11–18 Television phosphors as pixels (enlarged view).

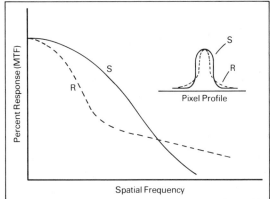

Figure 11–19 Response curves for two hypothetical television tubes. The information displayed by tube *S* will appear sharper but will have less resolution than that of tube *R*.

and playback of pictorial information as well as text information.

In new micrographic technology, laser beams focused to about one-half micron are used to imprint information on a light-sensitive optical disk as tiny pits or holes. These pixels can be packed closely enough to record the images of 10,000 letter-size documents on a 12-inch-diameter disk.

On a television screen the individual areas of the phosphors can be thought of as pixels. In the United States a conventional 525-line television screen consists of 49,152 pixels, 256 in the horizontal direction and 192 vertically (see Figure 11–18). (In Europe a 625-line screen is common. High-resolution screens have about 1,000 lines.) This provides the level of picture quality seen on broadcast television. A much better picture is available, however, with a TV monitor,

which consists of 128,000 pixels (640 × 200), or with higher-resolution monitors such as those with 262,144 pixels (512 × 512). In general, the more pixels per unit area of screen, the better the picture quality (Figure 11–19). There is presently no generally agreed-upon terminology for specifying the resolution of computer-generated images in terms of the number of discrete pixels. However, a qualitative grouping, as shown in Table 11–5, can be helpful. An illustration of the letter *A* on a low-, medium-, and high-resolution computer slide-generating system is shown in Figure 11–20.

Table 11–5 Arbitrary grouping of pixels in terms of resolution

Computer-generated Images Resolution		
Pixels	Total Pixels	Resolution Category
250 × 250[a]	62,500	Low
500 × 500	250,000	Medium
2,000 × 2,000	4,000,000	High
4,000 × 4,000[b]	16,000,000	Very high

[a]On a 19-inch diagonal TV screen, each pixel would have a diameter of about 1/16 inch.

[b]Kodak Ektachrome films can resolve about 3,000 × 4,500 pixels in 35 mm format (1 inch × 1½ inches) for a total of 13,500,000 pixels. Each resolved pixel would have a diameter of about 0.000,001 (one millionth of an inch).

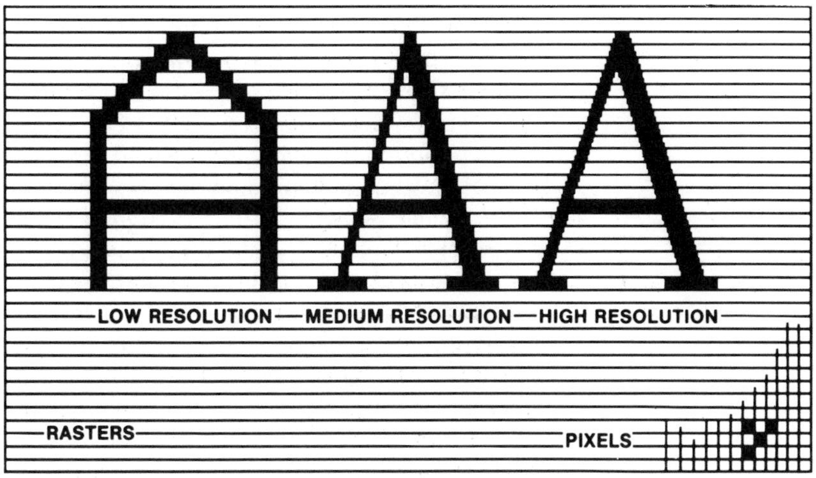

Figure 11–20 This illustration simulates the letter quality obtained by low-, medium-, and high-resolution computer slide-generating systems. (Illustration courtesy of Professor Deane K. Kayton, Audio-Visual Center, Indiana University.)

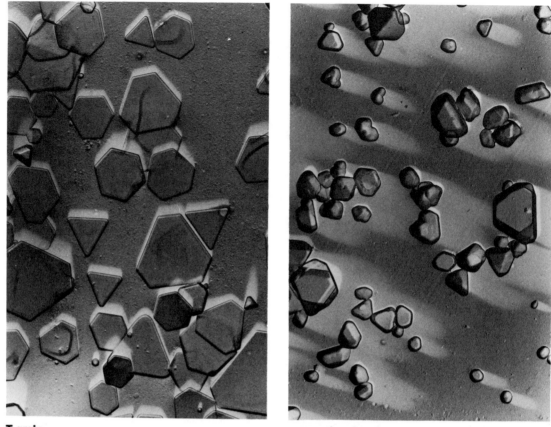

T-grain   **conventional grain**

Figure 11–21   Silver halide grains at 6,000× magnification. The T-grains appear flat and absorb light more efficiently than the conventional pebblelike grains to the right.

T-GRAIN EMULSIONS

Major advancements in the field of emulsion chemistry have led to films with significantly increased film speed and no loss of sharpness or increase in graininess. In 1982 the Eastman Kodak company announced a major breakthrough in emulsion technology with their new T-grain emulsion. Figure 11–21 compares a conventional-grain emulsion with the T-grain emulsion at a magnification of 6,000×. (An electron microscope was used, as optical microscopes are limited to magnification of about 850×.) Notice that the conventional grains are pebble-shaped whereas the T-grains are flat and present a larger surface per unit volume, which maximizes absorption of incoming incident light. The result is a much faster film with about the same amount of silver and no loss in image quality.

REVIEW QUESTIONS

1. The general term that includes sharpness, graininess, and detail is . . . (p. 258)
   A. acutance
   B. tone reproduction
   C. definition
   D. resolution
2. Graininess is most evident in the areas of a print having . . . (p. 262)
   A. a uniform light tone
   B. a uniform medium tone
   C. a uniform dark tone
   D. high-contrast fine detail
3. Some fine-grain silver images appear warm in color because the small particles of silver . . . (p. 263)
   A. transmit red light
   B. oxidize and turn reddish
   C. scatter short wavelength light

4. Granularity is determined by means of calculations based on . . . *(p. 263)*
   A. blending magnification
   B. blending distance
   C. microdensitometer densities
   D. the number of grains per square millimeter

5. Acutance is an objective term that corresponds to the subjective term . . . *(p. 266)*
   A. detail
   B. sharpness
   C. graininess
   D. definition
   E. localization

6. Resolving power is an objective term that corresponds to the subjective term . . . *(p. 267)*
   A. detail
   B. sharpness
   C. definition
   D. recognition
   E. detection

7. If a camera lens has a resolving power of 200 lines/mm and film has a resolving power of 100 lines/mm, the resolving power of the combination is . . . *(p. 268)*
   A. 50 lines/mm
   B. 67 lines/mm
   C. 100 lines/mm
   D. 200 lines/mm
   E. 300 lines/mm

8. Modulation transfer function graphs represent the relationship between . . . *(p. 272)*
   A. subject density and image density
   B. image exposure and image density
   C. subject contrast and image contrast
   D. subject resolving power and image resolving power

9. It is estimated that the retina of the human eye contains approximately 127 million pixels (rods and cones). In comparison, television screens in this country contain approximately . . . *(p. 275)*
   A. 5,000 pixels
   B. 50,000 pixels
   C. 500,000 pixels
   D. 5,000,000 pixels
   E. 50,000,000 pixels

12 Visual Perception

POST-STIMULUS PERCEPTION

Some of the light now reaching the earth from distant stars and galaxies originated millions and even billions of years ago, thus events being observed through powerful telescopes are not happening now but actually occurred long ago. Our visual system does not operate on real time even for events viewed at small distances. The delay between when light falls on the retina in the eye and the perception, although measurable, is extremely short and is of little importance. Perceptions that occur after the light stops falling on the retina, however, are of considerable importance. Post-stimulus perceptions can be divided into categories including persistent images, after-images, aftereffects, eidetic images, visual memory, and imagination imagery.

Laboratory studies indicate that *persistent images* last an average of approximately one-quarter second after the stimulus has been removed. If a slide of the letter *A*, for example, is projected on a screen in a darkened room and the presentation time is controlled with a shutter, the time cannot be made so short that the image cannot be seen, provided the luminance is increased proportionally so that the same total amount of light is available to the viewer. Bloch's law predicts that the visual effect will be the same for different combinations of luminance and time as long as the product remains the same. Like the photographic reciprocity law, it is not valid for high and low light values. Perception occurs with short exposure times because the persistent image keeps the image available to the viewer for somewhat longer than the actual presentation time.

Persistent images are essential to the perception of realistic motion-picture and television pictures. When motion-picture films are projected, it is necessary to darken the screen with a rotating shutter when the film is being moved from one frame to the next. The persistent image prevents the viewer from seeing the dark intervals, which would appear as flicker. Early motion pictures were sometimes referred to as "flicks" because they were shown at 16 frames per second and flicker was obvious. Even at the current 24-frames-per-second sound motion-picture speed, flicker would be objectionable with-

out the use of a chopper blade, which produces three flashes of light for each frame, or 72 flashes of light per second.

Anyone who has photographed a picture on a television receiver at a high shutter speed realizes that there is never a complete picture on the screen. Since the image is constructed sequentially with a scanning electron beam, the persistent image is necessary for the viewer to see a complete picture. Duration of the persistent image can vary depending upon several factors, including the image luminance level and whether removal of the image is followed by darkness or by another image.

It would be reasonable to speculate that increasing image luminance would produce a stronger effect on the visual system, which would increase the duration of the persistent image, but the effect is actually the reverse. The *increase* in persistent image duration with *decreasing* luminance can be explained by comparing the visual effect with exposure of film in a camera at low light levels. If the film does not receive sufficient exposure at a selected shutter speed and the lens is wide open, the exposure time must be increased in order to obtain a satisfactory image. The visual system in effect compensates for the lower image luminance by sacrificing temporal resolution and increasing the duration of the persistent image, which is the equivalent of increasing the exposure time.

We can illustrate this effect with motion-picture film being projected on a screen. If flicker of the intermittent image is just noticeable at a certain luminance level, reducing the luminance will increase the duration of the persistent image to include the time the screen is dark between flashes, and the flicker will disappear. In this example, it would not make any difference if the decrease in luminance were achieved by substituting a smaller projector bulb, increasing the projector-to-screen distance, substituting a gray screen for the white screen, or viewing the screen through a neutral-density filter.

Whereas a persistent image cannot be distinguished from the image seen while the stimulus is still present, *afterimages* are recognized by the viewer as a visual anomaly. Afterimages are most evident when the eye is exposed to an intense flash of light. Looking directly at a flashbulb or flashtube when

> Persistent images prevent us from seeing that motion-picture screens are dark when the projector moves the film from one frame to the next.

it is fired usually causes the viewer to see a vivid afterimage spot for some time. An afterimage can be formed with a stimulus having lower luminance, such as a white circle on a black background, by staring at it for a long time and then looking at a uniform surface (see Figure 12–1). One should not look at the sun, arc lights, UV sources, lasers, or other sources of radiation that can damage the retina.

Afterimages can be either negative or positive; but after looking at a bright stimulus such as a light bulb, the afterimage is typically dark (negative) if one looks at a white surface, and light (positive) if one looks at a black surface or closes and covers the eyes. Negative afterimages can be attributed to lo-cal bleaching of the visual pigments in the retina's receptors, positive afterimages to a continuation of the firing of the visual nerve cells. Negative afterimages of colored stimuli tend to be approximately complementary in color; for example, a yellow light bulb or other stimulus would tend to produce a bluish afterimage. Colored afterimages can also be seen after looking at a white light source. If the red, green, and blue visual pigments in the cones are not bleached and regenerated at the same rate, a sequence of different colors may result, usually in the order of blue, green, and red.

In contrast to persistent images, which contribute to the effectiveness of motion pictures by eliminating flicker, afterimages tend to interfere with subsequent perceptions. For example, an attempt to read immediately after looking directly at a flashbulb when it is fired can be frustrated by the afterimage. When a visual perception is altered by a preceding visual experience, the alteration is referred to as an *aftereffect*. Whereas the afterimage of a yellow light bulb tends to appear blue when the gaze is shifted to a gray surface, it tends to appear magenta when the gaze is shifted to a red surface. The perceptual result of mixing a blue afterimage with a red stimulus is similar to that produced by physically mixing blue and red light. If such an experiment were conducted to demonstrate an aftereffect, the viewer would be aware of the altered perception, but viewers are not commonly aware of aftereffects that occur in everyday life. Brightness adaptation to daylight alters the subsequent perception of the light level in a dimly lit interior, and it is only when the interior begins to lighten as the visual system adapts to the lower light level that the viewer is aware of the aftereffect.

Aftereffects can alter the perception of other subject attributes besides color and lightness. Watching continuous motion in one direction, such as a waterfall, can cause stationary objects to appear to move in the opposite direction due to the aftereffect. Similarly, a rotating spiral design that appears to be expanding will appear to shrink when the rotation is stopped after prolonged viewing. *Figural aftereffects* are changes in the size, shape, or orientation of a perception as the result of a preceding visual experience. Fixating on a set of curved lines for a minute or

Color afterimages can often be seen after looking at a bright white light source.

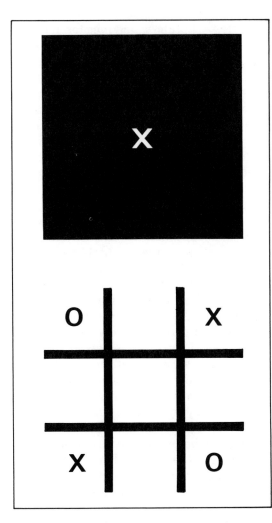

Figure 12–1 Afterimage. Stare at the isolated white X for about one minute. Then shift your gaze to the middle square and win the game of tick-tack-toe.

so will tend to make straight lines appear to be curved in the opposite direction. The reader can experience a figural aftereffect with the drawing in Figure 12–2. After looking steadily at the X above the curved lines for a minute or so and then shifting the gaze to the X above the straight lines, the straight lines tend to appear to be curved in the opposite direction.

Eidetic imagery, which is sometimes referred to as photographic memory, is an ability to retain a visual image for half a minute or longer after a stimulus has been removed from view. A 1968 study by Haber and Haber indicated that approximately 16 of 200 elementary school children experienced eidetic imagery. Since eidetic imagery is rare in mature adults, it is believed to decrease with increasing age. Eidetic imagery enables the viewer to read a printed page word for word and to see details in a complex picture for some time after removal of the printed page or picture. Whereas afterimages move with the eye, eidetic images remain stationary so that they can be scanned. Some persons who do see eidetic images are not aware of this capability because the experience seems normal to them and they assume that everyone has the same experience.

Visual memory shares many characteristics with other types of memory, but it consists of an ability to retain or to recall a visual image that is distinct from persistent images, afterimages, aftereffects, and eidetic images. Visual memories are usually divided into two categories—short term and long term. Just as a new telephone number can be remembered long enough to dial it by repeating it and keeping it in the conscious mind, so can a visual image be remembered for a short time by keeping it in the conscious mind. Experiments indicate that most people cannot retain more than five to nine independent pieces of information, such as numbers in random order, in short-term memory.

One should not expect to remember even for a short time all of the details in a complex picture. Instead, a person tends to remember the details that attract the most attention or hold the most interest. Short-term visual memory is used even as one scans a picture and fixates different parts of the image, and this enables the viewer to create a composite perception of the total picture.

Once the attention is allowed to go on to other things, *long-term memory* is required to revive an earlier visual image. If people are asked to indicate how many windows are in their home or apartment, they will use visual memory to form an image of the interior and then mentally move around it and count the windows. Some visual memories can be recalled easily, but those stored in the subconscious mind are more elusive and may require the use of special techniques such as hypnosis or association. Various procedures for testing visual memory include the use of recognition, where the subject selects the one picture of several that most closely resembles the original stimulus. Another procedure,

> According to one study, approximately one in twelve young children has photographic memory capabilities.

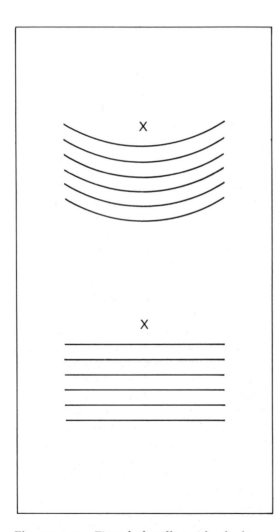

Figure 12–2 Figural aftereffect. After looking steadily at the X above the curved lines for a minute or so and then shifting the gaze to the X above the straight lines, the straight lines tend to appear curved in the opposite direction.

known as the recall method, is to ask the person to describe the original scene from memory or to draw a picture of it. Visual memory can be improved by several methods, such as using attention to produce a more vivid perception, keeping short-term memories in consciousness longer through rehearsal, and practicing making sketches from memory.

Except for simple stimuli such as a circle or a square, visual-memory images generally are not facsimile representations of the original scene, but rather they usually contain just the more important details. *Imagination imagery*, on the other hand, may bear no resemblance to any single earlier visual experience. Just as a writer can write a fictional story, so can a person create a fictional visual image that is as vivid as a visual memory of an actual object, scene, or event. Artists use imagination imagery when they draw or paint pictures of something that does not exist: a pocket watch flowing over the edge of a table in a surrealistic Salvador Dali painting is a dramatic example.

Photographers also use imagination imagery when they previsualize a finished picture, possibly even including a subject, props, background, arrangement, lighting effect, and colors (see Figure 12–3). Imagining impossible images has been recommended as an exercise for developing creative imagination, such as visualizing a large building jumping around like a bird, the water in Niagara Falls moving up rather than down, or a full moon shrinking in size and then disappearing.

Photographers make use of imagination imagery whenever they previsualize a finished picture.

Figure 12–3 Previsualizing a finished photograph, such as this one, before making it involves imagination imagery. (Photograph by Les Stroebel.)

PERCEPTION OF STIMULUS ATTRIBUTES

If one were required to describe precisely to another person the appearance of an object, using only words, it would be necessary to analyze the object's appearance in terms of attributes such as shape, color, and size. Different authorities on visual perception often disagree about the critical stimulus attributes for visual perception, the relative importance of each, and even the definitions of the terms. For example, the word *form* generally implies different object qualities to artists and photographers. Six attributes considered especially important to photographers will be discussed here: color, shape, depth, size, sharpness, and motion.

COLOR HUE

White, gray, and black are colors that are properly identified as neutral colors.

To accurately describe or identify a color, three different qualities of the color must be considered: hue, lightness, and saturation. Hue is the quality associated with color names such as red, green, blue, cyan, magenta, and yellow. White, gray, and black are also colors, but they are all neutral—without hue. Neutral colors are sometimes referred to as colors.

Since color hues are commonly associated with names, problems in communicating specific information about colors will occur unless everyone uses the same names for the same colors. Unfortunately, many children have learned that the names of the primary colors of water colors used for painting are blue, red, and yellow, rather than the correct names of cyan, magenta, and yellow. Advertisers have not helped matters by using exotic-sounding or unusual names for colors rather than more common and descriptive names. Color notation systems such as the Munsell system have done much to bring order to the identification of colors. The Munsell system uses five basic hues: red, yellow, green, blue, and purple. The complete Munsell hue circle, which includes subtle transitions between adjacent hues, has 100 hues—which is about the maximum that persons with normal color vision can distinguish in side-by-side comparisons.

Color photography is based on the subtractive primaries: cyan, magenta, and yellow.

Color television is based on the additive primaries: red, green, and blue.

Much of the discussion about color as it relates to color photography can be accomplished with combinations of the three additive primary colors: red, green, and blue. Combinations of pairs of these colors of light produce cyan, magenta, and yellow, which are called additive secondary colors or subtractive primary colors. The relationship of these six hues is often represented in the Maxwell triangle, shown in Figure 12–4.

Red, green, and blue are considered additive primary colors because by combining red, green, and blue *light* in different proportions it is possible to produce almost any color, including neutral colors. Cyan, magenta, and yellow are considered subtractive primary colors because dyes and other *colorants* in these hues absorb red, green, and blue *light*, respectively, from the white viewing light. Perceptually, however, primary colors are defined as those hues that appear to be pure rather than a mixture of other hues. In this sense, red, green, blue, and yellow are primary colors, and they are identified as *psychological* primary colors. Including the neutral pure colors of black and white increases the number of perceptual primary colors to six.

Persons having normal color vision are identified as *normal trichromats*, based on the three types of cones in the retina, which are sensitive to red, green, and blue light. Not all normal trichromats, however, respond to colors in exactly the same way. For this reason, scientific studies of color vision make use of the average response of a number of persons having normal color vision.

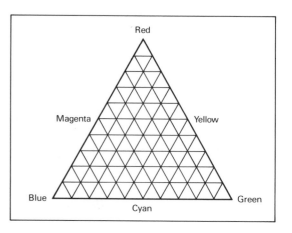

Figure 12–4   The Maxwell color triangle.

There are a number of types of defective color vision. A person who is missing one of the three types of cone pigments is known as a *dichromat. Monochromats* are missing two of the three cone pigments (or possibly have rhodopsin, the rod photopigment, in the cones). Dichromats have difficulty distinguishing between red and green, or more rarely between blue and yellow. There are very few monochromats, but their vision is the equivalent of black-and-white photography.

The red, green, and blue cone sensitivities suggest a simple trichromatic theory of color vision, but much of the experimental evidence supports the opponents' theory of color vision, whereby information from the red-, green-, and blue-sensitive cones is thought to be transmitted in combinations—specifically red-green, blue-yellow, and black-white—through three separate channels.

Other than missing one or two of the three types of cone photopigments, defective color vision can also be associated with reduced sensitivity of one or two of the three cone types, or with a shift in sensitivity along the spectrum for one or more of the cone types. People who have all three types of cone photopigments but who do not have normal color vision for either of the reasons cited are referred to as anomalous trichromats—as distinct from normal trichromats, dichromats, and monochromats.

A person with normal color vision may have inaccurate perception of stimulus colors under certain conditions. Some of these conditions are (a) when the image is formed near the periphery of the retina; (b) when the light level is very low or very high; (c) when the stimulus is very small in area; (d) when the stimulus is presented for a very short time; (e) when the stimulus is illuminated with other than white light; and (f) when the viewer is adapted to a different color. Accurate color identification requires normal color vision, standard viewing conditions, and the opportunity to make side-by-side comparisons with standard colors.

Heredity is responsible for most cases of defective color vision, although it can result from other causes such as the use of certain drugs, excessive use of alcohol, and brain damage. About 8% of Caucasian males and 0.4 percent of females of all races have some form of defective color vision. There is no cure for congenital defective color vision. Some people whose occupations require discrimination between certain colors have been helped by using filters of different colors over the left and right eyes. Photographers with defective color vision are able to make color prints that are acceptable to themselves, but the prints often do not appear correct to people with normal color vision.

Defective color vision can be detected with various types of tests. One pseudoisochromatic test contains simple geometric designs made up of colored circles against a background of gray circles (see Figure 12–5). People with defective color vision are unable to see some of the designs, which are visible to persons with normal color vision. Another type of test requires the subject to arrange color samples that vary in hue in the correct order to provide a consistent transition. This test can also determine how well people with normal color vision can make small discriminations between colors. With this test, defective color vision is revealed by large errors with certain hues, whereas low discrimination is revealed by randomly distributed errors with all hues.

COLOR LIGHTNESS

Luminance ratios as low as 100:1 in a photographic print can create the perception of a range of tones from white to black. This fact is misleading with respect to the ability of the visual system to detect differences in lightness (reflected light) or brightness (light sources) over a wide range.

Viewing transparencies or projected slides in a darkened room requires a somewhat greater luminance ratio to produce the perception of a range of tones from white to black—approximately 500:1 (see Figure 12–6). Since reversal color films typically produce a maximum density of over 3, it is possible to obtain a luminance ratio of 1,000:1 in the image, compared to a maximum density of approximately 2 in typical photographic papers. Placing one gray scale in direct sunlight and a second gray scale in the shade on a clear day when the lighting ratio

A glossy photographic print has a density range of about 2.0. This is a luminance ratio of 100:1.

Everyone is color blind under certain conditions, such as when the light level is very low.

To appear correct in contrast, slides and transparencies viewed in a darkened room must have higher contrast than photographs viewed under normal room illumination.

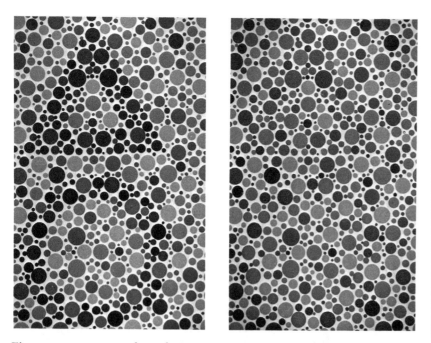

Figure 12–5   A pseudoisochromatic color vision test plate containing geometric designs of colored circles surrounded by gray circles. A person with normal color vision will be able to see the designs (left), whereas a person with defective color vision will not be able to see them (right).

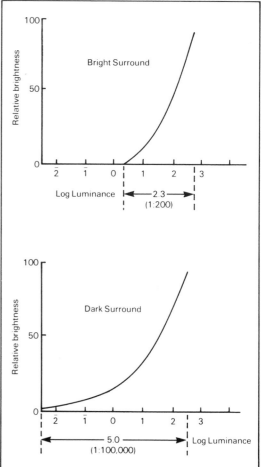

Figure 12–6   Curves representing the perception of brightness in relation to subject luminance with a bright surround (top) and dark surround (bottom). The difference in the log luminance ranges indicates that more contrasty photographic images are required for projection viewing in an otherwise darkened room than for viewing with normal room illumination.

is about 8:1 produces a luminance ratio of approximately 800:1 between the light end of the gray scale in sunlight and the dark end of the gray scale in the shade. A person may be able to see the separation between all steps on both gray scales due to *local adaptation*, whereby sensitivity of the visual system is increased in darker areas and decreased in lighter areas.

When *general adaptation* is taken into account, the ratio of luminances over which the visual system responds is truly amazing. A white object in sunlight has a luminance of approximately 100 million times that of a white object in starlight, and yet both can be seen quite easily when a person is fully adapted to each light level. Under optimum laboratory conditions a person can detect a flash of light having a luminance of only 1/100 that of white paper in starlight, and when adapted to bright light a person can detect luminance differences up to 1,000 times that of white paper in sunlight, which is approaching luminances that can be damaging to the retina. Thus the total response range of the visual system is approximately 10 trillion:1, or a log luminance difference of 13.

It is important to make a distinction between *luminance*, which is psychophysical and measurable with a light meter, and the perception of brightness (or lightness), which is influenced by physiological and psychological factors and is not directly measurable. The eye is not a dependable instrument for measuring luminance values. For example, it is difficult for a person to judge whether a black surface in direct sunlight is reflecting more or less light than a white surface illuminated with lower-level incandescent light indoors, because of two variables: reflectance of the two surfaces and the amount of light falling on them. The adaptation level of the visual system can affect perception in that a surface with a fixed luminance will appear lighter when the eye is dark adapted than when it is light adapted. Also, a gray tone appears lighter in front of a black background than in front of a white background, an effect known as lateral adaptation or simultaneous contrast.

The eye, however, is very good as a null instrument where very small luminance differences can be detected in side-by-side comparisons. Thus, visual densitometers can be quite accurate where the sample being measured is seen in a circle surrounded by a ring that can be adjusted to match the inner circle in brightness (see Figure 12–7).

It is more difficult to match the brightness or lightness of two areas if they differ in hue. If, for example, the inner circle in a visual densitometer is red and the outer ring is blue, the operator will have more trouble judging when the two match in brightness. When it is important to match the brightness of samples having different hues, a device called a flicker photometer can be used to present the two fields to the viewer alternately in rapid succession. When the two do not match closely in brightness, a flicker is seen. It is important for photographers to develop some skill in judging the lightness of subject colors so that they can anticipate whether there will be tonal separation in black-and-white photographs, such as in a photograph of a red object in front of a blue background made on panchromatic film. Viewing filters are used by some photographers in an effort to de-emphasize hues and thereby make lighting effects easier to see.

Equal amounts of light of different wavelengths do not generally appear equally bright. With light levels high enough for the retinal cones to function—*photopic* vision—the greatest sensitivity is at a wavelength of approximately 555 nm, which is usually identified as green yellow (see Figure 12–8). Since there is some variation among persons with normal vision, the luminosity function curve, or the American Standard Observer, is based on the average of a number of observers with normal color vision. With low light levels where only the rods function—*scotopic* vision—peak sensitivity shifts to a wavelength of about 507 nm. This change is known as the Purkinje shift, and can cause two colors, such as blue and red, that match in lightness when viewed in bright light to appear different in lightness when viewed in dim light, with the blue appearing lighter.

It is convenient to think of the limits of the visual system's response to electromagnetic radiation as being 400 nm and 700 nm. Although responses beyond these values are somewhat limited, 380 nm and 770 nm are more accurate limits, and responses have been detected as low as 300 nm and as high as 1,050 nm. The cutoff on the low end tends to increase with age as the transparent material in the eye becomes more yellow and absorbs more ultraviolet and blue radiation.

The human eye is a good measuring instrument only when making comparisons.

A Wratten 90 filter is used for visualization by Zone System photographers.

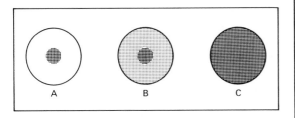

Figure 12–7 Field of view in a visual densitometer. (A) Dark small circle is the unknown density of the sample. The larger circle is the matching field set at 0 density. (B) The density of the matching field is increased but is insufficient to match the density of the sample. (C) More density is added to the matching field and it now matches the density of the unknown sample. The known density of the matching field now becomes the density of the sample.

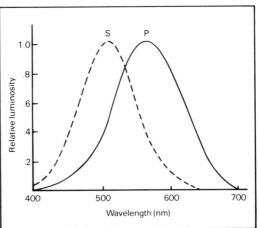

Figure 12–8 Photopic (P) and scotopic (S) response curves for the human eye.

Color saturation is the extent to which a color differs from gray.

COLOR SATURATION

Saturation is the third dimension of color. It is defined as the extent to which a color departs from neutral. Thus grays have no saturation, and spectral colors have high saturation. The saturation of pure blue light can be reduced by adding either white light or yellow (the complementary color of blue) light, and the saturation of a blue dye or other colorant can be reduced by adding a gray or yellow colorant.

It is difficult for a person to judge the saturation of a color seen in isolation except in general terms such as low, moderate, and high saturation. In side-by-side comparisons, where the eye functions as a null instrument, it is easy to detect small differences in saturation provided the two samples match in the other two attributes—hue and brightness/lightness. The smallest difference that can be detected in saturation, or any attribute, is called a just-noticeable difference (JND).

Whereas there are upper and lower limits to the ability of the eye to detect changes in brightness, and limits at both ends of the visible spectrum with respect to seeing different wavelengths of electromagnetic radiation or hues, there are no limitations of the visual system at the lower and higher ends of the range of saturations of stimulus colors. Indeed, it has been a problem over the years to obtain primary-color dyes, inks, and other colorants with sufficient saturation to reproduce subject colors satisfactorily in photographic, photomechanical, and other images. Color masks in negative color films and black ink images in four-color photomechanical reproductions are compensations for the limited saturation of the dyes and the inks. The most saturated color of a specific hue and lightness that can be reproduced with a given system is called a *gamut color. Gamut* is also used to identify the full range of colors that can be produced with different combinations of the primary colors with a given system.

The perceived saturation of a given color sample can vary depending upon a number of factors. A blue sample will appear more saturated when viewed in front of a yellow background than in front of a gray background, and it will appear less saturated when viewed in front of a blue background of higher saturation. Whereas large juxtaposed areas of

Under the proper conditions, hues such as blue and yellow can be seen when looking at a black-and-white image that is illuminated with white light.

yellow and blue enhance one another's perceived saturation—an effect known as *simultaneous contrast*—the opposite effect can result when small areas of the same two colors are closely spaced in a picture or design. In this configuration, the colors tend to neutralize each other, an effect known as *assimilation*. Viewing large areas of complementary colors, such as blue and yellow, in sequence rather than simultaneously will also increase the perceived saturations, an effect known as *successive contrast*. Prolonged viewing of a color sample will decrease the retina's sensitivity to that color and cause it to appear less saturated, an effect known as *chromatic adaptation*. Shifting the gaze to a neutral surface will tend to produce an afterimage that is approximately complementary in hue.

Color-temperature variations in viewing lights can produce either an increase or a decrease in the perceived saturation of colors. A decrease in the illumination level can produce a decrease in the appearance of saturation, and in dim light where only the rods function, even saturated colors tend to appear neutral. For the same reason, colors appear most saturated when viewed directly so that the images fall on the fovea of the retina, where the concentration of cones is highest.

It is also possible to induce the perception of hues with low to high saturation in neutral subjects. The color afterimage produced with prolonged viewing is one example; chromatic sensations can also be produced when black-and-white images are presented to a viewer intermittently at frequencies of about five per second; when certain black-and-white designs are rotated at appropriate speeds; and even with certain stationary black-and-white designs that are viewed continuously, where the involuntary small-scale nystagmus movements of the eyes produce the necessary interactions in the visual response mechanism. These perceptions of colors in neutral stimuli are known as *subjective colors* or *Fechner's colors*.

SHAPE

The word *shape*, as applied to an object, refers to its outline. Silhouettes emphasize shape and eliminate or deemphasize other attributes such as color, form, and texture.

We depend heavily upon the attribute of shape for the identification of many objects, and often that is the only attribute needed for a viewer to be able to recognize the object in a drawing or photograph.

Three-dimensional objects actually have many shapes because each can be viewed from many different angles. The choice of viewpoint selected by a photographer to provide the best shape for an object being photographed is important even when the object is lighted to provide detail, but it becomes critical in a silhouette. Silhouettes of people are commonly recognizable only in a full left or right profile view; thus profile views are normally used for the images of famous people on coins. Figure 12–9 shows an early American print in which George Washington is memorialized with an embedded profile between two trees.

Photographers can control the emphasis on object shapes by controlling the separation or contrast between object and background. The term *figure-ground* is commonly used to refer to the subject of a picture and the surrounding area. Figure-ground is an important concept in Gestalt psychology, where the emphasis is on the perception of the whole rather than an analysis of the parts. Experienced photographers have little difficulty separating an object from the background (figure from ground) in a photograph by means of choice of background, lighting, depth of field, etc. In military camouflage the objective is to conceal the shapes of objects so that they appear to be part of the background and therefore escape detection. In pictorial photography it is often just as important to de-emphasize shape in certain areas of a photograph as it is to emphasize it in other areas, and the principle of camouflage can be used for this purpose.

Although it is sometimes difficult to see the shape of an object clearly when it is not well separated from the background, we are seldom confused as to which is the object and which is the background when we can see both, either with actual objects or in photographs of objects. It is not difficult, however, to make simple drawings in which a given area can be seen alternately as figure and as ground. In a famous Rubin ambiguous picture, for example, the center area can be

Figure and ground can be seen to switch places in certain so-called "ambiguous" images.

Figure 12–9 The profile of George Washington, even though camouflaged, is easily recognized in this picture by Henry Inman. (Courtesy of the Metropolitan Museum of Art, New York City, New York.)

Figure 12–10 Rubin's figure, an ambiguous picture where the center area can be seen either as a vase or as a background for two profiles.

seen either as a vase (figure) or as a background for two profiles (see Figure 12–10).

It is usually unnecessary to see the entire outline shape of a familiar object in order to be able to identify it and to visualize its entire shape. Most people, for example, perceive the moon as being round, not only when there is a full moon but also when there is a half-moon or quarter-moon and the shadow side cannot be separated from the background (see Figure 12–11). One can conduct a simple experiment by looking at objects outdoors through Venetian blinds, starting with the slats in the full open position and then gradually closing them to determine how small the openings can become before encountering difficulty in identifying the objects. A small number of dots arranged in a circular pattern can easily be seen as representing a circle. In Gestalt psychology this effect is known as the principle of *closure*, where the viewer mentally fills in the spaces between the picture elements. A distinction should be made, however, between closure, and *fusion*, where the optical-retinal system in the eye cannot resolve the small discrete elements as in a halftone reproduction or a photographic image at a small to moderate magnification.

Reading provides an excellent example of how the mind fills in spaces between fixation points. The area of sharpest vision represented by the fovea of the retina is very small, so that when a person fixates one letter in a line of type, only a few letters on each side can be seen clearly. The reason it is possible to read rapidly with only two or three fixations per line is that the reader recognizes groups of letters as familiar words without examining each letter, and can understand the meaning of a sentence without examining each word. Printed material that contains unfamiliar words and a high concentration of factual information requires more fixations per line. Eye-motion studies have provided valuable information concerning reading, and how we look at photographs and other pictures. Viewers of pictures rarely scan them as thoroughly as one would a printed page, but rather fixate a few points in the picture and let the mind fill in the shapes and details between the fixation points.

The accuracy with which we perceive shapes is important in the study of visual perception, but it is not considered as critical in our daily lives as normal color vision and good acuity—which are usually tested before

Figure 12–11 Viewers have little difficulty visualizing Saturn as being a complete sphere, due to the principle of closure, even though there is no detail on the shadow side here. (This photograph was taken by Voyager 1 from a distance of 3.3 million miles on November 16, 1980.) (Courtesy of NASA.)

one can obtain a driver's license or be permitted to perform certain occupational tasks. It has been demonstrated that it is easy to deceive a viewer about shape under certain conditions; for example, a straight line can be made to appear curved, as shown in Figure 12–12. Under normal conditions, however, we are best able to detect changes in images having simple geometrical shapes such as straight lines, squares, circles, and triangles (see Figure 12–13). We are also better at making comparisons with superimposed or side-by-side images than with images that are separated in time or space, where memory becomes involved.

The perception of shapes is complicated by the fact that image shape changes with the angle and distance of the object relative to the eye or the camera lens. Parallel subject lines are imaged as converging lines except when viewed or photographed perpendicu-

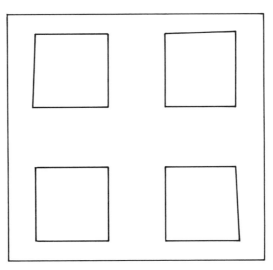

Figure 12–13 Which is the perfect square?

Shape constancy causes tilted circles to be perceived as tilted circles rather than as ellipses.

larly, and tilted circles are imaged as ellipses. Through experience we have learned that the parallel lines and circles do not change shape with a change in viewing angle, so we mentally compensate for linear perspective effects. *Shape constancy* refers to this stability of the perceived shape of objects as the viewing or camera angle changes.

Shape generalization is the tendency to perceive an irregular shape as a simpler shape—obtuse and acute angles seen in perspective may be perceived as right angles, and an ellipse seen in perspective may be perceived as a circle (see Figures 12–14 and 12–15). Memories of perceived shapes can also change with time. The simplification of irregular shapes due to memory is called *leveling*; and the exaggeration of a distinctive feature, such as a small gap in an otherwise continuous line, is called *sharpening*.

DEPTH

The perception of depth is important to photographers in two different contexts: how depth is perceived when viewing the three-dimensional world, and how depth is perceived when viewing photographs and other two-dimensional representations of the real world. Binocular vision, whereby objects are viewed from slightly different positions with the left and right eyes, is commonly given much of the credit for the perception of depth

Three-dimensional scenes do not suddenly appear flat when one eye is closed.

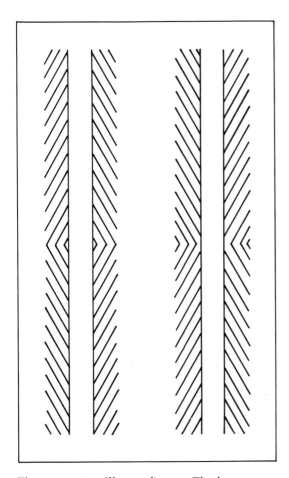

Figure 12–12 Illusory figures. The long straight lines tend to appear curved due to the influence of the diagonal lines.

Figure 12–14   Due to the principle of shape generalization, the ellipses in this image are generally perceived as perspective views of circles.

Figure 12–15   When the sheet of film in the top photograph is viewed obliquely, it is perceived as being rectangular, due to shape generalization, even though the perpendicular view in the bottom photograph reveals that the film has only one 90° corner.

Photographers tend to use their dominant eye when looking through a viewfinder.

in everyday life. The fact that a three-dimensional scene does not suddenly appear two-dimensional when we close one eye is evidence that there are other depth cues besides those provided by binocular vision.

When an object in a three-dimensional scene is fixated with both eyes, the images formed in the two eyes are for the most part identical. To the extent that they are identical, the mind can fuse them into a single image. The inability to fuse the two images in certain areas, because of differences in viewpoint of the two eyes, is referred to as *disparity,* and the disparity provides the mind with important depth information (see Figure 12–16).

Disparity can be demonstrated very easily by looking at a fairly close object, then covering the left and right eyes alternately and noting how objects in the background seem

to change position. If the background jumps when the right eye is covered but not when the left eye is covered, the indication is that the right eye is the dominant eye. Photographers usually feel more comfortable when they use the dominant eye for operations that can be done with only one eye, such as using a camera viewfinder or a focusing magnifier.

Disparity becomes so slight with objects at distances greater than approximately 450 feet that binocular vision contributes little to depth perception. Since the pupils in the eyes are separated, on the average, by 2.5 inches, at a distance of 450 feet the ratio of distance to separation is approximately 2,000:1. Conversely, at very close distances the disparity increases, and it is even possible to see opposite sides of one's hand when it is placed in a thumbing-the-nose position. When completely different images are presented to the two eyes with an optical device, so that no fusion is possible, the mind tends to reject

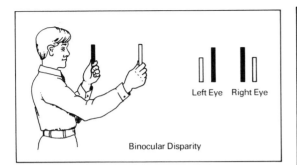

Figure 12–16 Binocular disparity.

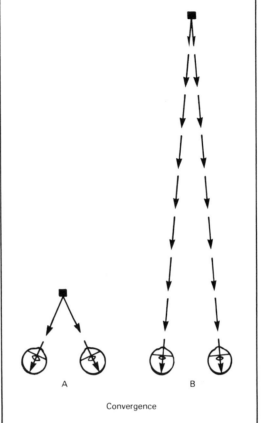

Figure 12–17 Convergence.

one image or the other, sometimes both on an alternating basis.

There is no disparity when one looks at a photograph or other two-dimensional image with both eyes. Thus the perception of depth in photographs must be due to cues other than binocular vision. On the other hand, the absence of disparity in two-dimensional pictures may remove a minor source of tension that makes some realistic pictures more satisfying to look at than the original scenes they represent.

Stereopsis, the perception of depth due to binocular vision, can be created in photographs. By taking two photographs of the same scene from slightly different positions, separated horizontally by about 2.5 inches, and presenting them so that each eye sees only the photograph taken from the corresponding position, the same disparities are produced as when looking at the original scene. Various methods have been used to present the two images to the appropriate eyes. With the stereoscope, the pictures are placed side by side and the lenses make it possible for each eye to view the corresponding image. The need for stereoscope lenses can be eliminated by viewing specially colored superimposed images with glasses that have different-color filters or polarizing filters that are rotated at right angles to each other. The two images can also be presented as alternating narrow strips covered with transparent lenticular embossings that present each set of strips to the appropriate eye. With holography, there is a single photographic image, but different interference patterns are presented to the two eyes so that the reconstructions constitute different images.

Convergence of the two eyes is stronger when viewing a closeup object than a distant object, which provides the mind with additional binocular information about depth (see Figure 12–17). Also, the focus of the eyes changes with object distance, although this is not a binocular function and the viewer is usually aware of the effect only when the object distance approaches the near point—the shortest distance the eye can focus on without strain. The near point varies among individuals, and increases with age from a minimum of approximately 3 inches to a maximum of approximately 84 inches.

Although the eyes have a limited depth of field like photographic lenses, the depth of field of the eyes tends to be relatively large because of the short focal length (17 mm). In addition, since the focus changes automatically as we look at different parts of a three-dimensional scene, and the angle of critical vision is very narrow, depth of field plays a much smaller role in the perception of depth in the real world than it does in photographs.

Other cues that contribute to the perception of depth in viewing both the real world and two-dimensional photographs include: (a) aerial perspective, where distant objects appear lighter and less contrasty than nearby objects; (b) overlap, where closer subjects obscure parts of more distant objects or the background; (c) linear perspective, where parallel subject lines converge in the image with increasing distance, and the images of objects of uniform size decrease with increasing distance; (d) lighting, where, for example, shadows provide a type of yardstick for measuring distances; and (e) color, where warm colors are known as advancing colors because they tend to appear closer than cool, receding colors (see Figure 12–18).

So far, the emphasis in the discussion of depth has been on the perception of distance, but the term *depth* properly includes two other important categories: form and texture. Form refers to the three-dimensional quality of objects, as distinct from the two-dimensional outline shape (see Figure 12–19). Form is the quality that can be determined through the sense of touch with the eyes closed, such as the spherical form of a baseball or the cubical form of a pair of dice. Effective representation of form in two-dimensional photographs depends largely upon the choice of an appropriate viewpoint and the use of light to reveal the different planes and curved surfaces with highlights, shadows, and appropriate gradations of tone.

Texture refers to the small-scale depth characteristics of a type that might be felt with the fingertips, such as the roughness of a wood file or the smoothness of window glass. Effective representation of texture in two-dimensional photographs depends largely upon using an appropriate scale of reproduction, as well as lighting that produces shadows in the recessed areas and highlights in the raised areas. Photographs made through optical and electron microscopes reveal that many surfaces thought of as being smooth, such as writing paper, appear to have a rough texture or even form and distance when magnified sufficiently. Conversely, the mountains on the moon appear to have a finger-touching type of texture when photographed from a distance with a small scale of reproduction.

The perceived size of an object has little relationship to the size of the image on the retina.

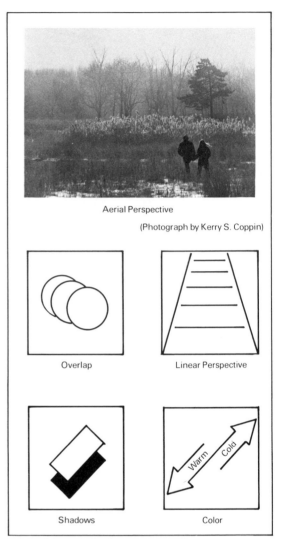

Aerial Perspective

(Photograph by Kerry S. Coppin)

Overlap

Linear Perspective

Shadows

Color

Figure 12–18   Some depth cues.

SIZE

The perceived size of an object has little relationship to the size of the image on the retina (or the size of the image in a photograph of the object). An automobile, for example, is judged to be about the same size when viewed over a wide range of distances—an effect known as size constancy. Experiments have demonstrated that the accuracy of judging the size of an abstract shape, such as a circle, depends greatly upon being able to estimate the distance. As distance cues are systematically eliminated, the accuracy decreases.

When the precise size of an object must be known, a ruler or other measuring instru-

Figure 12–19   Three-dimensional form in the original scene is represented by the different planes of the ceiling, wall, and floor; the bench; and even the protruding bricks that make up the triangular designs in the wall. The pattern of the protruding bricks is reduced in size sufficiently so that the viewer can visualize feeling the roughness of the pattern with the fingertips, a characteristic of texture.

ment is used for a direct side-by-side comparison. As with the perception of other object attributes, the eye is most precise when used as a null instrument in making comparisons between adjacent stimuli. It is sometimes necessary to include a ruler beside an object in a photograph when it is important to be able to determine the size of the object, as in some forensic photographs. In pictorial photographs, it is usually sufficient to include an object of known size with the unfamiliar object, and to provide good distance cues.

It is not difficult to deceive a viewer about the size of objects represented in photographs. Use of a short focal length camera lens tends to make foreground objects appear larger than normal and background objects appear smaller than normal when the photograph is viewed at a comfortable distance; and long focal length camera lenses have the reverse effect. Line drawings can also cause viewers to misjudge the relative length of lines or size of images. The Muller-Lyer arrow illusion and the Ponzo railway lines illusion both contain lines of equal length that are perceived as being unequal (see Figure 12–20). In nature, the moon is commonly perceived as being larger when it is near the horizon than when it is overhead. Various explanations of this illusion have been offered, but the most acceptable one is that the moon is thought of as being farther away when it is near the horizon, as an airplane would be, so that when the retinal image remains the same size as the moon moves toward the horizon from overhead it is perceived as being a larger object farther away. Emmert's law—that the size of an afterimage increases in proportion to the distance to the surface onto which the image is projected—supports this explanation.

The moon is commonly perceived as being larger when it is near the horizon than when it is overhead.

SHARPNESS

As an object of moderate or small size is moved farther away from a viewer and the retinal image decreases in size, a point is

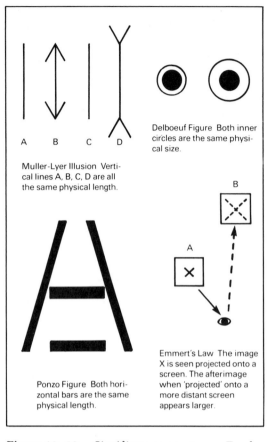

A

B

C

D

Muller-Lyer Illusion Vertical lines A, B, C, D are all the same physical length.

Delboeuf Figure Both inner circles are the same physical size.

B

A

Ponzo Figure Both horizontal bars are the same physical length.

Emmert's Law The image X is seen projected onto a screen. The afterimage when 'projected' onto a more distant screen appears larger.

Figure 12–20   Size/distance constancy. For familiar objects, size and distance are inseparable. For an object to be the same actual size at a farther distance it has to be larger, and therefore it appears to be larger. (The moon illusion is yet another example of this phenomenon.)

A person can detect movement of an object that is 90 degrees off to one side, but it might be difficult for the person to identify the color of the object.

reached where the viewer is no longer able to see the object. Visual acuity is a measure of the ability of a person to see detail. Visual acuity tasks are commonly classified as *detection, localization, recognition,* and *resolution,* with different types of test targets used for each task. A black dot or a black line on a light background can be used to measure detection. It is usual to measure the visual angle subtended by the appropriate dimension of the test target, rather than specify both the target dimension and the distance. A person with normal vision may be able to detect a black line having a width that represents a visual angle of only 1/120 of a minute. Visual acuity is the reciprocal of this angle in minutes, or 120. This value corresponds to a 1/4-inch black line, the thickness of a lead pencil, at a distance of 4/5 of a mile.

Localization is the ability to see where the opening in a Landolt C or the vernier displacement between two straight lines is located. Maximum acuity for the vernier lines is about 30, or about one fourth the acuity for the detection of a black line.

Snellen eye charts containing letters of the alphabet of varying size are commonly used for eye examinations to test vision. Maximum acuity for this visual task is about 2.

Resolution can be tested with a target having parallel black bars separated by spaces equal to the width of the bars, similar to targets used for measuring the resolution of photographic systems or components. The maximum acuity for this task is also about 2.

Visual acuity can be affected by other factors such as the illumination level, where acuity increases steadily with the illumination level over a wide range. Pupil size also affects acuity, with the maximum being obtained at an opening of approximately f/4. At the maximum opening of f/2, aberrations lower the acuity, and at the smallest opening of f/8 diffraction lowers the acuity.

MOTION

Motion refers to a change in position of an object, and the rate of motion is identified as speed. The speed of a moving object may be either too fast to be detected (a speeding bullet) or too slow (the moon, from moonrise to moonset). Slow movement can be detected more easily when the movement takes place in front of stationary objects. One experiment revealed that when a reference grid was removed, the rate of movement of a spot had to be increased 10 times to be detectable. In fact, a stationary spot of light in an otherwise darkened room is commonly perceived as moving—an effect known as the wandering-light phenomenon. Although the speed of a moving object can be measured in units such as miles per hour, viewers in general are not very accurate in estimating the rate of movement. Such ability improves with experience when subjective judgments can be compared to objective measurements; an experienced batter in a baseball game, for example, can detect the difference between an 80 mph pitch and a 90 mph pitch.

The human visual system is designed to detect motion over the entire area of the retina, so that moving objects in the periphery of the field of view, up to 90° to the side, can be detected. If identification of the moving object is important, a semireflexive movement of the eyes enables the viewer to position the image of the object on the fovea of the retina, where visual acuity is highest. Once the object is fixated, the visual system is very good in tracking the movement unless the speed is very high and the direction of movement is erratic. Strangely, it is almost impossible to pan the eyes smoothly except when tracking a moving object. Instead, the eyes tend to move in quick jumps, called saccades. By closing the eyes, one may be able to pan the eyes smoothly by visualizing tracking a moving object. Skill in tracking rapidly moving objects is important for both motion-picture and still photographers, as well as for participants (and spectators) in many sports and other recreational and occupational activities.

With present technology, it is impossible to accurately re-create motion photographically. In a single still photograph it is necessary to rely upon implied motion. A blurred image of a racing car is generally perceived as representing a rapidly moving object (see Figure 12–21). Similarly, a sharp image of a racing car with a blurred background implies motion. Motion can also be implied without blur, such as a sharp image of a skier going down a steep slope or suspended in air. With two or more still photographs viewed side by side, a change in position of an object is perceived as representing movement. A series of such still photographs viewed sequentially in the same position produces a more realistic perception of motion; and if the amount of change in position of the images and the rate of presentation of the still photographs is appropriate, the viewer accepts the images as representing smooth and continuous movement, as with contemporary motion pictures.

This perception of motion—when the image of an object appears in different positions in a sequence of still pictures presented in rapid succession—is known as the *phi phenomenon*. The phi phenomenon can be demonstrated in a darkened room with as few as two flashes of light, provided the separation

Figure 12–21   Blurring the image is one of several methods of representing motion in a still photograph. (Photograph by Robert Mulkern.)

in time and space are appropriate. In sound motion pictures, a projection rate of 24 frames per second produces generally acceptable results, although rapid panning of the motion-picture camera can produce an unrealistic jumpy effect. Early 8 mm motion pictures used a projection rate of 16 frames per second, which was increased to 18 frames per second for super-8 motion pictures.

The phi phenomenon can be demonstrated effectively by illuminating a moving object with a variable-speed stroboscopic light where the flash frequency can be altered from approximately one flash per second to a frequency that produces the perception of a continuous light source (see Figure 12–22). A black circle with a white radial line rotated on a motor will appear to be standing still when the light flashes once per revolution. Slowing down the frequency of the stroboscopic light so that the disk makes a little more than one complete revolution between flashes will make it appear to be rotating in the correct direction. Speeding up the frequency so that the disc does not quite complete a full revolution between flashes will make it appear to be rotating in the reverse direction. This explains why wagon wheels sometimes appear to be turning backward in cowboy movies.

Motion-picture photography offers the advantage over direct vision of being able to

The phi phenomenon accounts for the realistic perception of motion when looking at a series of still photographs in a motion picture.

Figure 12–22   A variable-speed stroboscopic light source was used to record 24 images of a dancer in a single photograph to produce a perception of implied motion. In a sound motion picture 24 still photographs are projected in rapid sequence to produce a perception of realistic motion due to the phi phenomenon. (Photograph by Andrew Davidhazy.)

Wagon wheels can appear to be turning backwards in motion pictures even though the wagon is moving forward.

speed up or slow down motion by altering the rate at which the individual pictures are exposed in the camera. In time-lapse photography, a flower can change from a bud to full bloom in seconds by exposing the film at a rate of about one frame per hour. Conversely, motion that is too rapid to be seen with the eye can be slowed down by using a high-speed motion-picture camera that can expose film at rates up to approximately 10,000 frames per second. If a film exposed at 10,000 frames per second is projected at a rate of 24 frames per second, the rate of movement is reduced to approximately 1/400 the original speed.

REVIEW QUESTIONS

1. Persistent images last . . . (p. 280)
   A. approximately 1/100 second
   B. approximately 1/4 second
   C. approximately 5 seconds
   D. as long as the viewers concentrate on the images
2. With respect to the perception of realism when viewing motion pictures, persistent images . . . (p. 280)
   A. contribute to the perception
   B. interfere with the perception
   C. have no effect on the perception
3. If a slight amount of flicker is apparent when viewing a motion picture, reduc-

ing the projector-to-screen distance to obtain a smaller and brighter image will . . . (p. 280)
   A. decrease the flicker
   B. increase the flicker
   C. have no effect on the flicker
4. The "spots" one sees after looking directly at a flash tube when it is fired are properly identified as . . . (p. 280)
   A. persistent images
   B. afterimages
   C. aftereffects
   D. eidetic images
   E. memories
5. To recall a visual image from a motion picture viewed an hour earlier would be identified as . . . (p. 282)
   A. an afterimage
   B. a short-term memory image
   C. a long-term memory image
   D. a rerun
6. The difference between the red and the blue in the U.S. flag is primarily in . . . (p. 284)
   A. saturation
   B. hue
   C. value
   D. chroma
   E. lightness
7. Persons having normal color vision are identified as normal . . . (p. 284)
   A. monochromats
   B. dichromats
   C. trichromats
   D. stereomats

8. The percent of Caucasian males who have defective color vision is approximately . . . *(p. 285)*
   A. 0.4
   B. 2
   C. 4
   D. 8
   E. 16

9. The maximum response range of the visual system to luminance variations is approximately . . . *(p. 286)*
   A. 100 to 1
   B. 10,000 to 1
   C. 1 million to 1
   D. 10 trillion to 1

10. The perceptual phenomenon of seeing yellow on a television screen in an area of red and green pixels is identified as . . . *(p. 288)*
   A. simultaneous contrast
   B. successive contrast
   C. local adaptation
   D. assimilation

11. The fact that the image of a tilted circle in a photograph is generally perceived as a tilted circle rather than its actual elliptical shape is attributed to . . . *(p. 291)*
   A. shape constancy
   B. shape generalization
   C. sharpening
   D. leveling
   E. closure

12. Binocular vision contributes most strongly to the perception of depth when viewing . . . *(p. 292)*
   A. photographs closeup
   B. photographs at a distance
   C. three-dimensional objects closeup
   D. three-dimensional objects at a distance

13. The moon tends to appear larger when it is near the horizon than when it is overhead due to . . . *(p. 295)*
   A. differences in refraction of light by the atmosphere
   B. differences in perceived distances of the moon
   C. size constancy
   D. size adaptation

14. The perception of motion when viewing a series of still photographs in a motion picture in rapid succession is identified as . . . *(p. 297)*
   A. a persistent image
   B. the gamma effect
   C. fusion
   D. motion adaptation
   E. the phi phenomenon

13 Filters

Dimitri Papadimitriou. *Mythology*. Copyright © 1988 by Dimitri Papadimitriou.

FILTERS

Light that falls on a surface can be *reflected, absorbed,* or *transmitted.* As a mnemonic device, these terms are commonly abbreviated as RAT (see Figure 13–1). The perception of the color of objects depends upon the spectral quality of the reflected or transmitted light. When illuminated with white light, an opaque white object appears white because the surface reflects a high proportion of the incident light, and a black object appears black because it reflects a small proportion of the incident light, *nonselectively.* In contrast, a red object appears red when illuminated with white light because it *selectively* absorbs most of the blue and green parts of the white light and reflects most of the red part. Although it will be necessary to consider what happens to the incident light wavelength by wavelength, for now it is sufficient to think in terms of the additive primary colors of light—red, green, and blue—and the additive secondary colors (or the subtractive primary colors)—cyan, magenta, and yellow.

Photographic filters function by removing part of the incident radiation by absorption or reflection, so that the transmitted radiation is of the desired spectral quality. Filters, of course, cannot add anything to the incident radiation—a red filter cannot transmit red light if there is no red light in the radiation falling on the filter. There are various ways of classifying filters, but with respect to the effect they have on the incident radiation, they remove some of the radiation (a) selectively by wavelength, as with color filters; (b) nonselectively by wavelength, as with neutral-density filters; or (c) selectively by angle of polarization, as with polarizing filters.

Filters work by removing some of the entering light.

A color filter transmits light of its own color and removes other colors of light.

COLOR FILTERS

Color filters are commonly identified by the perceived color of the transmitted light, viewed by white incident light (see Figure 13–2). Thus a red filter transmits red light and absorbs green and blue light. Color filters differ with respect to (a) which part of the visible spectrum is transmitted freely (identified by color hue or peak transmittance wavelength); (b) the width of the region transmitted and the sharpness of the cutoff (identified by a general description such as narrow, sharp-cutting passband or specific passband data); and (c) the degree of absorption of the unwanted colors (identified by general terms such as light yellow and dark yellow, or specific transmittance and density data).

Contrast filters are deep or saturated color filters that almost completely absorb the unwanted colors of light, and are commonly used in black-and-white photography to lighten or darken selected subject colors. Photographers need to be able to predict the effect that such filters will have on subject colors. One method is to look at the subject through the filter and note which colors are lightened and which are darkened. Accuracy of this process depends upon the film and the eye responding similarly to subject colors. Although panchromatic film and the eye do not respond identically, the responses are sufficiently close to make this a valid procedure. In Figure 13–3, steps of a gray scale were selected as the closest lightness matches with the adjacent subject color (identified with markers) without and with a filter be-

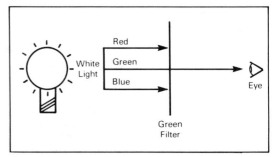

Figure 13–2 A green filter appears green to the eye because the filter absorbs the red and blue components of white light and transmits the green.

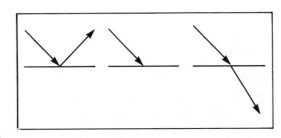

Figure 13–1 Light falling on a surface can be reflected (A), absorbed (B), or transmitted (C).

Figure 13–3   A white marker was placed on the gray-scale steps that most closely matched the lightness of the green background before the photographs were made with no filter (middle), a green filter (left) and a red filter (right). The closest visual matches were also the closest photographic matches.

fore making the corresponding photographs on panchromatic film.

The Maxwell triangle (containing the three additive primary colors—red, green, and blue; and the three additive secondary colors—cyan, magenta, and yellow) can be used to predict the effect of contrast filters on black-and-white photographs without looking through the filters (see Figure 13–4). The

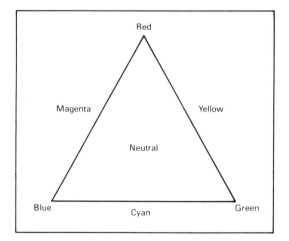

Figure 13–4   The Maxwell triangle can be used to predict the effect of contrast color filters on black-and-white photographs.

general rule is that a filter will lighten subject colors that are the same color as the filter or are adjacent to that color in the triangle. Thus a red filter will lighten red, magenta, and yellow subject colors in the print, and will darken the opposite color, cyan, and its adjacent colors, blue and green. Note that to lighten yellow and darken red simultaneously, it would be necessary to use a green filter (which is adjacent to yellow but is not adjacent to red in the triangle).

Since filters absorb light, the images of neutral-colored objects would be thinner in negatives if the same camera exposure settings were used with a filter as without (see Figure 13–5). The filter factor is a multiplying number that specifies the change in exposure necessary to obtain the same density in a neutral subject area with the filter as without. Since color filters commonly change the slope of the characteristic curve and the contrast of images, a filter factor cannot produce the same density for all steps in a gray scale exposed through a filter unless the development of the film is adjusted. It has been common practice when making color-separation negatives through red, green, and blue contrast filters to adjust developing times to

Filter factors are different for daylight and tungsten illumination.

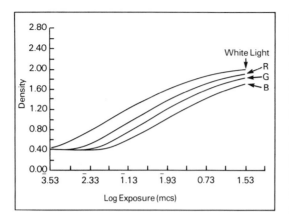

A filter factor of 16 corresponds to a 4-stop change in exposure.

Figure 13–5 Characteristic curves for film exposed in a sensitometer with white light and with red, green, and blue filters. Since no adjustment was made in the exposure, the filter negatives are thinner than the no-filter negative. Filter factors for this light source can be determined by finding the antilogarithm of the horizontal displacement of each filter curve from the no-filter curve at a specified density. Any effect the filter may have on image contrast of neutral subject tones can be determined by calculating the contrast index.

match curve slopes. This practice is less common when using contrast filters for pictorial black-and-white photography.

The dependence of filter factors on the degree of development is one reason manufacturers commonly indicate that published filter factors should be modified by the photographer if they do not produce satisfactory results. Two other aspects of the photographic process that must be taken into account in determining filter factors are the color quality of the illumination and the color sensitivity of the photographic material. Red filters, for example, have larger factors with daylight illumination than with tungsten illumination because they absorb a larger proportion of the bluer daylight illumination. Conversely, blue filters have smaller factors with daylight. Although red filters are not recommended for use with orthochromatic films, which have very low sensitivity to red light, an attempt to use this combination would require a very large filter factor compared to that required for the same filter with panchromatic film.

If the filter factor is applied to the exposure time, the exposure time without the filter is simply multiplied by the filter factor. For example, an initial exposure time of 1/60 sec-

ond modified for a filter having a factor of 2 becomes 1/60 x 2, or 1/30 second. If the factor is applied to the diaphragm opening, it is necessary to remember that opening the diaphragm doubles the exposure for each whole stop, so that a lens would have to be opened up by four stops for a filter having a factor of 16 (2-4-8-16). For factors that do not correspond to whole stops, the equivalent number of stops can be computed by dividing the logarithm of the filter factor by 0.3 (the logarithm of 2). Thus, a filter factor of 12 corresponds to 1.08/0.3, or 3.6 stops.

An advantage claimed by manufacturers of some cameras having behind-the-lens exposure meters is that the meter automatically compensates for the decrease in the amount of light that reaches the film when a filter is added. These claims are valid to the extent that the spectral response of the meter corresponds to the spectral sensitivity of the film being used. Some difficulty has been experienced in producing meters that have high overall sensitivity as well as the desired spectral response characteristics. A meter with high red sensitivity, for example, will give a false high reading, leading to underexposure when a red filter is used, and a false low reading, leading to overexposure when a blue filter is used.

A simple experiment can be conducted to determine the accuracy of a camera meter with filters. A neutral test card is metered with and without the filters, and the indicated changes in exposure are compared to the published filter factors. Or a gray scale is photographed with and without the filters as the camera meter indicates, and the densities of the resulting images are compared (see Figure 13–6).

Having noted that the objective of filter factors is to record neutral colors with the same density when a filter is used as without the filter, it is easier to explain why a filter lightens its own color and adjacent colors in the Maxwell triangle and darkens the opposite color and its adjacent colors (see Figure 13–7). If we think of white light as being composed of equal proportions of red, green, and blue light, a filter will lighten a subject color (on the print) when the filter transmits a larger proportion of that color of light than of white light from a gray scale or a neutral color object. In simple terms, a red contrast filter can

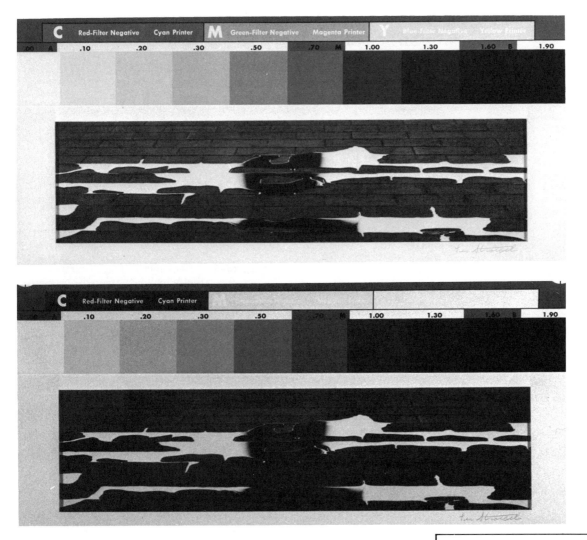

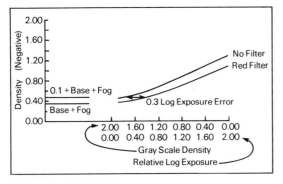

Figure 13–6 Copy photographs of a gray scale and a black-and-white print without a filter (top) and with a red filter (middle). Camera exposures were determined with behind-the-lens meter readings and the two negatives were printed exactly the same way. The difference in print density represents the meter error in reading through the red filter, due to high red sensitivity of the meter cell. (bottom) Characteristic curves based on the negative gray-scale densities. The two curves would be superimposed if the spectral response of the exposure meter matched the spectral sensitivity of the film. The 0.3 log exposure displacement represents an exposure error of one stop.

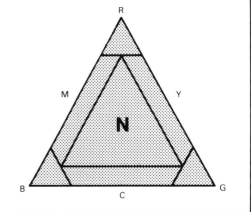

Figure 13–7 A triangular color chart containing the six basic colors—red, green, blue, cyan, magenta, and yellow—is used here to illustrate the effect of filters in lightening and darkening subject colors in black-and-white photographs. The lightness of the colors is such that without a filter they are all recorded with the same density as the neutral gray in the center.

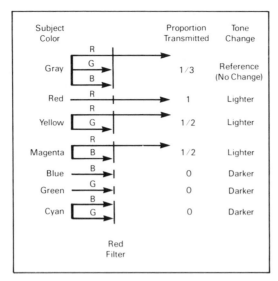

Figure 13–8   A red contrast filter is represented as transmitting one third of white light. The filter will lighten subject colors (in the print) when it transmits more than one-third of the light from the subject, and darken subject colors when it transmits less than one-third of the light. The visual effect is shown in Figure 13–9.

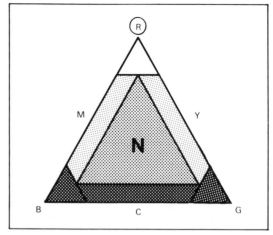

Figure 13–9   Since a red filter absorbs none of the red light, the red subject area will be dense in the negative and light in the print. The adjacent colors, magenta and yellow, will be lightened because the red filter absorbs less light from these areas (one-half) than from the neutral gray (two-thirds). The three remaining colors are darkened because the red filter absorbs essentially all cyan, blue, and green light.

be thought of as transmitting one-third of the white (R + G + B) light, all of the light from a red object, and one-half of the light from the adjacent yellow (R + G) and magenta (R + B) objects (see Figure 13–8). The red filter, therefore, will lighten its own color and the adjacent yellow and magenta colors (see Figure 13–9).

Similarly, cyan, magenta, and yellow contrast filters can be thought of as transmitting two-thirds of the white light but all of their own color and *all* of the adjacent colors (see Figure 13–10). A yellow contrast filter, for example, will transmit its own color and the adjacent colors red and green freely (lightening these colors on the print), will absorb all of the complementary color blue (darkening it considerably), and will transmit one-half (less than two thirds) of the colors adjacent to blue—cyan and magenta (darkening these colors slightly) (see Figure 13–11).

Even with contrast filters, the lightening and darkening effects in black-and-white photographs often are not as dramatic as photographers would like. The reason for this is that subject colors are seldom limited to such a narrow range of wavelengths that the light reflected from the subject is either com-

"Blue" skylight contains a significant amount of green light.

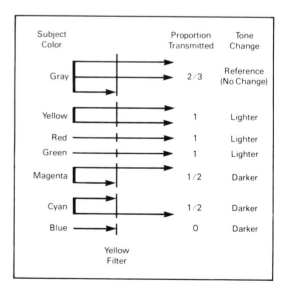

Figure 13–10   A yellow filter is represented as transmitting two-thirds of white light. The filter will lighten subject colors (in the print) when it transmits more than two-thirds of the light from the subject and darken subject colors when it transmits less than two-thirds of the light. The visual effect is shown in Figure 13–11.

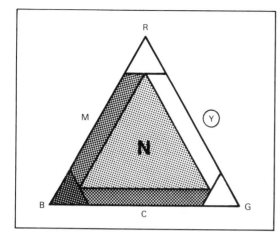

Figure 13–11 A yellow filter will lighten its own color and the adjacent red and green colors because it transmits these colors freely. The blue area will be darkened because the yellow filter absorbs blue light. Magenta and cyan will be darkened some because each of these areas is composed of one-half blue, which is a larger proportion than the one-third blue in the neutral gray.

ULTRAVIOLET AND INFRARED FILTERS

The silver halides in photographic films are inherently sensitive to ultraviolet radiation and they can be made sensitive to infrared radiation through the use of sensitizing dyes. When the objective is to make a photograph solely with ultraviolet or infrared radiation, a filter is needed that transmits the desired radiation but does not transmit unwanted radiation such as white light. Some UV and IR filters are therefore visually opaque, or nearly so. IR filters are usually used in front of the camera lens so that the film is exposed solely with IR radiation (see Figures 13–12 and 13–13). If the objective is to make surreptitious photographs at night, however, the filter can be used over the flash or other source of infrared radiation so that little or no light is transmitted.

In distant outdoor scenes, ultraviolet radiation often has a deleterious effect on photographs because short-wavelength radiation is scattered much more by the atmosphere than longer-wavelength radiation, creating the

Aerial haze results mostly from the scattering of light and ultraviolet radiation.

pletely absorbed or completely transmitted by contrast filters. Less saturated color filters are produced, however, for situations where more subtle tone changes are desired. For example, a red contrast filter will considerably darken blue sky on a clear day in a black-and-white photograph. A yellow contrast filter will not have as great an effect because the "blue" sky actually contains some green, which is not absorbed by the yellow filter. More subtle darkening can be obtained by switching from a contrast or deep yellow filter to a medium yellow or a light yellow filter.

Panchromatic films used without a filter tend to record blue sky as a little lighter than it appears to the eye due to the ultraviolet radiation in blue sky light, to which films are sensitive, and the slightly high blue sensitivity of panchromatic films with daylight illumination. Panchromatic films, therefore, produce a more realistic reproduction of the lightness of subject colors when used with a light yellow filter than when used without a filter. The recommended yellow filter is classified as a *correction* filter. With tungsten illumination, panchromatic films tend to record reds and blues too light, so a light green filter is used as a correction filter.

Figure 13–12 Both photographs were made on black-and-white infrared film. The photograph on the right was exposed through a red filter, which transmits red light and infrared radiation but absorbs blue light and green light, to which the film is also sensitive. The photograph on the left was made without a filter.

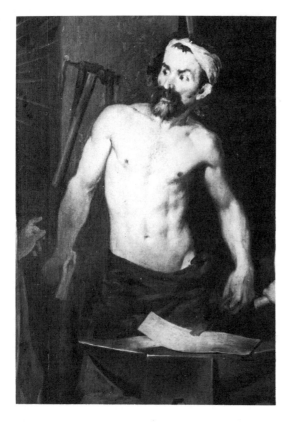

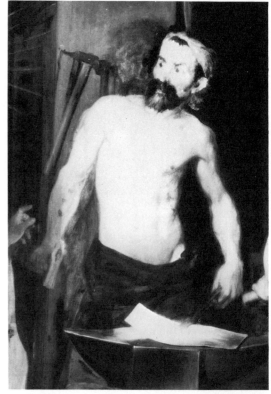

Figure 13-13   (Top left) Photograph of an oil painting made with panchromatic film. (Top right) Photograph made with black-and-white infrared film and an 87C filter over the camera lens, reveals a somewhat different original painting. The sketch on the right shows the visible image with dotted lines and the differences revealed by the infrared film with solid lines. (Photographs by Andrew Davidhazy.)

appearance of haze, which obscures detail. This Rayleigh scattering, as it is called, is inversely proportional to the fourth power of the wavelength of the radiation. To reduce the appearance of haze, a filter should be used that absorbs UV radiation, as well as blue and green light. Thus a red filter is effective in reducing the appearance of haze with panchromatic films, and the effect is even more dramatic when the photograph is made with IR film and IR radiation.

Ultraviolet radiation causes certain substances to fluoresce, whereby invisible UV radiation is converted to longer-wavelength visible radiation, revealing useful information about the material (see Figures 13–14 and 13–15). The fluorescent effect may be obscured, however, if the object is also being illuminated with white light from another source. It is common practice, therefore, in fluorescence photography to use two UV filters, one that transmits only UV radiation (known as an exciter filter) over the radiation source, and another that transmits light but absorbs UV radiation (known as a barrier filter) over the camera lens so that the film records only the visible fluorescence.

FILTERS FOR COLOR PHOTOGRAPHY

Red, green, and blue contrast filters can be used for color photography as well as for black-and-white photography, for example, to make separation negatives from an original scene with a camera or from a color transparency with an enlarger, and to do tricolor printing with color printing paper. More widely used, however, are less-saturated filters to alter the color temperature of the illumination and to alter the color balance of transparencies and prints. If one wants to use a tungsten-type color film (which is designed to be used with illumination having a color temperature of 3200 K) with daylight illumination having a color temperature of 5500 K, it is necessary to use an orange filter that will lower the color temperature of the light from 5500 K to 3200 K. Conversely, to use a daylight-type color film indoors with studio lights having a color temperature of 3200 K, it is necessary to use a bluish filter that will raise the color temperature of the light from

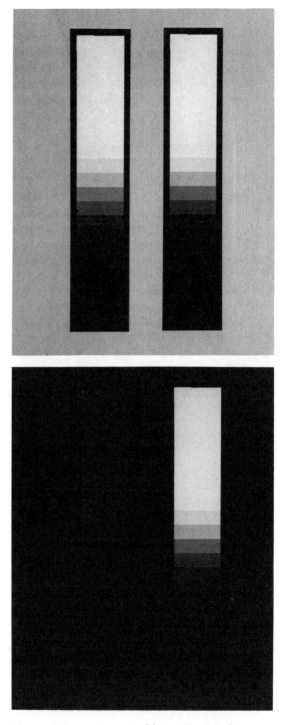

Fluorescent materials convert invisible UV radiation to longer wavelength light.

Figure 13–14   A step tablet was printed on a photographic paper containing a fluorescent brightener and on a non-brightener paper. Under tungsten illumination, which contains little ultraviolet radiation, the two prints appear similar (top). With ultraviolet radiation there is a dramatic difference in the appearance of the two prints (bottom).

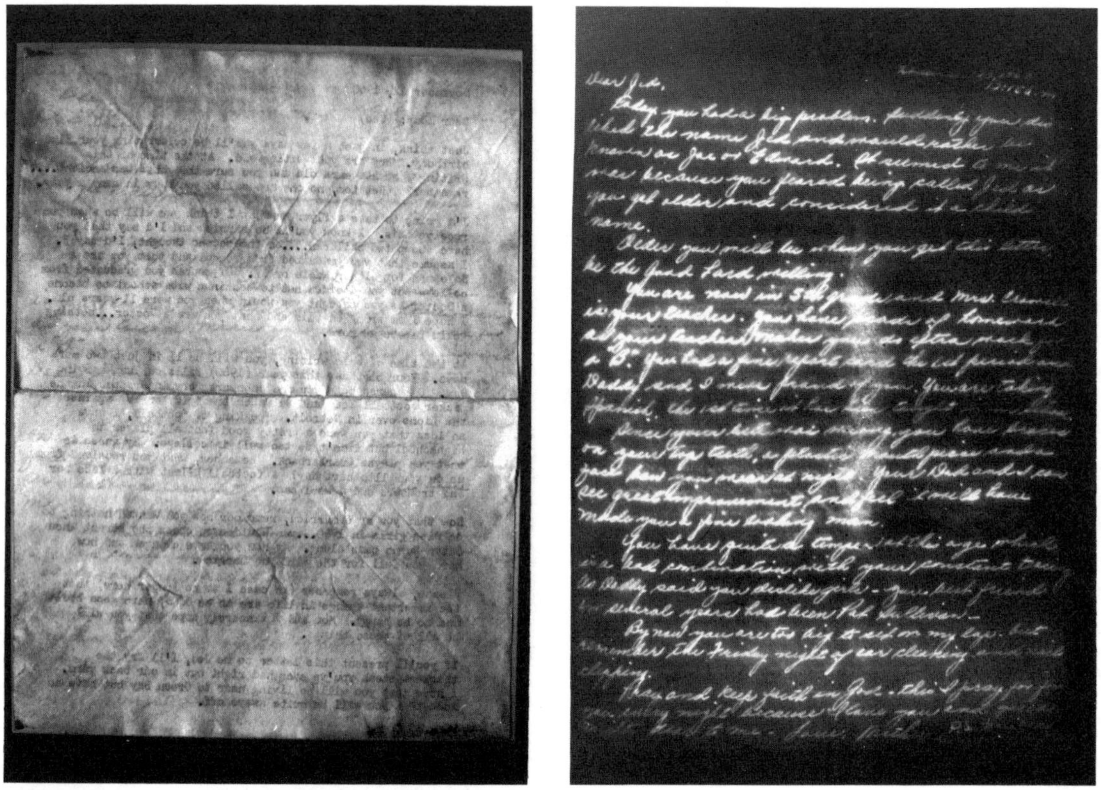

Figure 13–15   Ultraviolet and infrared radiation can sometimes reveal information that cannot be seen or photographed with white light. The water-damaged letter on the left was not legible under normal viewing conditions. The faded ink, however, fluoresced in the infrared region when illuminated with white light, producing a legible copy when photographed on infrared film with a filter. (Photographs by Andrew Davidhazy.)

3200 K to 5500 K. Such filters are identified as *conversion* or *light-balancing* filters.

Calibration of conversion and light-balancing filters presents a problem because the orange filter noted above that lowers the color temperature from 5500 K to 3200 K, a decrease of 2300 K, will not lower the color temperature by the same amount if used with illumination having a different color temperature. For this reason, such filters are usually calibrated in terms of *mired shift value*. As noted in Chapter 1, the mired value is found by dividing a constant—1,000,000—by the color temperature. The mired shift value is the difference between the two mired values. Thus the mired value for 5500 K is 182 and the mired value for 3200 K is 312; therefore an orange filter with a *positive* mired value of 130 (312 − 182) is required to use

tungsten-type film with daylight illumination (see Figure 13–16). This filter will produce the same mired shift with other light sources having either higher or lower color temperatures.

Color-compensating filters are designed to absorb any of the six additive and subtractive primary colors—red, green, blue, cyan, magenta, and yellow. Thus a red color-compensating (CC) filter will absorb green and blue light, whereas the complementary color cyan filter will absorb only red light. Color-compensating filters are used on cameras and enlargers to control the color balance of color photographs, for the purpose of compensating for variability in any part of the process including the photographic material, the illumination, and the processing; or to create mood or other special effects. CC filters are

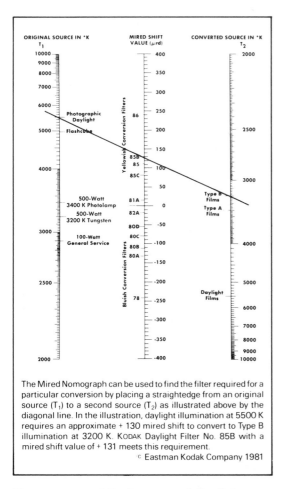

ORIGINAL SOURCE IN °K
T₁

MIRED SHIFT
VALUE (μrd)

CONVERTED SOURCE IN °K
T₂

The Mired Nomograph can be used to find the filter required for a particular conversion by placing a straightedge from an original source (T₁) to a second source (T₂) as illustrated above by the diagonal line. In the illustration, daylight illumination at 5500 K requires an approximate + 130 mired shift to convert to Type B illumination at 3200 K. KODAK Daylight Filter No. 85B with a mired shift value of + 131 meets this requirement.
ⓒ Eastman Kodak Company 1981

Figure 13–16   Mired nomograph for light source conversion. The mired nomograph can be used to find the filter required for a particular conversion by placing a straightedge from an original source (T₁) to a second source (T₂) as illustrated above by the diagonal line. In the illustration, daylight illumination at 5500 K requires an approximate $+130$ mired shift to convert to Type B illumination at 3200 K. Kodak Daylight Filter No. 85B with a mired shift value of $+131$ meets this requirement. (© Eastman Kodak Company 1981.)

produced in a range of saturations in each of the six hues, with the saturations calibrated in terms of peak density. Thus a CC30M filter is a magenta color-compensating filter with a density of .30 (the decimal is omitted in the designation). Filters are available in densities ranging from .025 to .50.

The density of CC filters is based on the maximum absorption, not the proportion of white light that is absorbed. For this reason, exposure corrections are not derived directly from the density but instead are obtained by trial and error, from the manufacturer's data sheet, or by measuring the transmitted light with a meter having appropriate response characteristics. A manufacturer's data sheet, for example, suggests two-thirds-stop exposure increase for CC30 filters in all hues except yellow, which is one-third stop.

Even though contrast filters are never sandwiched, because theoretically no light would be transmitted through most combinations, such as red and green, CC filters can be combined to obtain almost unlimited control over the color quality of the image-forming light. When two or more CC filters of the *same* hue are combined, the densities are additive. The combined density of a CC20M filter and a CC30M filter is the same as the density of one CC50M filter, although it is necessary to increase the exposure by approximately 10% for each additional filter to compensate for light lost due to surface reflection.

The situation is different when filters of different hues are combined. Sandwiching any two of the three subtractive primary hues—cyan, magenta, and yellow—produces the same effect as a single filter of the resulting additive primary hue and the same density. Thus, CC20M plus CC20Y filters are equivalent to one CC20R filter. Combining equal densities of all three of the subtractive primary hues results in neutral density. Thus, CC20M plus CC20Y plus CC20C is equivalent to a neutral-density filter with a density of 0.20. Combining complementary hues produces the same result; for example, CC20Y plus CC20B is equivalent to a neutral-density filter with a density of 0.20. When three or more CC filters are combined, as in color printing, it is usually better to eliminate any neutral-density combinations unless the resulting shorter exposure time would be inconveniently short.

It should be noted that combining two of the *additive* primary hues is not exactly equivalent to using one filter of the corresponding subtractive primary hue. For example, combining equal densities of red and green results in a greater absorption of blue, as with a single yellow filter, but there is also a neutral-density effect because the red filter absorbs blue and green, and the green filter absorbs blue and red.

CC20M + CC20M = CC40M

CC20M + CC20Y = CC20R

SPECTROPHOTOMETRIC ABSORPTION CURVES

Spectrophotometers measure the transmittance of filters wavelength by wavelength throughout the spectrum.

The general type of filter data discussed above is usually sufficient to meet the needs of most photographers. However, when more specific information is required for technical and scientific applications, spectrophotometric curves that show the absorption characteristics of filters throughout the spectrum are useful. In the spectrophotometer, white light is dispersed to form a spectrum. A narrow slit permits the measurement of the transmittance of a filter to a narrow band of wavelengths as the spectrum is scanned from the short-wavelength ultraviolet radiation through the visible region to the long-wavelength infrared radiation. Since the transmittance is based on a comparison of the transmitted beam to an unfiltered reference beam, the specific response characteristics of the sensor are unimportant as long as it responds to the full range of wavelengths of interest.

Spectrophotometric curves typically represent wavelengths on the horizontal axis and transmittance and density values on the vertical axis, with the baseline representing 100% transmittance, or a density of 0. As illustrated in Figure 13–17, the transmittance scale on the vertical axis is a *ratio* scale, which makes it difficult to estimate values between the marked values. In contrast, the density scale is an *interval* scale, which provides uniform values for the smaller divisions.

Curves for red, green, and blue tricolor contrast filters are shown in Figure 13–17. The maximum transmittance (minimum density) is in the wavelength region that identifies the hue of the filter. It can also be seen that the curves do not neatly divide the visible spectrum into thirds (it is impossible to obtain colorants that have perfectly sharp cutting characteristics at the desired wavelengths) and that all of the filters transmit infrared radiation freely. Infrared radiation transmittance is of little consequence with respect to tone reproduction if the photographic material is not sensitive to infrared radiation.

Furthermore, the curves reveal that some filters do not transmit their own colors freely. The red filter comes the closest to the ideal with a maximum transmittance of more than 90% in the red region beyond 630 nm. The blue filter has a maximum transmittance of approximately 50% to blue light, and the green filter has a maximum transmittance of only 40% to green light. It is difficult to obtain colorants with all of the desired characteristics for use in filters and in color films. Color masking, which is used in negative color films and in some color printing processes, is necessitated by the unwanted absorption of certain colors of light by the cyan and magenta image-forming dyes.

Color-compensating filter curves for the set of yellow filters from CC025Y to CC50Y are

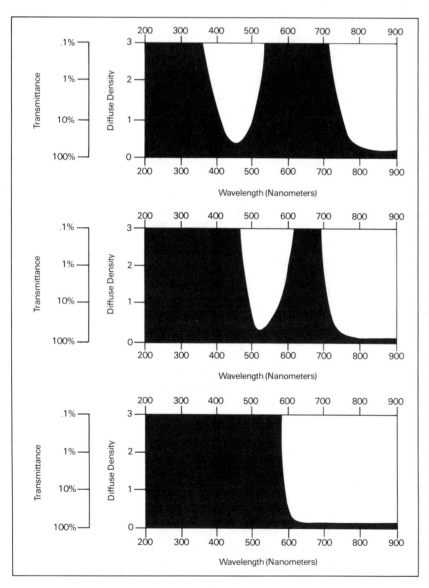

Figure 13–17   Spectrophotometric curves for blue, green, and red (top to bottom) tricolor contrast filters.

shown in Figure 13–18. All of the curves peak in the blue region of the spectrum, and all of the filters are quite transparent to green and red light.

NEUTRAL-DENSITY FILTERS

Neutral-density (ND) filters are used to reduce the illuminance by a known factor in cameras and other optical systems. The intent is to have the same transmittance for all wavelengths of radiation within specified limits. ND filters are usually calibrated in terms of white-light density. A filter with a neutral density of 0.3 has an opacity and filter factor of 2 (the antilog of 0.3) and a transmittance of 1/2 (the reciprocal of the opacity). Some ND filters have been calibrated in terms of stops, where each 0.3 in density corresponds to one stop. ND filters offer an alternative method of controlling exposure when the conventional aperture and shutter controls are inadequate. With highly sensitive photographic materials, a combination of the smallest diaphragm opening and the highest shutter speed may result in overexposure at high illumination levels. There are also situations in which the diaphragm opening is selected to produce a certain depth of field and the shutter speed is selected for a certain action-stopping capability rather than for exposure considerations. With some motion-picture cameras, the shutter setting is not a useful exposure control.

ND filters provide a convenient method of obtaining accurate small decreases in exposure. A 0.1 ND filter reduces the illuminance by the equivalent of one-third stop, 0.15 corresponds to one-half stop, and 0.2 corresponds to two-thirds stop.

In black-and-white photography an ND filter need not be entirely neutral with respect to the absorption of the different colors over the visible part of the spectrum. The relatively low-priced gelatin ND filters, which use carbon and dyes as light-absorbing materials, absorb somewhat more in the blue region (see Figure 13–19).

A more dramatic deviation from neutrality is noted in the ultraviolet region below 400 nm, where the density increases, and in the infrared region above 700 nm, where the density decreases. Clear gelatin and glass absorb considerable ultraviolet radiation. Thus the gelatin in the ND filter and in the photographic emulsion, and the glass lens in the camera or enlarger, absorb some (but not all) UV radiation. When photographs are to be made entirely with UV radiation, special precautions must be taken to minimize this absorption in the optical system, such as by substituting a quartz lens and avoiding use of gelatin and glass filters.

The gelatin in gelatin filters absorbs UV radiation but transmits light and infrared radiation freely.

0.30 neutral density = 1 stop
0.15 neutral density = 1/2 stop
0.10 neutral density = 1/3 stop

Pinhole cameras have no lenses and can be used to record ultraviolet radiation.

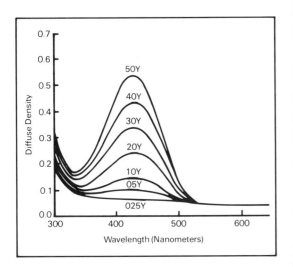

Figure 13–18 Yellow color-compensating filter spectrophotometric curves.

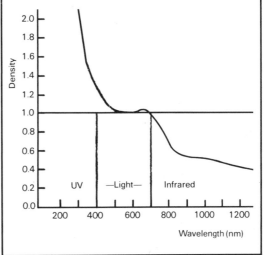

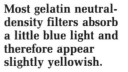

Figure 13–19 Gelatin neutral-density filters containing dyes and finely divided carbon absorb too much blue light and ultraviolet radiation and too little infrared radiation.

Most gelatin neutral-density filters absorb a little blue light and therefore appear slightly yellowish.

On the other hand, when UV radiation interferes with obtaining the desired effect, as in color printing, a special filter may be required to completely absorb the unwanted UV radiation. The excess IR radiation transmitted through ND filters will have no effect on photographic emulsions that are not sensitive to these wavelengths, but it can introduce exposure errors if the exposure is based on meter readings made with a meter that is sensitive to IR radiation.

Some of the absorbing materials other than dyes that are used in ND filters include carbon, finely divided particles of photographic silver, and thin layers of metallic alloys deposited on a transparent base by evaporation in a vacuum. Colloidal carbon used in gelatin ND filters is somewhat yellowish, requiring the addition of dyes to reduce the deviation from neutrality. Larger carbon particles are more nearly neutral but they scatter the light more, so these filters are not suitable for use in imaging systems.

The color of finely divided particles of silver depends upon their size. Most black-and-white negatives appear quite neutral in color,

but negatives made on fine-grained film developed in a fine-grain developer appear somewhat warmer in color. Even smaller particles of colloidal silver function as an efficient yellow filter in the Carey-Lea filter layer, used in color films to prevent blue light from reaching the green-blue and red-blue sensitive layers.

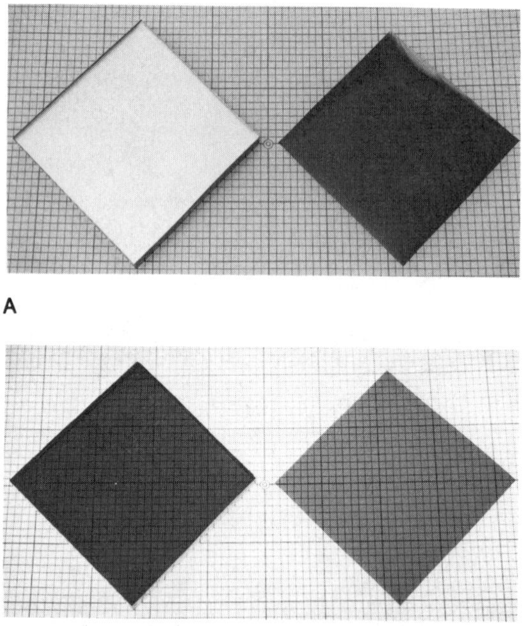

INTERFERENCE FILTERS

Traditionally, photographic filters have consisted of colorants that absorb the unwanted part of the incident radiation and transmit the desired part suspended in transparent materials. Interference filters typically consist of thin layers of metallic alloys separated by thin layers of transparent dielectric materials. In these layers the refractive indexes and thicknesses are controlled so that constructive and destructive interferences separate the wanted and unwanted components of the incident radiation by means of reflection and transmission (see Figure 13–20).

POLARIZING FILTERS

A ray of ordinary light can be visualized as consisting of waves that undulate in all directions perpendicular to the line of travel. Under certain conditions the light can become plane polarized so that there is only one set of waves, and the undulation is restricted to a single plane—like that produced in a rope when one end is snapped up and down. Polarization occurs, for example, when light is reflected as glare at an appropriate angle from most nonmetallic surfaces. The angle at which the polarization is at a maximum varies somewhat with the reflecting material. For precision work, this angle—called the Brewster angle or the polarizing angle—can be determined from Brewster's law, which states that the tangent of the angle equals the index of refraction of the reflecting material. For example, if the glass in a window has an index of refraction of 1.5, the tangent of the polarizing angle equals 1.5, and the angle equals the inverse tangent of 1.5 or 57°.

In optics, angles of incidence and reflection are measured to the normal, a line per-

Reflections viewed at an angle of 35° to nonmetallic surfaces can be eliminated with a polarizing filter.

Interference filters reflect, rather than absorb, unwanted light.

A

B

Figure 13–20   A dichroic filter (left) and a gelatin filter (right) were placed on a translucent grid on an illuminator. With the room lights on and the illuminator off, the dichroic filter looks like an opaque mirror (top left). With the room lights off and the illuminator on, the transparency of both filters is evident (bottom).

pendicular to the surface. When eliminating reflections in practical picture-making situations, it may be more convenient to measure the angle of reflection to the surface, which in this example would be 90° − 57° = 33° (see Figure 13–21). Since the polarizing angle does not vary greatly with the different reflecting materials ordinarily encountered, and the polarizing effect decreases gradually with deviations from the polarizing angle, an angle of 35° to the surface is commonly used.

An artificial method of polarizing light is to pass ordinary light through a polarizing filter, which contains very small crystals all oriented in the same direction. The crystals transmit light waves oriented in the same direction and absorb light waves oriented at right angles; thus the transmitted light is plane polarized. Placing a polarizing filter in front of the eye or camera lens and rotating it to the proper position will absorb the polarized light of glare reflections from nonmetallic surfaces at an angle of approximately 35° to the surface. Therefore, the detail that had been obscured by the reflection can be seen and photographed (see Figure 13–22).

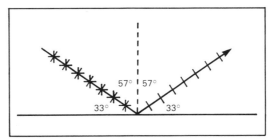

Figure 13–21 Maximum polarization of light reflected from nonmetallic surfaces occurs when the angle of incidence and the angle of reflection are approximately 57°.

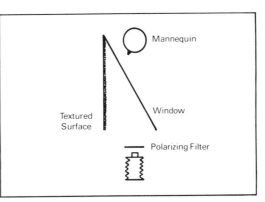

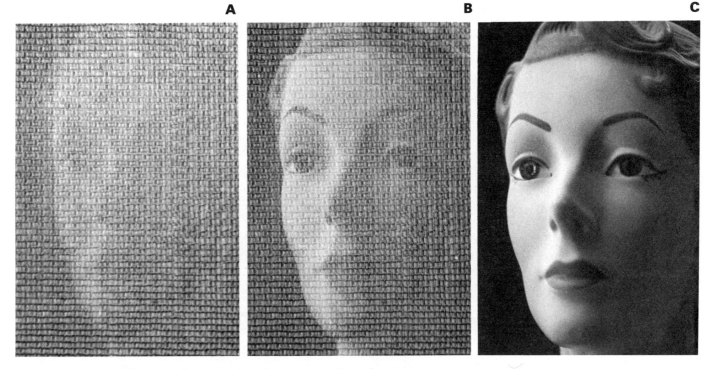

Figure 13–22 A reflection of a textured surface on a window obscures the object behind the window (A). Rotating a polarizing filter in front of the camera lens provides control over the reflection, which is reduced in photograph B and almost entirely eliminated in photograph C.

If the filter absorbed only the polarized light from the glare reflection, no increase in camera exposure would be required. Theoretically, the polarizing filter also absorbs half of the ordinary (unpolarized) light from the scene. In practice, a little more than half of the light is absorbed, and a filter factor of approximately 2.5 is generally recommended. The fact that the ordinary light from the scene becomes polarized by the filter is of no consequence as far as the quality of the image is concerned, as long as the exposure adjustment is made.

In addition to obvious glare reflections on water, glass, and other shiny surfaces, there are often more subtle veiling reflections on plant leaves and other surfaces that may be removed or reduced with a polarizing filter, producing the appearance of more saturated subject colors. Whereas polarizing filters are supposedly neutral in color, in practice some filters deviate sufficiently from neutrality to produce a noticeable shift in the color balance of color slides and transparencies. Polarizing filters can also be used with color films to darken blue sky, which contains considerable polarized light. The maximum effect is obtained at a right angle to a line connecting the camera and the sun.

Under controlled lighting conditions indoors, glare reflections can be removed from both metallic and nonmetallic surfaces with the camera at any angle to the surface, provided the subject is illuminated with polarized light (by placing a polarizing filter in front of the light source) in addition to using a polarizing filter on the camera (see Figure 13–23). With polarizing filters over the light sources, surface glare reflections consist entirely of polarized light and thus can be removed with the camera filter; the polarized light that penetrates the surface, however, becomes depolarized and thus can be used to record the detail under the reflections. Polarized light is commonly used in copying textured surfaces, such as oil paintings. The filter on the camera must be rotated at a right angle to the direction of rotation of the filters in front of the light sources.

Crossed polarizing filters can be used to reveal stress patterns in transparent plastic items by placing them between the two filters, using transillumination. The stress patterns are revealed as areas of varying colors

Polarizing filters are used in pairs to reveal stress patterns in plastic materials and some crystals.

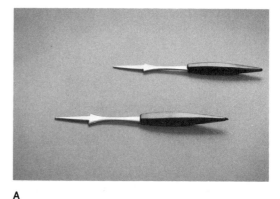

A

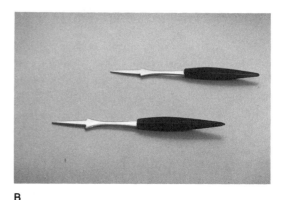

B

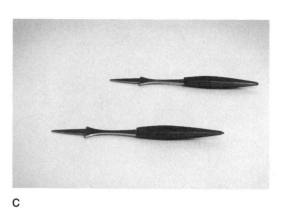

C

Figure 13–23   A single polarizing filter on the camera lens can eliminate the reflections on the black plastic parts of these objects because the camera angle to the reflecting surfaces is approximately 35°, but the filter has no effect on the reflections on the metal parts (B). When a polarizing screen is placed in front of the light source producing the reflections so that the reflection on the metal consists of polarized light, the filter on the camera lens can remove it (C).

and densities (see Figure 13–24). A similar technique is used in photomicroscopy to reveal the structures of crystalline materials.

SAFELIGHT FILTERS

Whereas camera and enlarger filters are used mostly to control tone-reproduction characteristics—including density, contrast, and color balance—of photographic images, safelight filters are used to permit the photographer to see in an otherwise darkened room without altering the images. Generally, safelight filters are intended to absorb the colors of light to which the photographic film or paper is sensitive. A yellow filter, for example, absorbs blue light, so it is recommended for use with printing papers that are sensitive only to blue light. Prolonged exposure of photographic materials to safelight illumination, however, even with the recommended safelight filter, will generally produce fog (see Figure 13–25). This means

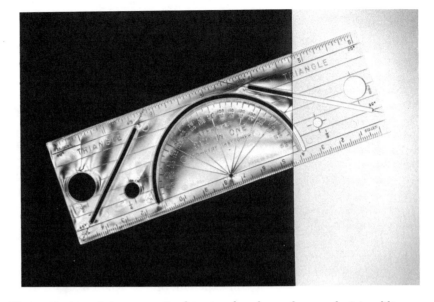

Figure 13–24   A transparent ruler was placed on a large polarizing filter on an illuminator. A second polarizing filter was rotated 90° and placed on top of the ruler, leaving the right end uncovered. Stress patterns in the ruler are revealed as areas of varying colors and densities.

either that the safelight does not absorb all of the light it is intended to absorb or that the photographic material has some sensitivity to colors to which it was assumed to be insensitive. Thus it is important not only to use the recommended filter for a given photographic material, but also to observe the recommendations concerning bulb wattage, minimum working distance, and maximum time of exposure, and to remember that filters should be replaced when they fade.

Safelight fog affects print picture areas more than the unexposed border.

Figure 13–25   A safelight fog test. A uniform picture exposure was made in complete darkness. The paper was then exposed to safelight illumination for zero, one, three, and seven minutes. Fog is evident in the picture area with the one-minute safelight exposure but is not evident in the border until the paper has received seven minutes of safelight exposure.

REVIEW QUESTIONS

1. A yellow object appears yellow when viewed with white light because the object . . . *(p. 302)*
   A. reflects 580 nanometer wavelength radiation
   B. absorbs light nonselectively
   C. reflects blue light
   D. absorbs blue light
   E. absorbs yellow light
2. A practical use for the Maxwell triangle is to . . . *(p. 303)*
   A. determine how many different hues there are in white light

B. determine how many primary colors there are in white light

C. predict the effect filters will have on photographic exposure

D. predict the effect filters will have on contrast index

E. predict the effect filters will have on color reproduction

3. In terms of the final black-and-white photographic print, a red filter used on the camera should darken . . . *(p. 303)*

A. cyan, magenta, and yellow objects

B. red, yellow, and magenta objects

C. white, gray, and black objects

D. cyan, blue, and green objects

E. red, green, and blue objects

4. A filter factor of 8 corresponds to a change in exposure of . . . *(p. 304)*

A. one stop

B. two stops

C. three stops

D. four stops

E. five stops

5. The maximum reduction in the appearance of haze occurs when the image is recorded entirely with . . . *(p. 309)*

A. ultraviolet radiation

B. blue light

C. green light

D. red light

E. infrared radiation

6. The color of the recommended filter for use with tungsten-type color film and daylight illumination is . . . *(p. 309)*

A. orange

B. blue

C. green

D. cyan

7. Superimposing a CC05R filter and a CC30R filter is equivalent to a single . . . *(p. 311)*

A. CC15R filter

B. CC25R filter

C. CC35R filter

D. CC150R filter

E. None of the above.

8. Superimposing a CC20Y filter and a CC20M filter is equivalent to a single . . . *(p. 311)*

A. CC10R filter

B. CC20R filter

C. CC40R filter

D. ND 0.20 filter

E. ND 0.40 filter

9. To eliminate reflections when copying an oil painting, with the camera axis perpendicular to the surface of the painting, it is necessary to use . . . *(p. 316)*

A. two polarizing filters in front of the camera lens

B. a polarizing filter in front of each light source

C. a polarizing filter in front of each light source and a polarizing filer in front of the camera lens

10. A neutral density filter used with color film could cause a shift in the . . . *(p. 313)*

A. red direction

B. green direction

C. blue direction

D. yellow direction

14 Color

John Fergus-Jean.

INTRODUCTION

The visual experience of perceiving color does not depend upon language or numbers, but the communication of information about the experience does. It is important that we use an agreed-upon and established vocabulary to talk about color. In the past, various disciplines including photography, art, and graphic arts have had difficulty communicating with each other due to the absence of an agreed-upon common color language. The Munsell Color System and the Standard CIE System of Color Specification have emerged as two of the more successful attempts to organize the variables of color into a universally accepted system. In addition, the Inter-Society Color Council has brought together representatives of the various disciplines in an effort to lower the barriers to communication. The stated aims and purpose of the organization are "to stimulate and coordinate the work being done by various societies and associations leading to the standardization, description and specification of color and to promote the practical application of these results to the color problems arising in science, art and industry."

Before examining color systems, however, it is necessary to establish a definition of color. Strictly speaking, color is a visual experience. It is a perception, and therefore the word *color* refers to physiological and psychological responses to light in much the same way that taste, smell, touch, and sound are responses to other physical stimuli. In everyday language, color is associated with words such as *red, green, blue,* and *yellow.* Actually, these words refer to only one of three attributes of color—namely hue. The other two attributes are brightness (or lightness) and saturation.

Clarity of language is achieved when we distinguish between using color terms to refer to the visual perception of an object and the physical characteristics of the object. For example, it can be argued that we should not refer to a "red" apple as being red. We see it as red because the surface of the apple has physical properties that reflect certain wavelengths and absorb other wavelengths of the white light that illuminates it (see Plate II). If a "red" object is illuminated with "blue" light, most of the light will be absorbed and the object will be perceived as being black. Thus it is understood that when we identify an object as being red, we mean that it will generally be perceived as being red by persons with normal color vision when it is illuminated with white light and viewed under normal viewing conditions.

Colorant is a general name for colored substances, including pigments and dyes used to alter the color of objects or other materials. The colors we see in color photographs are a result of mixtures of three colorants in the photographs: cyan, magenta, and yellow dyes. The purpose of the three dyes, which are present in different proportions in various areas of the photographs, usually is to create the same perception of color as when the original scene is viewed directly—even though a wavelength-by-wavelength comparison of the light entering the eye from the scene and from the photographic reproduction might reveal entirely different compositions. In other words, the photograph is intended to create the illusion that it has faithfully reproduced the reflection characteristics of the original scene, without actually doing so.

In a sense we live in two different worlds, the physical world and the psychological world. The physical world lends itself rather easily to measurement. We use instruments such as thermometers, clocks and rulers to measure temperature, time, and distance.

SPECTROPHOTOMETERS

The most fundamental physical instrument for measuring color is a spectrophotometer. It is a specialized photometer (light meter), as the name suggests; *spectro* refers to the fact that the instrument is capable of measuring light of different hues or wavelengths in the color spectrum from blue to red (measurements can also be made of radiation in the infrared and ultraviolet parts of the electromagnetic spectrum with some spectrophotometers). One can think of a color densitometer that measures red, green, and blue densities as a limited spectrophotometer capable of measuring only three broad parts of the color spectrum according to the particular type of red, green, and blue filters used.

A red object illuminated with blue light appears black.

Spectrophotometers measure the reflectance or transmittance of color samples wavelength by wavelength throughout the spectrum.

A spectrophotometer is a much more sophisticated instrument capable of measuring the density (or transmittance or reflectance) of a given sample at very precise and narrow intervals of color (wavelengths). The intervals are specified in terms of bandwidth. Light, for example, covers a bandwidth of about 300 nm (400 nm to 700 nm). Red, green, and blue light can each be thought of as covering a bandwidth of roughly 100 nm. A spectrophotometer has a dispersing element (prism or diffraction grating) that can separate white light into a full color spectrum. Then, through the arrangement of an optical network of lenses and slits, narrow bandwidths of color 10 nm or less can be isolated and measured after they pass through or are reflected from a sample (see Figure 14–1). This is done at 10 nm intervals over the entire spectrum of colors to be measured.

SPECTROPHOTOMETRIC CURVES

Spectrophotometric curves, such as those shown in Figure 14–2, can then be plotted from the data, or plotted directly by some instruments. The curves provide a contour or envelope that describes the reflection or transmission characteristics of a sample, such as an image area in a color print or color transparency. It is important to understand that such curves describe the physical characteristics of the colored samples, *not the colors* that would be perceived by a person viewing the samples. The curves for skin colors in Figure 14–2, for example, show that human skin has the highest reflection in the red region of the spectrum and the least in the blue region. Light skin has the highest reflection at all wavelengths, as one would expect, and the curve is bumpier. The reflectance for any specific wavelength is easily obtained from the curves. At 700 nm, for example, light skin has a reflectance of 70%, dark skin 45%, and very dark skin 15%. There are many different skin colors. Each would have its own curve shape but, in general, would be similar to those shown in Figure 14–2. Depending upon how the spectral response is specified, spectrophotometric curves can be referred to as spectral trans-

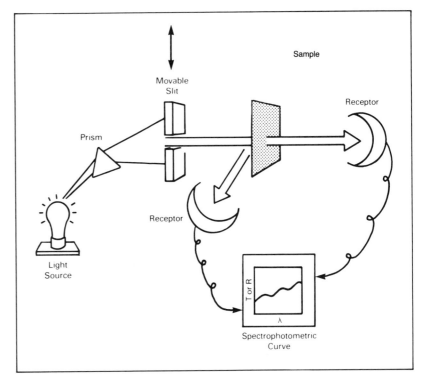

Figure 14–1 A simplified illustration of a spectrophotometer in transmission and reflection modes.

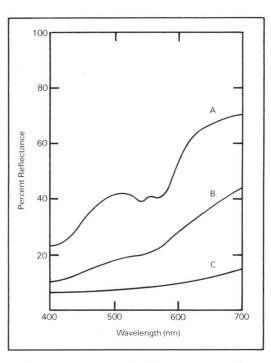

Figure 14–2 Spectral reflectance curves for human skin: (A) light skin; (B) dark skin; (C) very dark skin. (Evans, R., *An Introduction to Color.* New York: John Wiley & Sons, 1948, p. 88.)

Spectrophotometers also are used to measure the spectral quality of light emitted by various sources.

The word "spectrum" resulted from the belief in early times that this phenomenon was a phantom of light or a specter.

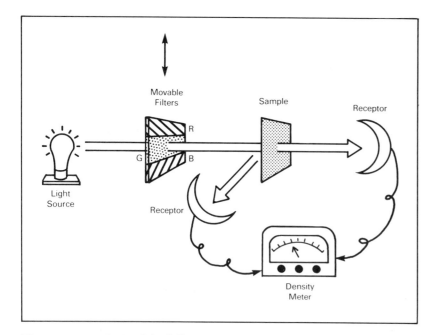

Figure 14–3   A simplified illustration of a color densitometer in transmission and reflection modes.

reflectance or transmittance of a color sample. Whereas a spectrophotometer uses a prism or diffraction grating to spread the light and a slit to isolate narrow spectral bands of light ranging between 1 nm and 10 nm, a densitometer uses red, green, and blue filters to isolate broad bands of light of about 50 nm (see Figure 14–3). The result is not a curve showing the spectral transmittance or reflectance of a color sample at each wavelength, such as in Figure 14–2, but simply three numbers, expressed as density (the log of the reciprocal of the transmittance or reflectance). The recommended red, green, and blue filters for use in measuring most color transparencies and prints are the Wratten 92, 93, and 94 filters, with bandwidths of approximately 60 nm, 30 nm, and 35 nm, as shown in Figure 14–4.

Color densitometers can be classified as either visual, if the human eye serves as the receiver, or physical, if some other light-sensitive device such as a photodiode is used. Densitometers can be further classified as direct reading, if the receiver output indicates the density, or null type, if the receiver indicates when two beams of light are balanced or equal. Visual densitometers are always of the null type based on the equality of light passing through the sample being measured and light passing through an area of known density.

mittance, spectral reflectance, or spectral density curves.

COLOR DENSITOMETERS

An essential difference between a color densitometer and a spectrophotometer is in the bandwidth of light being used to measure the

COLOR DENSITIES

Color densitometers measure the density of a colored layer to broad spectral bands of light as determined by the choice of filters. In photography, color densitometers are used to measure the density of cyan, magenta, and yellow dye layers to red, green, and blue light. Ideally, cyan dye would absorb only red light and transmit blue and green light freely. In practice, cyan dyes absorb some blue and green light. Therefore, the combined density of the three dye layers in a color transparency would be different if the layers were peeled apart and measured than when they are measured in the normal superimposed position. If all three layers are measured as one unit, the results are called *integral* color densities. When each layer is measured separately, the results are called *analytical* color densities.

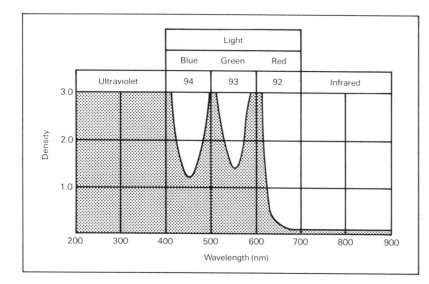

Figure 14–4   Wratten 92, 93, and 94 filters are recommended for color density measurements of photographic color transparencies and prints. (All gelatin filters absorb UV but transmit IR.)

Integral density measurements are typically used for the control of photographic color processes and to determine the printing times required for color negatives. Analytical densities are used primarily for research and development to improve existing color products and to develop new products. Figure 14–5 shows a spectrophotometric curve (expressed as spectral density) for a typical color film. The three lower curves represent the spectral analytical densities for each layer separately, while the upper curve represents the spectral integral densities of the cyan, magenta, and yellow layers combined to form a neutral.

Integral densities made with a color densitometer would give three separate values: density to red light (D_r), density to green light (D_g), and density to blue light (D_b). If the bandwidth of the three filters in the densitometer coincided with the three peaks of the upper curve in Figure 14–5, one could speculate that D_r would be a little less than 1.0, while D_g and D_b would be higher. The difference between the peak analytical densities and the integral densities at the same wavelengths represents unwanted absorption of blue and green light by the cyan dye, unwanted absorption of red and blue light by the magenta dye, and unwanted absorption of red and green light by the yellow dye.

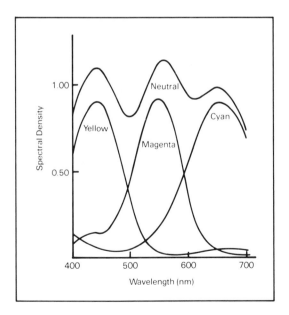

Figure 14–5 Spectrophotometric curves for a visually neutral sample of color reversal film.

Color density measurements are highly dependent upon the choice of red, green, and blue filters used to make the measurements and the spectral sensitivity of the phototube. Even when the same filters and phototubes are used, one should not be surprised to discover differences. And since filters and phototubes age and change, it is good practice to maintain control charts for color densitometers.

PHYSICAL, PSYCHOPHYSICAL, AND PSYCHOLOGICAL MEASUREMENTS

Spectrophotometric curves describe the physical reflection or transmission characteristics of colored substances. We use our own perceptual systems, however, to make psychological measurements of hue, saturation, and brightness. Since psychological measurements are personal and subject to great variability, specialized instruments have been developed that provide, under rigidly specified conditions, physical measurements that correlate well with psychological visual perceptions. Such measurements are called psychophysical since they combine both worlds.

Figure 14–6 provides a comparative listing of terms to describe the attributes of light when the measurements made are physical, psychophysical, or psychological. Thoughtful description of these attributes will distinguish the intended meaning and facilitate clarity. For example, a spectrophotometer will measure the *spectral distribution* of light reflected from an object or transmitted by an object. This is strictly a physical measurement. An instrument called a *colorimeter* (to be discussed later) has a response characteristic similar to the human eye and specifies the *chromaticity* of the spectral distribution. This is a psychophysical measure that describes the quality of color relative to a neutral color in terms of dominant wavelength (hue) and purity (saturation), and allows it to be positioned on a color map (CIE diagram to be discussed later). How that particular chromaticity is sensed by a human observer falls into the realm of psychological measurement and is called *chromaticness* (or

Color analyzers used in color printing darkrooms are essentially color densitometers that indicate the filtration that should produce an appropriate color balance in the print.

Although instruments can measure the physical characteristics of light, the measurements cannot accurately predict the color perceptions that will be produced by the light.

PHYSICS	PSYCHOPHYSICS	PSYCHOLOGY	
Light Source	Source x Eye	Color Perception	
Radiant energy Spectral composition 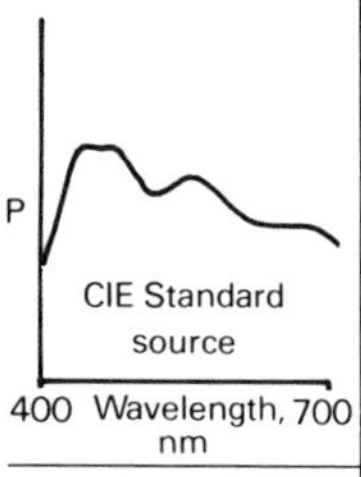	Luminous energy 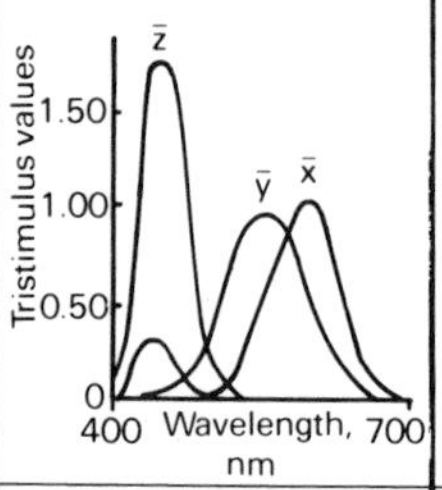	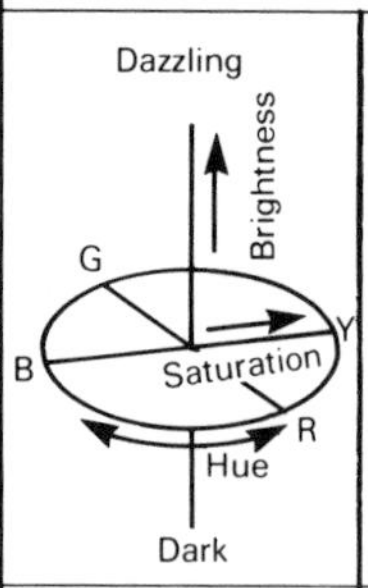 Dazzling	
(graph: P, CIE Standard source, 400 Wavelength 700 nm)	*(graph: Tristimulus values 1.50, 1.00, 0.50, 0; z̄, ȳ, x̄; 400 Wavelength 700 nm)*	*(diagram: Brightness, Dazzling, Dark, G, B, Y, R, Saturation, Hue)*	
Characteristics of Radiant Energy	Characteristics of Luminous Energy	Attributes of Color	Corresponding Modes of Appearance
Radiant flux	Luminous flux	Brightness	Aperture
Radiance	Luminance		Illuminant
Irradiance	Illuminance		Illumination
Radiant reflectance	Luminous reflectance		Object modes: Surface
Radiant transmittance	Luminous transmittance		Volume
Spectral distribution	Chromaticity	Chromaticness	Attributes of modes of appearance:
(Relative spectral composition, quality)	Dominant wavelength (or complementary)	Hue	1. Brightness (or lightness)
Radiant purity	Purity	Saturation	2. Hue
	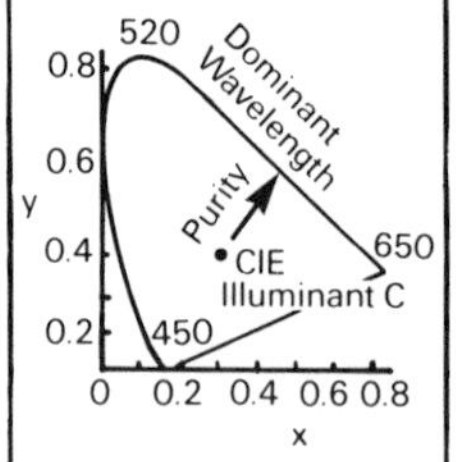 *(chromaticity diagram: 520, 650, 450, CIE Illuminant C, Dominant Wavelength, Purity, y axis 0.2–0.8, x axis 0–0.8)*		3. Saturation
			4. Size
			5. Shape
			6. Location
			7. Flicker
			8. Sparkle
			9. Transparency
			10. Glossiness
			11. Luster

Figure 14–6 Adapted from the system of nomenclature as given by the Committee on Colorimetry of the Optical Society of America 1943. (Sheppard, J., 1958, p. 11)

Computers can generate and display over one million physically different colors.

chrominance). Chromaticity and chromaticness represent only two of the three attributes of color; the other is luminance (brightness/lightness).

Color, like time and space, is a distinctive property of our daily experiences. Different individuals' concern with color will depend to a large extent upon their chosen professions. A chemist may be concerned with the molecular dye structure of the colorants; a colorimetrist with precise psychophysical measurements and specification; a psychologist with repeatable and predictable psychological measurement; a painter with the esthetics; and a photographer with the esthetics, accuracy of reproduction, archival qualities, and communication of color.

COLOR CHAIN

The perception or measurement of color follows a chain of events as shown in Plate III. There are four major links in the chain. The characteristics and interactions of the first three—light source, color stimulus, and receiver—determine the response, which can be described in psychological, physical, or psychophysical terms. Some of the terms used to describe the characteristics of the various links are:

1. *Light source:* illuminant, color temperature, correlated color temperature, spectral energy distribution, tungsten, sunlight, daylight.
2. *Color stimulus:* colorant, dye, ink, pigment, color object, spectral reflectance or transmittance.
3. *Receiver:* human eye, photocell, photographic film.
4. *Response:* Munsell system (psychological), spectrograph (physical), CIE system (psychophysical).

COLOR MODE

The characteristics of the color stimuli can be further expanded to include the particular mode of appearance or context in which colors are experienced. There are five such modes generally recognized, which influence the response to the color stimuli: *surface, volume, aperture, illumination,* and *illuminant.* When photographing color it is helpful to recognize and distinguish between these various modes (see Plate IV).

The most common is the surface mode or surface color, which refers to light reflected from and modified by a surface. It is seen as belonging to a surface. Examples are painted walls, color photographs, fruits, vegetables, human skin, and all such objects that present reflecting surfaces.

Volume color refers to color perceived when one looks into or through a uniformly transparent substance such as a liquid. Examples would be the bluish color of water in a swimming pool, yellowish color of cooking oil in a bottle, pinkish color of a rosé wine, and the like. A liquid filter used to change color temperature is yet another example. The strength of the volume color is determined by the concentration of the liquid and the thickness or depth.

Aperture color refers to color perceived in space and not recognizable as an object. Examples include the bluish color of the sky and the reddish color of a sunset or sunrise, which seems to fill the "emptiness" above us. Light seen in an opening, such as the color lights projected into an aperture of a colorimeter for color matching, is a prime example of an aperture color mode. A more practical example would be a virtual image that one could see at the film plane of a view camera when the ground glass is removed. (Aperture color is sometimes referred to as film color, which can be confusing to a photographer.)

Illumination color refers to the perceived color of light falling on an object. A familiar example in color photography occurs when we are taking photographs outdoors under sunlight conditions and we find that the shadows in the white snow (or on light sand or concrete) tend to have a bluish cast. This is a result of fill light from the blue sky, which casts a diffuse blue light over the entire landscape. Such bluish shadows are recorded objectively by the film as illumination color, but rarely by the eyes, which tend to compensate.

We may not perceive them as such but the colors we see reflected from a movie screen can be thought of as illumination colors. A projector has a white light source that is modified by the color film in the projector. The various colors of light passing through the film fall on a white screen surface and are reflected. We could, but generally do not, see them as illumination colors. In a studio setting we usually use white lights to obtain surface colors. If we used a warm (reddish) light to represent light from a fireplace, however, the colors would be modified. We would then attribute the altered appearance of the color objects to illumination color.

Illuminant color is distinguished from illumination color in that it is the color of the light source viewed directly. Instead of looking at the illumination color on the movie screen, we could turn around and look directly at the projector and the colors of light passing through the lens. More common experiences of illuminant colors are red, green, and yellow traffic lights, Christmas tree lights,

and neon lights. Another example of illuminant color is the viewing of a television picture or television display from a microcomputer or word-processing machine. The television screen (cathode ray tube) consists of thousands of dots of red, green, and blue phosphors that emit light as they are energized. When we view color pictures or color graphics on a television screen, we are looking directly at the light source. This system of color reproduction is uniquely different from a photographic system. (When working with video color systems it is important to realize that there is a decided difference between displaying pictures of three-dimensional objects on a television screen and using a computer to generate pictures and graphics on a television screen. Important to our perception of color are texture and line, which are often inadequate with computer-generated graphics.)

PRIMARY COLORS

Surprisingly enough, the minimum number of colors needed to form most other colors, by mixing, is three. They must be independent of each other; i.e., two of the three colors when mixed must not form the third color. For subtractive systems of color, such as photography and printing, the primaries are called cyan (blue-green), magenta (blue-red), and yellow (red-green) *colorants* (inks, pigments, dyes). In additive systems of color, the primaries are red, green, and blue (Plate V). Television is an additive system, which uses small dots of red, green, and blue phosphors that are too small to resolve as separate dots, although they can be seen with a magnifier. (Avoid doing this for any length of time since television tubes emit some potentially harmful radiation.) An early example of an additive system used to produce photographic color transparencies is Dufaycolor, which used a mosaic of minute red, green, and blue filters and was introduced in 1934.

Since the subtractive color primaries—cyan, magenta, and yellow—are complementary to the additive color primaries—red, green and blue—they are sometimes referred to as minus-red, minus-green, minus-blue. In color printing it is especially helpful to think of cyan, magenta, and yellow filters in this

One of the problems in making water and sky look natural in a color print is that the print can only represent volume color and aperture color as surface color.

"Color possesses me. I don't have to pursue it. It will possess me always . . . Color and I are one."
— *Paul Klee*

Although all color films contain cyan, magenta, and yellow dyes, they will not produce identical color images.

way. It is also helpful to remember that a neutral can be formed by combining complementary colors (see Plate VI).

Although the selection of the three primary colors for the subtractive as well as additive systems is somewhat arbitrary, the choice is usually based on a number of factors: ability to produce the colorants, ability of the colorants to reproduce the greatest number of secondary colors, stability, availability, cost, and convenience.

COLOR ATTRIBUTES

When we describe the colors we experience, we use such familiar words as *red*, *green*, *yellow*, *orange*, *blue*. Further, we might distinguish one red from another by saying it is darker or lighter, stronger or weaker. This is because colors have three distinct attributes: hue, saturation, and brightness (or lightness). When we say lipstick is red we refer to hue. Saturation describes how strong or intense the red hue is, and brightness or lightness how light or dark the red hue is (see Plate VII). We shall see later that the specific terms used to describe these attributes distinguish the different systems for specifying color.

Neutrals are sometimes called achromatic colors. This expression is an oxymoron—a contradiction of terms. White, gray, and black are colors that have lightness but no hue or saturation.

NEUTRALS

When mixed in the proper proportions, complementary colors produce neutral colors. Colors such as cyan and red, magenta and green, yellow and blue will, when mixed, form neutrals having various levels of lightness or brightness (see Plate VI). Neutrals are distinguished from other colors in that they have no hue or saturation, so they are sometimes referred to as achromatic colors. A scale of such neutrals ranges from black to white and is called a gray scale. It is wrong, as many of us have been taught in early schooling, to think of black or grays as the absence of color. Grays appear neutral because they do not alter the color balance of the incident white light illumination. They appear dark or black when little light reaches the eye, and they appear light or white when much light reaches the eye.

One of the most critical tests of a photographic color material is its ability to reproduce a scale of neutral colors. Small variations in color balance can be detected more easily with neutral colors than with saturated colors. To the eye, a neutral color is either neutral or not, whereas a color such as red can be modified considerably by the addition of blue (magenta-red) or yellow (orange-red) and still be accepted as red.

In testing the ability of a color film or paper to reproduce neutrals it should be noted that the results are often more dependent upon the user than the manufacturer. Regardless of how good the product is, if it is improperly stored, exposed to the wrong color temperature light or to mixed lighting, improperly processed, or viewed under poor lighting conditions, one cannot expect quality results.

COLOR-VISION THEORY

As early as 1666, a 23-year-old instructor at Cambridge University demonstrated that white sunlight is made up of all colors of light. He did this by passing sunlight through a prism to create a spectrum of colors. This demonstration gave visual evidence of how light can be separated into different colors. It was not until 1704, however, that Sir Isaac Newton put forth a hypothesis to explain the process by which we see colors. He speculated that the retina contains innumerable light receptors, each of which responds to a specific color stimulus. This was rejected by Thomas Young nearly a hundred years later, in 1801. In a terse 300-word statement, Young hypothesized that there are only three different kinds of light receptors in the retina, each responding to one color of light—red, green, or blue. His theory was in turn rejected and ignored by his contemporaries. Some 50 years later, in 1850, Young's theory was rediscovered by Maxwell and Helmholtz.

Although there are many theories that attempt to explain our response to color, there remain many unanswered questions. Some theories, however, are more useful than others. The oldest, most efficient, and most persistent theory of color vision is the *Young-Helmholtz three-component theory*. This theory postulates that there are three kinds of receptors in the retina that react selectively. One can think of these as red, green, and blue receptors that somehow combine in our visual system and produce the perception of other colors.

Some unanswerable questions have been raised regarding this three-color theory for human vision. For example, the eye will distinguish not three but four fundamental or primary colors—colors that are distinct and have no trace of other colors. The colors are red, yellow, green, and blue and are called the psychological primaries. The Hering theory of color vision takes this into account.

A somewhat more complex theory is the *Hering opponent-colors theory.* Whereas the Young-Helmholtz theory is based on three color stimuli, the Hering theory is based on the response to pairs of color stimuli. It assumes that there are six basic independent colors (red, yellow, green, blue, white, and black). Rather than postulating special and separate receptors for the six colors, Hering proposed that the light absorbed by the red, green, and blue sensitive receptors in the retina starts a flow of activity in the visual system. Somehow this flow is channeled into three pairs of processes with the two components of each opposing one another (opponents). The opposing pairs are blue-yellow, green-red, white-black. For example, a color may look bluish or yellowish but never both at the same time. Blue would oppose and cancel yellow, and vice versa.

A variation on the Hering theory that provides for quantification is the *Hurvich-Jameson quantitative opponent-colors theory.* This theory assumes a two-stage process. The first stage is the excitation stage, located in the cones of the retina and consisting of four light-receiving cells that contain combinations of three photo chemicals. The second stage is an associated response stage located beyond the retina in the visual nerve center. It has three paired-opponent processes: blue-yellow, green-red, and white-black. This opponent-response theory represents differences in the neural response to the stimulation that originated when the cones in the retina were excited. A single nerve transmits two different messages by one of the pairs of opponent colors raising the neural firing rate above the normal rate (excitation) and the other lowering the firing rate below the normal rate (inhibition).

OBJECT-COLOR SPECIFICATION SYSTEMS

Five different modes of perceiving color were mentioned earlier. The most common mode is object color or surface color, since it is the light reflected from the surfaces of objects that we normally see; a red tomato, green grass, blue shirt, brown pants, skin color, and so on. Systems for color specification of object colors consist of a large variety of colored chips from which a person can choose. The choice provides information on *matching colors* or *mixing colorants.* To determine which colorants need to be mixed to produce a certain color, a color chip is chosen. This provides information on the percentages of different inks, pigments, or paints needed to produce that chip color. Matching a color sample to a particular color chip provides a color-standard color name and notation for the sample. Color-matching and color-mixing information provide an agreed-upon means of communication. There are three different ways in which object-color specification systems can be designed.

The first is called *colorant mixtures;* a limited number of colorants such as dyes and pigments are used to generate a range of colors on some surface. This is done by systematically varying the proportions of colorants. One such system was developed from eight basic paints, including black and white. These were mixed in measured proportions and put on 1,000 different cards. This system is useful in preparing specific colors of paints or inks by known mixtures of other basic paints or inks.

The second system for object-color specification, *color-mixtures,* is similar to the colorant-mixture system in that a limited number of colorants are used to produce a range of colors. It is different in that it is an *optical mixture* of colorants. The colors of light reflected from specially prepared surface colorants are blended in systematically varied proportions to generate a range of colors (see Plate VIII).

Optical mixtures can be obtained by *spatial integration* or by *temporal integration.* An example of spatial integration is an integrating sphere as used in a colorimeter. An example of temporal integration is a mixture of colors of light that result from a rapidly

When making color prints, photographers sometimes use color matching to determine the correct filtration and exposure.

Viewing color television pictures, composed of red, green, and blue primary colors, is an example of spatial color integration.

revolving sector wheel having different colorants. This was the basis for producing the colors for the Ostwald system. Object-color samples obtained by color mixtures more closely approximate the way object colors are seen in our daily experiences. A photographic example of spatial integration is the measurement of the color balance and illuminance of the integrated light transmitted by an enlarger lens to determine the filtration and exposure time when printing color negatives and transparencies. The human eye also integrates colors when it cannot resolve the individual colors in an object or reproduction composed of fine detail, as when viewing dots of colors in photomechanical halftone reproductions of color photographs. Even the original color photographs are made up of minute clumps of colors that conform to the grain structure of the emulsion.

The third system for object-color specification is the *appearance* system, in which the color chips or samples are spaced so that they appear to have uniform intervals of hue, saturation, and lightness. This requires judgment by a standard observer under a standard set of viewing conditions. One of the more comprehensive examples of the appearance system is the *Munsell Color System*, which consists of hundreds of object-color samples in a *Munsell Book of Color*. The book contains more than 1,200 individually removable chips having either a mat or glossy surface. The Munsell Color System is very useful to practicing photographers, artists, and designers.

Object colors can be analyzed in terms of primary colors or in terms of wavelengths. In a colorimeter, a color sample is matched with a mixture of red, green, and blue primaries. The proportions of red, green, and blue light necessary to make a match specify the color sample being analyzed. A spectrophotometer provides a wavelength-by-wavelength analysis of a color sample in terms of spectral reflectance or transmittance. A densitometer analyzes a color sample in terms of densities measured with red, green, and blue light.

MUNSELL SYSTEM OF COLOR SPECIFICATION

In 1905 Munsell published his first edition of *A Color Notation*. One of the Munsell sys-

tem's strengths is that it provides a common international notation for color in which a person first identifies a color visually (nonverbally) and then uses language to describe it and communicate it. In order to accomplish this, Munsell had to first prepare a large variety (gamut) of painted color chips. Imagine hundreds of such chips having different *hues*, which Munsell described as the "name of the color"; different *values*, which he described as "the lightness of color . . . that quality by which we distinguish a light color from a dark one"; and *chroma*, which is "the strength of a color . . . that quality by which we distinguish a strong color from a weak one; the degree of departure of a color sensation from that of a white or gray; the intensity of a distinctive hue; color intensity." In sorting out the gamut of color chips, one would first group them according to hue (red, yellow, green, blue, and the in-between hues). Then for each grouping of hue they would be arranged according to their values of lightness or darkness. A red hue grouping, for example, could be arranged so that all the red chips increased in lightness from dark to light. The final arrangement, more difficult, would be to group all of the chips having the same hue and lightness according to their color intensity or chroma. On one end would be red chips that are near neutral, on the opposite end would be a red chip that one might call an intense or vibrant red—one that was a highly saturated red.

This type of ordering is essentially what Munsell did with some refinement—the most important refinement being that he ordered the colored chips so that the interval between adjoining chips would be *visually equal in hue, value, and chroma*, quite a monumental task. Once this visual structuring was completed, he had to communicate it to others so they could also visualize the arrangement and use a common language to describe the colors. Since color has three attributes (hue, value, chroma), it occupies a three-dimensional space or volume. Munsell described this concretely in terms of a color tree. "The Munsell color tree has a vertical trunk which represents a scale of *values*, branches which represent different *hues*, and leaves extending along the branches which represent *chromas*. The hues change as one walks around the tree, the values increase as one climbs up

"Music is equipped with a system by which it defines sound in terms of its pitch, intensity, and duration, without allusion to the endless varying sounds of nature. So should color be supplied with an appropriate system."
—*Albert Munsell*

the tree, and the chromas increase as one moves out along the branches" (see Figure 14–7).

A more abstract representation of this color system is shown in Plate IX. The trunk of the tree, representing values from black to white, is now the center post of a color sphere; the branches representing hues are now vertical sheets of plastic attached to the center post; and the leaves across the branches now become chips of color extending from the center post outward in all directions. Changes occur as one circles the Munsell color sphere. Value changes vertically, and for a specific hue and value, chroma increases from the center post outward.

The three attributes of color (hue, value, and chroma) are visually represented in the Munsell tree as a color *space* or volume. The overall shape of the color space is not symmetrical. This is because the range of chromas is not the same for all hues. In the extreme it is possible to have a yellow hue of high value and high chroma but not one of low value and high chroma. Similarly, a blue hue of high value and high chroma is not attainable. The three-dimensional asymmetrical color solid represents colorants now available, not including fluorescent colors.

Of the 100 hues that can be represented in the Munsell system a select sample of 10 are actually shown. The others would fall in between, and all 100 hues would be equally spaced in terms of visual differences as illustrated in Plate X. The five basic hues are red, yellow, green, blue, and purple.

Plate X shows how the 10 major hue names can be used for qualitative notation. The 100 numbers in the outer circle provide more precise notation. The numbers make it easy to use the Munsell system for statistical work, cataloging, and computer programming. The combination of numerals with hue initials is considered the most descriptive form of designation. This combination is shown in the inner circle: 5R, 7.5R, 10R, 2.5YR, 5YR, 7.5YR, 10YR, etc. *The Munsell Book of Color* shows the actual colors of these 40 constant hues.

For a particular hue, say one of the 10 major hues represented in the Munsell tree, there are a number of different values and chromas, and therefore different *colors*, of that hue. Imagine the 5 Red hue segment removed from the tree.

Figure 14–7 The Munsell color tree. (From *Munsell, A Grammar of Color*, edited by Faber Birren, © 1969 by Litton Educational Publishing, Inc. Reprinted by permission of Van Nostrand-Reinhold Co.)

The 5R notation designates that this hue is midway between a 5RP and a 5YR on the hue circle. The 5R by itself designates only the particular hue, *not* the color. As indicated in Figure 14–8, the 5R hue can be further specified in terms of value and chroma numbers or described in terms agreed upon by Inter-Society Color Council and the National Bureau of Standards (the ISCC-NBS).

In the Munsell system, color is specified in the alphanumeric sequence Hue Value/ Chroma (H V/C). For example, a 5R hue with a value of 8 and a chroma of 4 would be called a Light Pink and designated 5R 8/4 or simply R 8/4 (for hues having positions other than 5 on the hue circle the position must be indicated—for example, 3R 8/4). R 5/12 translates to a Strong Red having a value of 5 and a chroma of 12. At the extreme left of the dia-

"Hue is the name of a color. Value is the lightness of a color. Chroma is the strength of a color."
—*Albert Munsell*

In other systems chroma is called saturation, and value is called lightness or brightness.

"Viewing of color in a particular situation is, at best, a perculiar mixture of attention, intention, and memory."— *Ralph Evans*

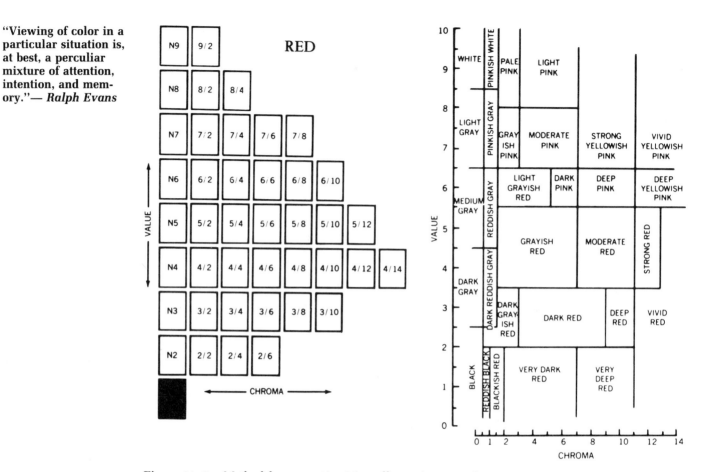

Figure 14–8 Method for converting Munsell notations to color names using The ISCC-NBS (Inter-Society Color Council and National Bureau of Standards) *Method of Designating Colors, and a Dictionary of Color Names.* (Available as circular 553 from the Superintendent of Documents, U.S. Government Printing Office, Washington, D.C. 20402.)

grams (center of the Munsell tree) the colors are neutral and are so represented. The notation for a neutral (achromatic) color is written NV/; for a middle gray (18% gray) the notation is N5/. Since the chroma is zero, it is usually omitted but could be included (N 5/0). (For those interested in Zone System notation a black, Zone 0, would be about N1/; a Zone V gray, N 5/; and a Zone IX white, N 9/.)

Near neutrals having chromas less than /0.3 are usually treated as neutrals. If more precision is needed, the form N V/(H,C) is used (H would be represented by one of the 10 select hues). For example, a light gray that is slightly yellowish might be noted as N 8/Y,0.2 or, using the regular H V/C form, Y 8/0.2. The form N V/0 can be reserved for absolute neutrals.

Although the spectrum is commonly divided into regions having seven different hue names (red, orange, yellow, green, cyan, blue, and violet), the Munsell system identifies 10 major hues and 100 distinguishable hues.

PANTONE® SYSTEM OF COLOR SPECIFICATION

The Pantone Matching System is an object-color system widely used in the graphic arts industry. (Pantone® is Pantone, Inc.'s check-standard trademark for color reproduction and color reproduction material.) It differs from the Munsell system in that it is not based on equal visual differences in color, and in that the colorants used are inks common to graphic-arts printing. Using colorants similar to those used for printing provides a closer match between the colors selected and specified by the client and the colors produced by the printer. The system comprises more than 500 standard Pantone® colors by blending various proportions of eight basic chromatic colors, plus black and transparent white. To produce a Pantone® 292 color similar to a Munsell 5PB 5/10, for example, would require 2 parts Pantone® Reflex Blue, 2 parts Pantone® Process Blue, and 12 parts transparent white. The reproduction of a gray card can be specified by matching it first to a Pantone® color such as Pantone® 424 and then giving that information to the printer. The printer knows that the gray-card color can be reproduced with an ink mixture of 3 parts Pantone® Black, 1 part Pantone® Reflex Blue, and 12 parts transparent white.

The colors used in Plate VIII were chosen by the authors from the Pantone® Color Formula Guide (see Plate XII). The Pantone numerical designation for each color specified was communicated to the book's graphic designer, who in turn used the same notation to communicate the desired colors to the printer. In this way, author, designer, and printer shared a common color language, which facilitated communication and reduced error. Any standard system of color notation provides this. In communicating colors to others, however, it is essential to know which system they are using. (For more information on Pantone® Matching System write to Pantone, Inc., 55 Knickerbocker Road, Moonachie, NJ 07074.)

CIE SYSTEM OF COLOR SPECIFICATION

The letters CIE stand for Commission Internationale de l'Eclairage (The International Commission on Illumination). In 1931 the CIE system of color specification became an international standard for colorimetry.

Color specification with the CIE system differs from that of the Munsell in that a mixture of red, green, and blue light is used to match a given sample rather than the selection of a color chip. The Munsell system is based on *appearance*, whereas the CIE system is based on *stimulus-synthesis* (the additive mixture of three color primaries). Both systems have advantages, and specifying a color sample with one system allows conversion and notation to the other. The obvious advantage of the Munsell system is its simplicity and its directness. The selection of one of 1,490 color chips provides a color and color notation directly in terms of hue,

value, and chroma (H V/C). It is a physical match easily made and easily understood. The CIE system is somewhat abstract and is based on mathematical conventions. It requires instrumentation but has the advantage that any color can be matched, numerically specified, and positioned on a CIE diagram or map. The CIE system provides psychophysical measurements of color, while the Munsell system, based on color appearance, provides psychological or perceptual measures.

CIE CHROMATICITY MAP

Chromaticity describes two attributes of color: hue and chroma. These two qualities, called dominant wavelength and purity in the CIE system, are plotted on a horseshoe-shaped map, shown in Figure 14–9. Around the periphery of the curved section are wavelengths of light scaled from 380 to 770 nm. The x and y axes are used to plot the position of the chrominance of a particular color. The color is always plotted relative to a specific daylight or tungsten light source. The daylight source is located in the lower middle of the map, where it serves as a neutral reference point. A green filter is identified by the coordinates x = .22 and y = .52. Not shown is

Munsell notations can be converted to CIE notations and vice versa.

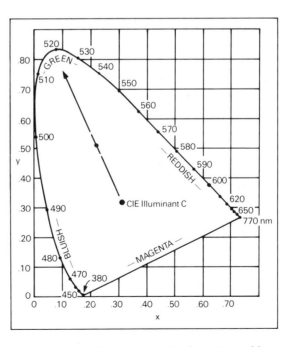

Figure 14–9 Chromaticity plot for a Green filter relative to a daylight source.

the luminance (brightness) of the green. This value is obtained separately from the data used to calculate the x and y coordinates. The farther a color plots from the reference neutral, the greater the purity (chroma). The dominant wavelength (hue) is determined by drawing a line from the reference point through the plotted point to the curved wavelength line. This green filter has a dominant wavelenth of 520 nm.

REVIEW QUESTIONS

1. The most appropriate definition of color from the following choices is that color . . . (p. 320)
   A. is the wavelength attribute of light
   B. is the attribute of light associated with names such as red, green, and blue
   C. is the quality of all light that is not white
   D. is a perceptual response to light
2. With respect to the relationship of pigments, dyes, and colorants, all . . . (p. 320)
   A. colorants are dyes
   B. pigments are colorants
   C. dyes are pigments
   D. colorants are pigments
3. Of the light-response systems listed, the one that is most similar to a spectrophotometer in principle is . . . (p. 320)
   A. an exposure meter
   B. the human eye
   C. a color densitometer
   D. black-and-white panchromatic film
   E. color film
4. In addition to the color of light that cyan dyes are supposed to absorb, they also absorb considerable . . . (p. 323)
   A. red light and green light
   B. red light and blue light
   C. blue light and green light
5. The mode of the blue seen in blue sky on a clear day is classified as . . . (p. 325)
   A. surface mode
   B. volume mode
   C. aperture mode
   D. illumination mode
   E. illuminant mode
6. White, gray, and black are . . . (p. 326)
   A. chromatic colors
   B. subtractive primary colors
   C. not colors
   D. neutral colors
7. The Munsell term *value* corresponds to the more general term . . . (p. 328)
   A. color
   B. lightness
   C. saturation
   D. harmony
   E. darkness
8. The five principal hues in the Munsell system are . . . (p. 329)
   A. red, green, blue, yellow, and gray
   B. cyan, magenta, yellow, white, and black
   C. yellow, red, green, blue, and cyan
   D. red, green, blue, yellow, and purple
9. A color designated as 5R 8/2 is a . . . (p. 329)
   A. light red
   B. dark red
   C. medium value red

15 Color Reproduction

Ellen Current.

OBJECTIVES OF COLOR REPRODUCTION

Generally, the objective of color photography is to produce either an accurate color photograph or a pleasing color photograph.

When considering the properties required of color photographic materials, it is necessary to include the purpose for making the photograph. Generally, the objective of color photography is to produce either an accurate color photograph or a pleasing color photograph. Accurate color reproduction implies a certain correspondence of the colors in the photograph to those of an original scene. If it is a photograph of an object or scene with which the viewer is familiar, it would at least be required that the viewer would not name any of the colors in the picture differently from those in the original scene, based upon memory. With a much more demanding form of accurate color reproduction, the photograph is held in the hand and critically compared directly to the original scene or object. Except with very simple subject matter, there are no color photographic systems capable of fulfilling this definition of accurate color reproduction.

There are at least three major reasons why color photographic systems are unable to achieve accurate color reproduction; they are physical, physiological, and psychological in nature. The most troublesome of the physical reasons is that the dyes employed to construct the photographic image are not the same colorants that existed in the original subject. This will affect the range of colors that can be reproduced. Furthermore, the dyes have unwanted absorptions, which limit their ability to accurately simulate the real-life colors.

The human visual system adapts its color sensitivity to different visual environments, whereas color films do not.

Among the physiological factors is the fact that the human eye records the original subject colors in a fashion far different from that of a color film. Since the spectral response of color film differs from that of the eye, the various wavelengths of light reflected from the subject will be encoded (analyzed) in a different way. Additionally, since the eye adapts its sensitivity to the current visual field, its spectral response changes frequently.

Finally, it is obvious that a photograph is not an original scene, but rather something derived from it. It is two-dimensional, typically much smaller than the original scene, and viewed under significantly different lighting conditions than the original scene. Thus the perceptual conditions are significantly different when viewing a photograph than they are when viewing the original scene. In other words, the dyes in the photograph may be physically identical to those in the original subject and present in the same quantities, but still not give an accurate color reproduction because of these psychological differences.

Most often the goal is pleasing color reproduction, which can also be defined in a variety of ways. For example, it is possible to differentiate between acceptable color reproduction and excellent color reproduction. Acceptable color prints can be defined as those containing a reasonable resemblance to the colors in the original scene. The color prints produced by amateur photofinishing labs using automated equipment and the color prints from instant cameras are examples of generally acceptable color photography. Every color photographic process has certain defects and limitations, which may work strongly against a particular subject color. However, for every subject, a most pleasing print can be made by whatever process is used.

To obtain such an excellent print without substantial retouching requires a very flexible process such as the dye-transfer process and expert manipulation of its variables. This is the nature of the work performed by professional photographers and custom color-printing laboratories, and these are examples of what is termed excellent color photography. Obviously the differences between acceptable and excellent color reproduction are a matter of opinion, and as such are highly dependent upon the audience. Thus the definition of pleasing color reproduction will ultimately be determined by the user.

It is safe to say that if accurate color reproduction were the most important objective in color photography, the wide proliferation of color photographic systems would never have occurred. The majority of photographers require only pleasing color reproduction, which is the principal objective of most color photographic endeavors. Cost and convenience may be as important to users as the quality in selecting a color system.

ADDITIVE AND SUBTRACTIVE COLOR FORMATION

All color reproduction systems operate on the assumption that the human eye contains three different types of color receptors. Many experiments show that almost all colors can be produced through the appropriate mixture of red, green, and blue light. Although many theories of color vision have been proposed and used to describe various visual phenomena, the trichromatic theory of color vision offers the most satisfactory explanation of color perception as it relates to color photography. Thus in order for a color photographic system to be successful, it must be capable of recording and controlling the red, green, and blue components of the light being reflected, transmitted, or emitted by the subject. There are two basic methods for controlling the color of the light produced in a color reproduction system: additive color mixture, which involves combining red, green, and blue light; and subtractive color mixture, which involves colorants that selectively absorb red, green, and blue light.

In the additive system it is possible to produce a wide variety of colors by mixing various amounts of red, green, and blue light. This can be illustrated most directly by considering three identical projectors, each equipped with a different-colored filter (red, green, and blue) and aimed at a projection screen. When the projector with the blue filter is turned on, this causes blue light to reach the screen; and if the projector is equipped with a variable aperture, the illuminance of the blue light can be increased or decreased, producing lighter or darker colors of blue. If the second projector, employing a green filter, is turned on, green light will reach the screen and can be made to partly overlap the light from the blue projector, allowing many additional colors to occur. In addition to the scale of greens that can be produced where only the green light is striking the screen, a series of blue-greens (cyans) occurs where the two projector beams overlap. This latter series of colors could be made not only to differ in lightness by increasing or decreasing the size of both apertures simultaneously, but also in greenness and blueness

by changing the relative amounts of green light and blue light.

The addition of the third projector, utilizing a red filter, provides the red light on the screen, which again can be made to partially overlap the other two projected areas. Where the red light overlaps the blue light a new blue-red (magenta) color is produced, which likewise can be altered in lightness and blueness-redness. Where the red light overlaps the green light the color yellow is produced, which can also be manipulated by controlling the individual projectors.

Such a triple-projection system is illustrated in Plate XIII. Each of the three beams of light is falling individually on the screen and also overlaps the other beams. The color cyan appears where the blue and the green beams overlap. The overlapping of the red and blue beams produces the color magenta. That cyan is formed where the blue and green overlap is not surprising, nor is the formation of magenta from the combination of blue and red light. Visually, the colors produced are consistent with the contributions made by the primary colors. However, for the mixture of red light and green light to appear yellow is a somewhat amazing result, since yellow in no way resembles the red light and the green light. This phenomenon is related to the theory of color vision. Where the beams of the three projectors overlap, the color white is produced. The area of the screen where no direct light from any projector is falling appears black. This illustrates the fundamental method of additive color formation.

The word *additive* implies that different colors are produced by adding together different amounts of the three colors of light. All additive color reproduction systems require the use of the three colors represented in this example: red light, green light, and blue light. Consequently, these colors are referred to as the additive primaries. Table 15–1 lists the basic facts of additive color formation using these three additive primaries.

By effectively controlling the operation of the three projectors, practically any color can be produced. Thus the attributes of color—hue, lightness, and saturation—can all be controlled by changing the amounts of red, green, and blue light falling on the screen. About the only colors that can't be produced

Both additive and subtractive color reproduction systems are based on controlling red, green, and blue light.

There are various ways by which red, green, and blue light can be combined, including superimposing, spatial blending, and temporal blending.

Table 15–1   Basic facts of additive color formation (the mixing of lights)

Lights	Produce
Blue light plus green light	Cyan
Blue light plus red light	Magenta
Green light plus red light	Yellow
Blue light plus green light plus red light	White
No light	Black

are spectral colors, since they are too saturated to be simulated with this method.

There are two additional ways to employ the principles of additive color formation. A single projector can be used to project three colored images in rapid succession with red, green, and blue following each other so rapidly that the eye cannot distinguish the individual projected colors. Thus the mixing of the three lights is achieved through the perceptual phenomenon known as persistence of vision.

This technique is referred to as temporal color mixture. In an alternative process, small transparent bits of red, green, and blue are placed side by side to form the picture. The bits are so small that the eye cannot readily distinguish them as colored spots. This is the method used in color television; the picture is composed of small disks of red, green, and blue light from the phosphors coated on the screen. Observed from the proper viewing distance, these discs blend to form the color image, and with their intensities electronically controlled they reproduce the colors of the original subject. This method is often referred to as spatial color mixture.

In summary, additive color formation is characterized by the mixing of lights. Since the human color visual system has three different types of color receptors (red, green, and blue), the colors of the three lights used must be red, green, and blue. The additive mixture of lights can be achieved through either temporal fusion or spatial fusion, as well as by physically mixing the lights.

The alternative to additive color formation is subtractive color formation. It is characterized by the mixing of colorants (dyes, pigments, paints, and inks). Although subtractive synthesis involves a significantly different principle, such a system can produce nearly the same range of colors as the additive sys-

tem. In order for a subtractive system of color formation to be successful, the colorants must be capable of controlling the red, green, and blue light components of the white viewing light. As a result, there are only three colorants that can meet this requirement: cyan, magenta, and yellow. Cyan is used because it absorbs red light and reflects (or transmits) blue and green light, creating its blue-green appearance. Magenta is used because it absorbs green light and reflects (or transmits) blue light and red light, causing its blue-red appearance. Yellow is used because it absorbs blue light and reflects red and green light. As a result, the subtractive primaries of cyan, magenta, and yellow can be summarized as follows:

$$cyan = minus\ red$$
$$magenta = minus\ green$$
$$yellow = minus\ blue$$

Plate XIV illustrates the results of mixing various amounts of the cyan, magenta, and yellow colorants. Initially the paper is illuminated by white light, which contains nearly equal amounts of red, green, and blue light. Since the paper is white, it reflects white light, again consisting of nearly equal amounts of red, green, and blue light. Where the yellow colorant is mixed with the cyan colorant, the area of overlap produces the color green. In other words, the yellow colorant absorbs the blue portion of the white light while the cyan colorant absorbs the red portion of the white light, leaving only the green to be reflected back to our eyes. Similarly, where the yellow and magenta colorants overlap, the color red appears. This is the result of the blue light absorption of the yellow paint and the green light absorption of the magenta, leaving only the red light to be reflected to our eyes. Finally, where the cyan and magenta colorants overlap, the color blue appears. This occurs because of the red light absorption by the cyan colorant and the green light absorption by the magenta, leaving only the blue light component of the original white light to be reflected to our eyes. In this fashion, a wide variety of hues can be reproduced by mixing varying amounts of these three colorants.

Furthermore, as seen on the righthand side of Plate XIV, the three subtractive primaries of cyan, magenta, and yellow can be mixed

Cyan, magenta, and yellow subtractive primary colorants respectively absorb red light, green light, and blue light.

in the proportion required to produce a scale of neutrals from black to white. Table 15–2 summarizes the information contained in Plate XIV. The use of the term *subtractive color formation* for this system is appropriate, since it depends upon the use of substances that remove varying amounts and colors of light from an initially white source of light.

It is important to point out here that artists and photographers often use different names for the same colorants. Artists and printers often refer to their three primary colors as red, blue, and yellow. However, if the range of colors to be produced is to be as complete as possible, the red paint (or ink) must actually be magenta, and the blue paint (or ink) must actually be a blue-green or cyan. It is unfortunate that the terms *red* and *blue* have been used to describe *magenta* and *cyan*, because their use in this context has undoubtedly limited the understanding of the principles of subtractive color mixture.

THE RECORDING OF COLOR (ANALYSIS STAGE)

All color photographic systems consist of two major stages: analysis and synthesis. First, the colors of light from the subject must be recorded in terms of primary colors—typically, red, green, and blue—which is referred to as the analysis stage. Second, when the photographic image is reconstructed, the process must be able to control the red, green, and blue light that will ultimately reach the viewer of the image. This is referred to as the synthesis stage.

The oldest and perhaps most fundamental method of achieving the first stage is to use red, green, and blue filters with black-and-white panchromatic film (which is sensitive to red, green, and blue light). Plate XV illustrates this approach to color analysis. The subject is first photographed through each of the filters on three separate sheets of film. After processing, the negatives reveal three separate records of the subject and are referred to as separation negatives. The red separation negative is dense in the areas where red light was reflected from the subject. Similarly, the green and blue separation negatives are records of the green and blue content

Table 15–2 Basic facts of subtractive color formation (mixing of pigments)

Colorants	Absorbs	Produces
Cyan	Red	Blue-green (cyan)
Magenta	Green	Blue-red (magenta)
Yellow	Blue	Green-red (yellow)
Cyan plus magenta	Red and green	Blue
Cyan plus yellow	Red and blue	Green
Magenta plus yellow	Green and blue	Red
Cyan plus magenta plus yellow	Red, green, and blue	Black
No colorant	Nothing	White

of the subject. Notice that the yellow patch of the subject is recorded as a dense area in both the red and the green separation negatives. This is because a yellow object absorbs only blue light and reflects both red and green light. Therefore, both the red and green negatives record light from the yellow patch and produce a heavy density.

The exact density information recorded in this tricolor analysis is determined by various factors including the color quality of the illumination; the spectral characteristics of the subject; the transmission characteristics of the red, green, and blue filters; and the spectral sensitivity of the photographic emulsion. A wide variety of tricolor separation filters and photographic emulsion characteristics have been used for tricolor photography. One combination consists of Eastman Kodak Wratten No. 25 (red), No. 58 (green), and No. 47B (blue) filters with Eastman Kodak Super XX black-and-white film. Figure 15–1 illustrates the spectral density curves for these filters and the spectral sensitivity curve of Super XX. Figure 15–1 shows the way in which this combination of filters and film would respond to the red, green, and blue components of the light reflected from the scene.

The color-analysis method used with modern color films is the multilayer approach, whereby three emulsions are coated one above the other on a single base, as illustrated in Figure 15–2. The emulsions are typically separated by interlayers of clear gelatin or colored layers that act as filters. The color analysis is achieved by limiting the emulsion layer sensitivities to the three (red, green, and blue) spectral bands required.

The spectral sensitivities of the multilayer camera speed film are shown in Figure 15–3.

Process red is actually magenta, and process blue is actually cyan.

Analysis and synthesis are involved in all color reproduction processes.

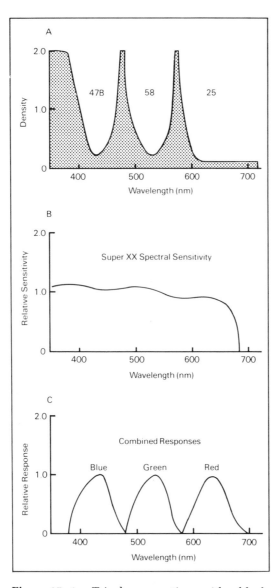

Figure 15–1   Tricolor separations with a black-and-white film. (A) Spectral density curves for tricolor separation filters 47B (blue), 58 (green), and 25 (red). (B) Spectral sensitivity of Eastman Kodak Super XX black-and-white film. (C) Resulting spectral responses when tricolor filters are used with Super XX film.

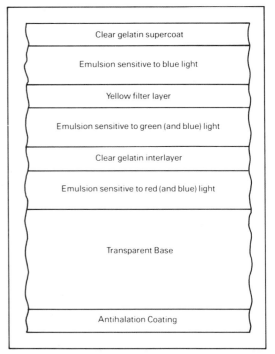

Figure 15–2   Cross section of a typical integral tripack color film (not drawn to scale).

Since all three emulsion layers of conventional color films are sensitive to blue light, a yellow filter must be used to prevent the blue light from reaching the two bottom emulsion layers.

On the left are shown the inherent sensitivities of the blue-, green-, and red-sensitive layers, where it is shown that the green- and red-recording layers also have blue sensitivity. This problem is minimized through the use of a yellow filter layer above the green- and red-sensitive layers. The top emulsion is sensitive only to blue light, and the green and red light pass through it without effect, al-

lowing the blue light alone to be recorded in that layer. The yellow filter layer immediately below the top emulsion layer absorbs the blue light and prevents it from reaching the two lower (green- and red-sensitive) emulsion layers. Thus the effective spectral sensitivities of the three layers are shown on the left of Figure 15–3.

Consequently, multilayer films possess three more or less independently sensitized emulsions. Such a process requires only a single camera exposure to produce the red, green, and blue record negatives. The three negatives will be stacked on top of each other and, consequently, are inseparable.

At this point it is important to understand what each of the separation negatives represents. The red record negative must contain a record of only one kind of information; the densities in this negative must be present only where there was red light being reflected from the subject, and it must not record the blue or the green light. Similarly, the green and blue record negatives must record only their respective colors and no others.

THE PRODUCTION OF COLOR (SYNTHESIS STAGE)

Once the subject colors have been analyzed, separated into their red, green, and blue components, they must be put back together by some process that re-creates the appearance of the original colors. This step is referred to as the color synthesis stage and may be achieved by either the additive or the subtractive methods of color formation previously described.

In all color photographic processes the viewer is interested in seeing the picture as it originally appeared—as a positive image. This can be achieved by making three positive silver images, one from each of the three silver separation negatives. Since the densities in the red record negative are formed where red light is reflected from the subject, the densities in the positive will have an inverse relationship with the red content of the scene. The red record positive has the greatest density in those areas where the least red light was reflected from the subject. Similarly, the green and blue record positives will have the greatest density in the areas reflecting the least green and blue light, respectively.

The reconstruction of the subject colors by using these three positives is illustrated in the portion of Plate XV labeled the synthesis stage. Here, three projectors are equipped with red, green, and blue filters. The red record positive is placed in the projector equipped with the red filter. Since the red record positive contains the most density in the areas that reflected the least red light, it will in those same areas prevent red light from reaching the screen. This same positive is thin in the areas that reflected much red light from the subject, and will for those areas permit red light to reach the screen. Therefore the red light is reaching the screen in approximately the same relative amounts as red light was reflected from the subject.

Similarly, the green and blue record positives are placed in the projectors equipped with the green and blue filters. The three images are projected on the screen so that they are exactly superimposed. When this procedure is carefully performed, a full-color reproduction of the original subject will be formed on the screen. Since this color for-

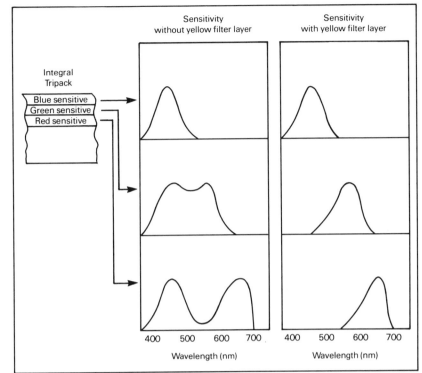

Figure 15–3 The spectral sensitivity of an integral tripack color film with and without the yellow filter layer.

mation is the result of mixing lights, it is defined as an additive color reproduction system.

The relationship between the densities of the separation negatives and the densities of the separation positives is summarized in Table 15–3 for the simple subject matter shown in Plate XV. Notice that for the yellow pot the absence of density in the red and green record positives allows red and green light to reach the screen, where they mix together to produce the color yellow. For the black base, no light reaches the screen from any of the projectors. For the white background, all three projectors produce light on the screen, creating the color white.

If the color analysis stage involved a single exposure through a color mosaic screen, the separation negatives will exist side by side in the emulsion. A positive image must be obtained with reversal processing or by producing a negative and contact printing it onto another piece of film. The positive is then placed in register with the original color mo-

The densities in a red record separation negative indicate where red was present in the original subject.

Table 15–3   The relationships between separation negatives and separation positives in terms of density

Subject	Densities of Separation Negatives			Densities of Separation Positives		
	Blue Negative	Green Negative	Red Negative	Blue Positive	Green Positive	Red Positive
Red flower	Low	Low	High	High	High	Low
Green leaves	Low	High	Low	High	Low	High
Yellow pot	Low	High	High	High	Low	Low
Black base	Low	Low	Low	High	High	High
White background	High	High	High	Low	Low	Low

With reversal color films a red image is obtained by removing blue and green from the white viewing light with yellow and magenta dyes.

saic screen, and the combination is viewed by transmitted light. Such a system is illustrated in Plate XVI. Again, the synthesis stage of this process involves a mixture of red, green, and blue light transmitted through the filter mosaic; therefore it is described as an additive color reproduction system.

The color analysis method that employs a multilayer film as the recording medium contains a set of separation negatives stacked vertically on top of each other. Because the red, green, and blue records are superimposed and inseparable, it is impossible to employ the active system of mixing red, green, and blue lights to form the color reproduction. Consequently, for multilayer materials an alternative approach to color formation must be used.

In the additive color synthesis, silver positives were used to absorb the appropriate amounts of the red, green, and blue light in the three-projector system. The alternative approach is to substitute colors (dyes), superimposed in the emulsion, for the three silver positives. In the red record it is necessary to use a dye that absorbs only the red part of the spectrum. A properly formed cyan dye image is nearly as effective in absorbing red light as is the red record silver positive image. The cyan dye will allow the blue light and green light to be transmitted. In the green record positive it is necessary to use a dye

that absorbs only green light. Thus magenta dye is used since it absorbs green light and transmits blue and red light. In the blue record positive, yellow dye absorbs only blue light and allows red light and green light to pass.

Therefore, if the multilayer separation positive consists of cyan, magenta, and yellow dyes, the image can be projected using a single projector. This arrangement is illustrated in Plate XVII. It is successful because the dye images, unlike silver, absorb only one color of light, and allow the other two to be transmitted. Therefore, by combining the three dye images in exact registration a tripack image is produced, necessitating only one source of light for viewing. The red portion of the white light is controlled by the density of the cyan dye image alone, while the green portion of the white light is being controlled by the magenta dye image alone. The blue portion of the white light is controlled by the yellow dye image. In this fashion the cyan, magenta, and yellow dye images act as light valves, absorbing appropriate amounts of red, green, and blue light. Since this system involves the removal of red, green, and blue light from the initially white viewing light, it is defined as a subtractive color reproduction process.

Table 15–4 summarizes the effects of the three dye positives for the simple subject considered in Plate XVII. The process affected some of the subject areas in these ways:

1. *The green leaves.* These were recorded only in the green record negative as a heavy density while the other two negatives were thin in the area representing this portion of the subject. When the positives were made, no dye was formed in the magenta positive and a large amount of dye was formed in both the cyan and yellow pos-

Table 15–4   Analysis of the yellow, magenta, and cyan dye separation positives in terms of the amount of dye formed

Subject	Yellow Positive	Magenta Positive	Cyan Positive
Red flower	Large	Large	No
Green leaves	Large	No	Large
Yellow pot	Large	No	No
Black base	Large	Large	Large
White background	No	No	No

itive images. In the final reproduction shown in Plate XVII, the cyan and yellow dyes are superimposed, producing the color green. Since the cyan dye absorbs the red light and the yellow dye absorbs the blue light, the combination absorbs all but the green from the white viewing light.

2. *The yellow flower pot.* This is recorded as a heavy density in both the red and the green record negatives, but as a thin area in the blue record negative. Since the positives are reversed in density, only the blue record positive will contain any dye. The yellow dye absorbs blue from the white viewing light and this area appears yellow (minus blue).

3. *The black base of the flower pot.* This part of the scene was recorded as a thin area in all three negatives, and as a dense area in all of the positives. This area was therefore recorded as a heavy deposit of all three dyes, which when superimposed absorb red, green, and blue light, leaving very little light to reach the viewer.

4. *The white background.* This was recorded as a heavy density in all of the negatives and as a thin area in all of the positives. Since no dye was formed in any of the positives, practically none of the white viewing light is absorbed.

Even with a subject as simple and containing as few colors as this, such a process of subtractive color reproduction can produce a tremendous variety of subject colors. Furthermore, the use of the subtractive process allows for the production of both transmission and reflection color images.

SPECTRAL SENSITIVITY

For a color photographic system to reproduce colors approximately as the eye perceives them, the responses to red, green, and blue light must bear the same relationship to each other as do the responses of the eye to these colors. For example, if a reversal color film has relatively too much sensitivity to blue light, blue objects in the scene will appear too light in the color reproduction—assuming the exposure is correct for the green and the red objects—and white objects will appear bluish. Thus one of the requirements for accurate color reproduction is that the red, green, and blue sensitivity of the initial color recording (analysis stage) must match those of the human eye.

Matching the color sensitivity of the three receptor systems involved in human vision is a very complex task, since the human visual system is constantly changing in its response to light. The perceptual phenomena of color adaptation and color constancy are impossible to duplicate with a color film. The human color visual system can adapt to many different forms of color balance, while a color film is limited in its response to a single adaptation level, and therefore has a given color balance that is determined at the time of manufacture. In the negative/positive process, it is relatively easy to adjust the color balance during the printing operation. However, reversal color films provide less opportunity for change, since the image exposed in the camera will be that which is viewed as the final positive.

Since a color film can record the light reflected from the subject with only one particular set of sensitivities, the optimum reproduction of color in a reversal film can be obtained only when the illumination is of the particular color quality for which the film is balanced. As a result, film manufacturers market multilayer camera films of three different spectral responses. Daylight-type color films are designed to be exposed to light with a color temperature of 5500 K, which is the mixture of sunlight and sky light most commonly encountered outdoors. Type A films are designed to be used with light sources operating at 3400 K, which is produced by special photoflood lamps. Type B films are balanced for light sources operating at 3200 K, which is the most commonly encountered type of tungsten lamp in photographic studios.

The difference between color films balanced for 5500 K (daylight) and 3200 K (tungsten lamps) is most easily seen by inspecting the spectral sensitivity curves of the films in Figure 15–4. Such curves indicate the sensitivity to light, wavelength by wavelength, for each of the film's three layers. The principal difference between the responses of the two films is in the red-sensitive (cyan-dye-forming) layer, where the response is much lower in the tungsten film than in the day-

Daylight type color films have lower blue sensitivity and higher red sensitivity than tungsten type color films.

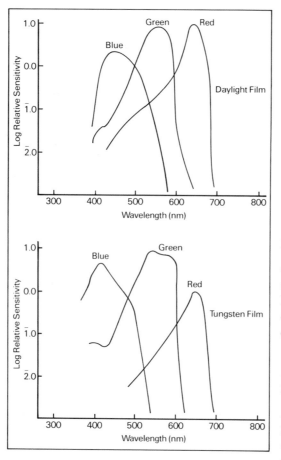

Figure 15–4 Spectral sensitivity curves for daylight-balanced and tungsten-balanced color reversal films of the same speed without the yellow filter that is beneath the blue-sensitive emulsion. This filter prevents blue light from reaching the blue-and-green sensitive and the blue-and-red sensitive emulsions.

Cyan dyes absorb considerable blue light and green light even though they should transmit both colors freely.

resulting transparencies would show a strong bluish cast. Similarly, if the daylight-balanced film were used with a tungsten source without filtration, the resulting color transparencies would have a reddish cast to them. Therefore, when the color film is to be exposed to light of a color quality other than that for which it was balanced, light-balancing filters must be used. Even when appropriate light-balancing filters are used over the camera lens, errors in the rendering of the colors of certain objects may result, and only a practical test of the film, filters, and light source will determine if the results are suitable. Also, notice that there are no color films balanced for fluorescent lighting. Consequently, all color films invariably require color-correction filters for fluorescent lights. Proper filter use depends upon the film's original color balance and the type of fluorescent lamp used.

DYE CHARACTERISTICS

As discussed previously, subtractive color formation involves the mixture of the three dyes: cyan, magenta, and yellow. The purpose of the dyes is to control the amount of red, green, and blue light that will be reflected (or transmitted) to the viewer. Therefore, each of the dyes should absorb only one color of light: the yellow dye should only absorb blue light; the magenta dye should only absorb green light; and the cyan dye should only absorb red light. In other words, each dye should only absorb its complementary color and allow the other two primary colors of light to be transmitted freely. However, even the best available dyes, pigments, and printing inks absorb some colors of light that they should transmit: (1) Yellow dye absorbs a small amount of green light; (2) magenta dye absorbs a considerable amount of blue light and a smaller amount of red light; (2) cyan dye absorbs significant amounts of green light and blue light.

Perhaps the easiest way to define the dye properties is to examine the spectral dye density curves of a typical set of cyan, magenta, and yellow dyes. Such a graph illustrates the wavelength-by-wavelength density of each dye. In Figure 15–5, graph A illustrates the spectral dye density curves for a theoretically

light film. This is because tungsten sources produce relatively large amounts of red light compared to blue light, whereas daylight has more nearly equal amounts of blue light and red light. Thus the low red sensitivity of the tungsten-balanced film compensates for this high proportion of red light in tungsten sources, producing a correctly balanced color image.

The blue sensitivity of the green (magenta-forming) layer and the red (cyan-forming) layer of both films is essentially eliminated by the yellow filter interlayer that is coated between the top blue-sensitive layer and the bottom two layers in the multilayer films. If the tungsten-balanced film were used with daylight illumination without filtration, the

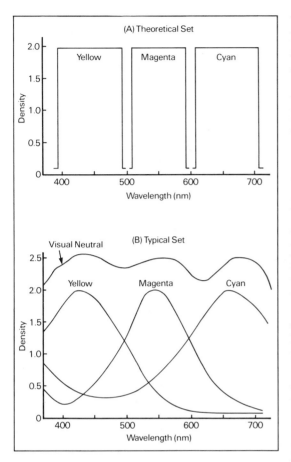

Figure 15–5 Theoretical and typical spectral dye density curves of a color film. (A) Theoretical set. (B) Typical set.

perfect set of dyes. It is shown that each dye should absorb only one-third of the spectrum and allow the other two-thirds to be transmitted; and, therefore, provide for an independent control for each of the red, green, and blue light regions.

Graph B shows the spectral dye density curves for a typical set of cyan, magenta, and yellow dyes. The yellow dye shows a high density to blue light (its primary or "wanted" absorption), a small amount of unwanted absorption in the green region, and practically no absorption in the red region. The magenta dye shows a high density in the green region (its primary absorption) and a significant amount of unwanted absorption in the blue and the red regions. The cyan dye shows a high density to red light (its primary absorption) and a significant amount of unwanted absorption in the green and blue regions. Based upon the amount of unwanted absorp-

tions, it appears that the yellow dye is the purest and the cyan dye is the least pure, giving the greatest amount of unwanted absorptions.

In order to understand the effects of these unwanted absorptions, consider the problems of attempting to form a neutral (gray) color. Assume that an initial deposit of cyan dye, which has a primary absorption of red light, is produced in the film. Since the cyan dye has unwanted green light absorption, a lesser amount of magenta dye is required than would otherwise have been used. The magenta dye has significant unwanted blue light absorption, requiring that a lesser amount of yellow dye be formed than would otherwise have been necessary. Therefore, the neutral is formed from unequal amounts of the three dyes because of their unwanted absorptions. Although it is possible to produce a visual neutral in this fashion, the reduction in the magenta and yellow dyes of the process will influence the rendering of other colors such as red, since it is formed from magenta and yellow dyes; green, which is formed by mixing yellow and cyan dyes; and blue, which is formed by mixing cyan and magenta dyes. In other words, these colors and others will be reproduced inaccurately.

This illustrates the fact that the nature of the dyes makes it impossible to obtain accurate reproduction of all colors. Fortunately, accuracy in color reproduction is seldom required; pleasing color reproduction is the most obvious goal. Color films are invariably designed so that the dyes produce the most pleasing color renditions of such important subjects as skin, grass, the sky, and others familiar to memory. Since color processes are adjusted to produce pleasing reproductions of these memory colors, it is not surprising that slight departures from neutrality occur in the rendering of grays.

This compromise in color reproduction, which is the result of unwanted dye absorptions, does not seriously degrade the quality of a color print or transparency that is to be viewed directly. However, if the color print or transparency is to be duplicated again with the photographic process, the errors in the original are compounded by the errors due to the colorants used in the duplicate reproduction. Consequently, as the chain of color photographic stages gets longer (a duplicate from a duplicate from a duplicate, etc.), undesirable changes in color rendering as well as contrast increase in magnitude.

COLOR DENSITIES

Neutral gray areas in color transparencies do not transmit equal amounts of all wavelengths of the white viewing light.

The measurement of density for colored photographic materials is complicated by the nature of the dye images formed. The spectral dye density curves for the three dyes of a typical color process, together with the curve representing the visually neutral image formed when these dyes are superimposed, are shown in Figure 15–5B. As noted previously, each dye has appreciable density to all wavelengths of light. Therefore, every measurement of the density of a color image (either negative or positive) is simultaneously affected by all three dyes. Furthermore, the visual neutral represented by the top curve, which is produced by mixing these three dyes, departs from a "perfect" neutral, which would yield the horizontal straight line. Since the combined densities of the three dyes is not constant at all wavelengths, density measurements of color images will be greatly influenced by the color characteristics of the light source as well as the color response of the receptor in the densitometer.

Since the structure of the dye clouds that comprise the color image is much finer than the grains of silver that comprise the image in a black-and-white material, problems arising from light scattering in color materials are much less serious. Consequently, the differences between specular and diffuse densities for dye images are quite small, with the Q factor approaching 1.0.

Color films consist of three dye layers, each with different spectral characteristics. Since it is possible to generate either one, two, or all three of the dyes in a color image, it is necessary to distinguish between such images. If a color image has all three dyes present, it is referred to as an integral image. If a color image contains only one of the three dyes, it is referred to as an analytical image. The differences in the nature of these images are the source of the following two fundamental classifications of color densities:

1. *Integral densities.* Integral densities are measures of the combined effect of the three superimposed dye layers. Since each dye has significant absorptions to all wavelengths of light, no information about individual contributions of the dyes can be obtained.

2. *Analytical densities.* Analytical densities are measures of the individual contributions of the dye layers to the total image and indicate the composition of the image in terms of the amounts of the dyes present.

In practice, integral densities are far more useful, since they describe the performance of the image. Therefore, integral density measurements are usually satisfactory for the consumers of photographic materials. However, since they give no direct information about the image composition, integral densities are usually insufficient for the manufacturer of color materials. During their manufacture, color emulsions are made and coated separately, and thus the manufacturer must have insight into the nature of each layer. Analytical images can be obtained by exposing and processing individual coatings of each layer. An alternative method is to expose a multilayer emulsion through a very saturated narrow-band filter, thereby producing dye in only one layer.

The density obtained with a blue-filter densitometer reading of a neutral area on a color transparency is identified as an integral density since it represents the combined densities of all three dye layers with blue light.

If all three dyes are present in the image, it is defined as an integral image, and any density measurement made on it is termed an integral density. Likewise, when only one dye layer is present, it is defined as an analytical image and any measurement made on it is termed an analytical density.

Each of these two types of color densities can be classified in terms of the information supplied by the measurements. The following sub-classifications are typically used:

1. *Spectral densities.* These measure the wavelength-by-wavelength density of the image, and the results are typically displayed as a graph. If all three dyes are present in the image, the resulting data are termed integral-spectral densities, and if only one of the three dye layers is present, the results are termed analytical-spectral densities. Referring to Figure 15–5B, the uppermost curve would be termed an integral-spectral density plot, and the three curves below would be defined as analytical-spectral densities since they relate

to each of the three dyes. Such information is fundamental to studies of the dyes used in color photographic materials.

2. *Wide-band three-filter densities.* These densities are based on the use of three arbitrarily chosen red, green, and blue filters. The densities are termed *wide-band* since the filters have band widths of 30 nm to 60 nm. Thus the data that are derived from a color densitometer describe the red-, green-, and blue-light-stopping abilities as dictated by the filters that were used in the instrument.

In addition, the spectral response of the densitometer's photo detector can have a significant effect on the resulting measured values. Therefore, for the purposes of standardization, the American National Standards Institute has defined the following two conditions for color densitometry:

1. *Status A densitometry:* A set of red, green, and blue response functions for the measurement of color photographic images intended for direct viewing (either transmission or reflection). The standard response functions for status A are illustrated in Figure 15–6. In order to achieve these aim response functions, densitometer manufacturers must select a light source, a filter set, and a photodetector whose combined response will approximate

(within 5%) these response functions. This is most commonly achieved through the use of a tungsten lamp operating at 3000 K, an S-4 photo-detector, and the use of Eastman Kodak Certified AA red, green, and blue filters. The peak responses of these functions correspond fairly closely to the peak densities for each of the three dyes in a typical color-positive material. Furthermore, the response functions are narrow enough to allow fairly high densities (above 2.0) to be measured with good integrity (no cross-talk between the spectral bands). Thus a densitometer equipped with such a set of response functions provides color densities capable of detecting small changes in the densities of the image. It is important to note, however, that such red, green, and blue densities are not a direct indication of the appearance of the colors in the image.

2. *Status M densitometry:* The standard used when reading the densities of color images that are to be printed. The red, green, and blue aim response functions for status M densitometry are illustrated in Figure 15–7. Again, densitometer manufacturers must select a combination of lamp, filter set, and photo-detector response that approximates these response functions. The use of a tungsten lamp operating at 3000 K, the S-4 photo-detector, and Eastman Kodak Certified MM filters meets this re-

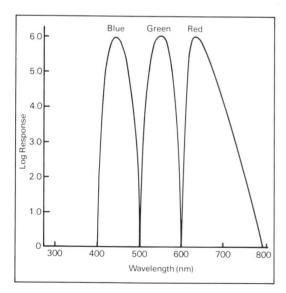

Figure 15–6 Red, green, and blue aim response functions for status A densitometry.

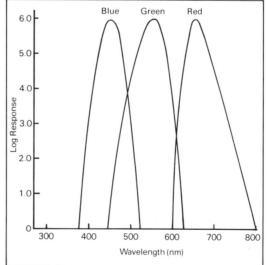

Figure 15–7 Red, green, and blue aim response functions for status M densitometry.

quirement; and densitometers so equipped are said to read "Status M Density."

Notice that the response functions for status M show considerable overlap when compared to the status A response functions (i.e., they have greater cross-talk). This is intended to simulate the red, green, and blue spectral response functions encountered in color print emulsions. Thus densitometers equipped with these response functions give densities that are descriptive of the printing properties of the image.

The benefit of using densitometers with these standardized response functions is principally a reduction of interinstrument variability. The result is that densitometers from the same manufacturer and densitometers from different manufacturers tend to show better agreement when measuring the same image. As noted above, if the image being measured consists of all three dyes superimposed, the densities are referred to as integral; and if only one of the three dyes is present, such wide-band densities are termed analytical. This type of color densitometry is more commonly employed on integral images to obtain information about the performance of the color materials. Further, such readings are often used in conventional process monitoring where it is necessary to determine significant changes in the processing conditions by using standardized control strips.

Regardless of the use of the readings and the filter set actually employed, it is necessary to check the instrument's behavior periodically by using statistical control methods. Periodic readings should be made on standard control patches to detect unwanted changes in the densitometer's response.

Figure 15–8 is a simplified diagram of the basic components of a transmission color densitometer. The basic parts are (1) a light source that emits light in all regions of the visible spectrum, typically a tungsten lamp operated at 3000 K; (2) a set of wide-band red, green, and blue filters plus a fourth filter that adjusts the instrument response to match that of the eye (visual density); and (3) a photoelectric receiver that senses the light transmitted by the sample being measured. Most commercial color densitometers allow for interchangeable filter sets, so that status

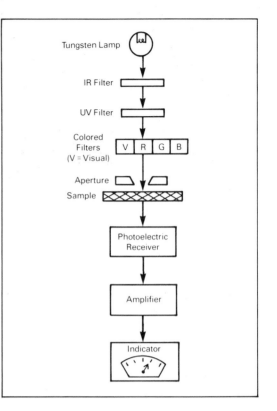

Figure 15–8   Major components of a photoelectric transmission densitometer.

A, status M, or nonstatus three-filter color densitometry can be used.

The measurement of reflection color densities requires a significantly different design, as illustrated in Figure 15–9. A tungsten lamp is used to illuminate the sample at an angle of approximately 45°, and the light reflected from the sample is collected over a 5° angle around the perpendicular to the sample. A mirror in the center reflects the light to the receiver after passing through the filter (red, green, or blue) that has been placed in front of the receiver.

Such a design simulates the optimum condition for viewing a reflection image, which is with the illumination striking the image at about a 45° angle with the light that is reflected on the perpendicular being received by the viewer's eye. In practice, however, light reaches the image over a very wide angle as a result of light reflections from the walls, ceilings, floors, etc. Since it is impossible to simulate these conditions in a reflection densitometer, the values that result do not accurately represent most viewing conditions

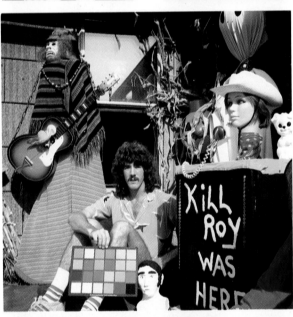

PLATE I *(See legend on following page.)*

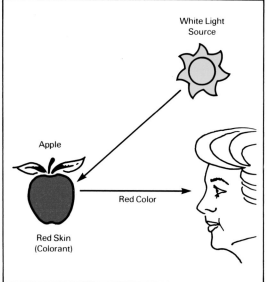

PLATE II A "red" apple is seen as red because the surface of the apple acts as a filter absorbing the green and blue components of white light and reflecting the red component.

PLATE III A simplified chain of events showing how color can be measured.

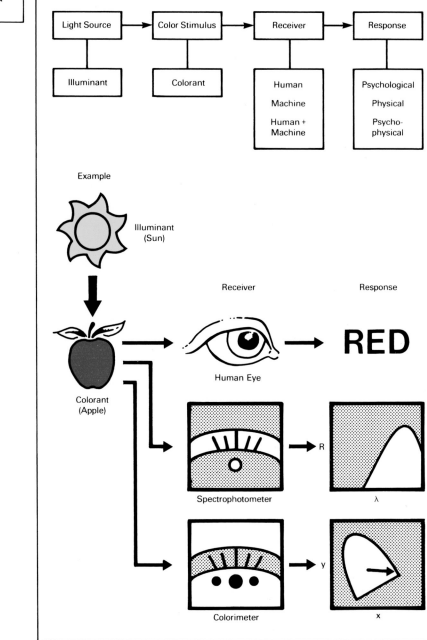

PLATE I *(Preceding page.)* Reciprocity effect for Ektacolor Printing Paper. All the prints received the same total exposure but the exposure times for each print increased by a factor of two (4, 8, 16, 32, 64, 128 seconds). Although there is a decrease in density as print exposure times are extended to one minute, there is no significant shift in color balance. (Photographs by Gary Hadlock.)

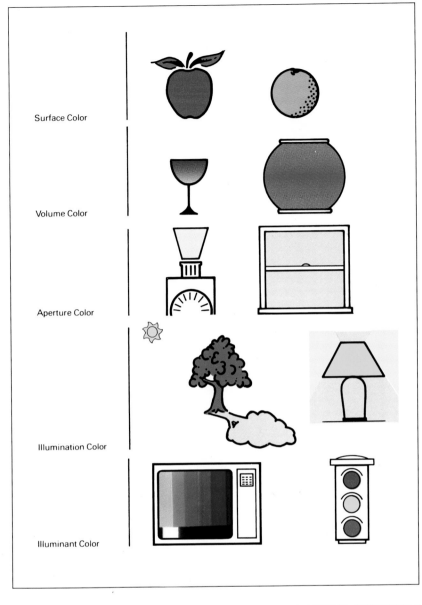

Surface Color

Volume Color

Aperture Color

Illumination Color

Illuminant Color

PLATE IV Modes of color perception.

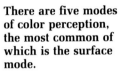

There are five modes of color perception, the most common of which is the surface mode.

PLATE V The four psychological primaries are distinct from any other color. The three additive primaries are colors of light that can be mixed to form other colors. The three subtractive primaries are colorants that can be superimposed to form most other colors.

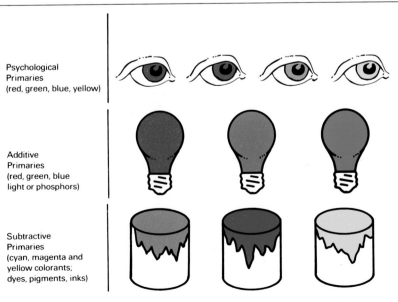

Psychological Primaries
(red, green, blue, yellow)

Additive Primaries
(red, green, blue light or phosphors)

Subtractive Primaries
(cyan, magenta and yellow colorants; dyes, pigments, inks)

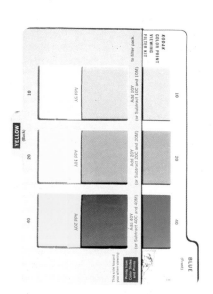

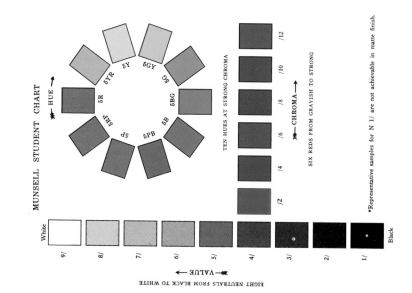

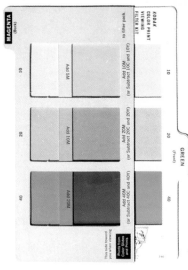

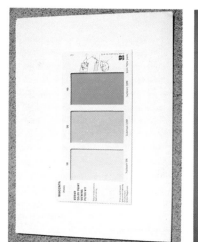

PLATE VI When mixed in proper proportions, complementary colors form neutrals or near neutrals.

PLATE VII The three attributes of color: hue, saturation (chroma), and brightness (lightness or value).

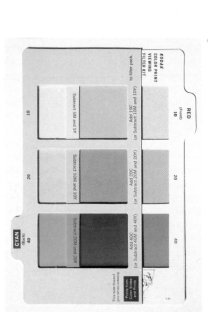

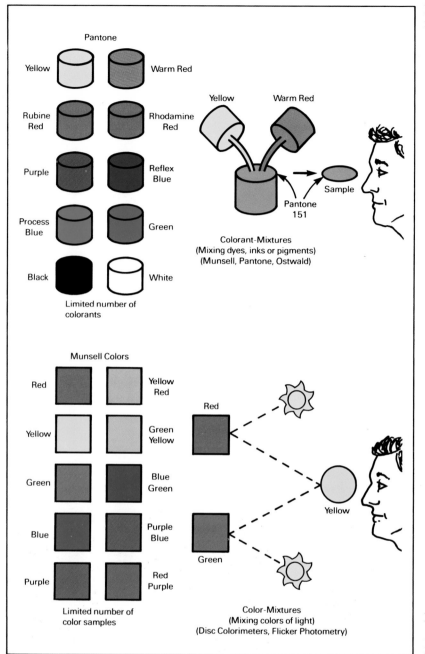

Pantone

Yellow | Warm Red

Rubine Red | Rhodamine Red

Purple | Reflex Blue

Process Blue | Green

Black | White

Limited number of colorants

Yellow | Warm Red

Sample

Pantone 151

Colorant-Mixtures
(Mixing dyes, inks or pigments)
(Munsell, Pantone, Ostwald)

Munsell Colors

Red | Yellow Red

Yellow | Green Yellow

Green | Blue Green

Blue | Purple Blue

Purple | Red Purple

Limited number of color samples

Red

Green

Yellow

Color-Mixtures
(Mixing colors of light)
(Disc Colorimeters, Flicker Photometry)

PLATE VIII   A comparison of two systems for producing a gamut of object-color samples.

Color matching consists of adjusting the components of color mixtures until they match a color sample, a process that is used by manufacturers of paints, dyes, and other colorants.

PLATE IX   This Munsell color tree contains 10 constant hue charts on clear plastic leaves. (It may be purchased from Macbeth, Munsell Color, 2441 N. Calvert St., Baltimore, MD 21218.)

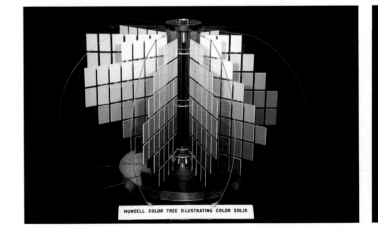

MUNSELL COLOR TREE ILLUSTRATING COLOR SOLID

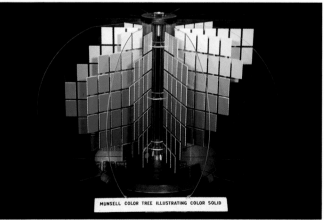

MUNSELL COLOR TREE ILLUSTRATING COLOR SOLID

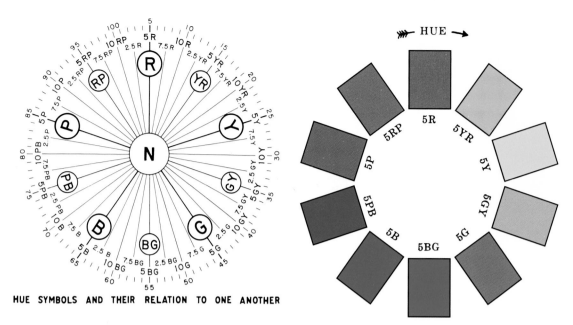

HUE SYMBOLS AND THEIR RELATION TO ONE ANOTHER

PLATE X The right diagram shows the 10 major hues of the Munsell system. The left diagram displays the names of these hues as letters in 100 divisions representing 100 different hues.

PLATE XI The Macbeth ColorChecker Color Rendition Chart contains an array of representative colors and a gray scale.

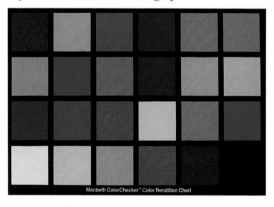

Macbeth ColorChecker™ Color Rendition Chart

PLATE XIII Additive mixture of blue, green, and red lights.

PLATE XII Pantone® Matching System.

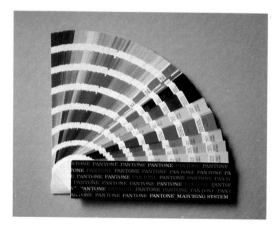

PLATE XIV   The subtractive mixture of yellow, magenta, and cyan pigments.

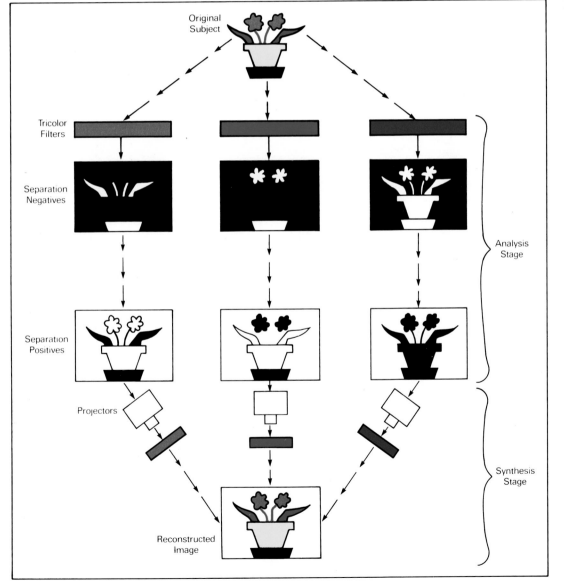

PLATE XV   The additive system of color photography with both analysis and synthesis stages illustrated.

Original Subject

Tricolor Filters

Separation Negatives

Separation Positives

Analysis Stage

Projectors

Synthesis Stage

Reconstructed Image

PLATE XVI The reproduction of color through the use of the mosaic screen process.

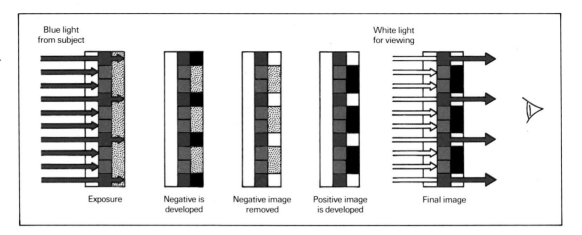

PLATE XVII The subtractive system of color photography with both the analysis and synthesis stages illustrated.

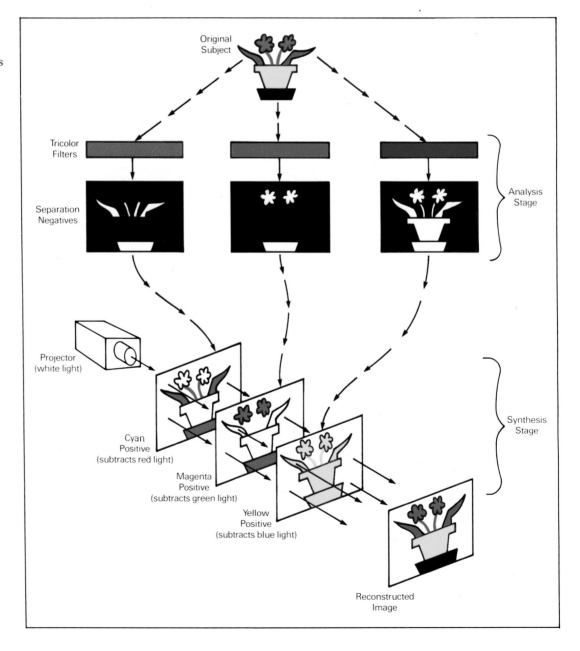

and, as a result, represent an acceptable compromise.

Reflection densitometers can provide useful information about the light-modifying characteristics of reflection photographic images. Additionally, red, green, and blue filters can be used over the receptor to obtain information about the color characteristics of the image.

By far the most commonly used means of determining color densities, integral three-filter densities must be interpreted with great care. For example, it is often assumed that a color film will give a good visual neutral if the red, green, and blue densities are equal. In fact, color images with equal red, green, and blue densities usually show a slight color cast. There are at least two reasons for this discrepancy. First, the three dyes that comprise the image each have unwanted absorptions. Second, the overall red, green, and blue response characteristics of the densitometer are considerably different than those of the human eye.

SENSITOMETRY OF COLOR PHOTOGRAPHIC MATERIALS

The determination of the sensitometric properties of color photographic films and papers essentially involves the same procedures that were discussed in Chapter 4: exposure, processing, image measurement, and data interpretation through the use of characteristic curves. When exposing the film, the sensitometer must be equipped with a light source that produces the color temperature for which the color film was designed. Any attenuators (neutral-density filters and step tablets) must not depart significantly from neutrality, or they will alter the color balance of the light reaching the film.

Figure 15–10 shows the spectral density curves for a variety of neutral attenuators. Notice that the Wratten neutral-density filter shows a higher density at the blue end of the spectrum than the green or red end. Furthermore, photographic silver step tablets also show a slightly higher blue density. When extreme neutrality is required, as when following ANSI standard methods, M-type carbon attenuators are preferred. These at-

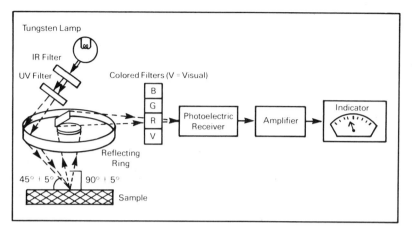

Figure 15–9   Major components of a photoelectric reflection densitometer.

tenuators are composed of carbon particles dispersed in gelatin and coated on an acetate base. Additionally, the Inconel coated materials (iron alloys evaporated onto glass) provide excellent neutrality in the visible region.

It is important to note that all four of these attenuators depart considerably from neutrality in the ultraviolet region of the spectrum. Furthermore, only the Inconel coated material shows good neutrality in the infrared regions of the spectrum. Thus when selecting filters for specialized work involving either ultraviolet or infrared energy, extreme care must be taken in selecting the attenuators.

The exposure time used in the sensitometer must simulate those times that typically will be used with the material in practice. This is to avoid the problem of rec-

> **Equal red, green, and blue densities in an area of a color photograph do not guarantee that the area will appear neutral.**

> **Gelatin neutral density filters tend to absorb more blue light than green light or red light, and therefore are not entirely neutral.**

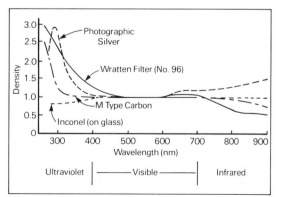

Figure 15–10   Spectral density curves for a variety of neutral attenuators.

iprocity failure. The processing of the sensitometric samples involves either standardized methods (ANSI methods) or techniques used in practice. The manufacturer of the sensitized material is likely the only one to use the standardized methods since this approach would provide information primarily about the emulsion properties. The photographer, however, requires information about the film and its corresponding processing in order to control the quality of the resulting photographic images.

In both cases, the resulting color photographic image consists of a set of superimposed cyan, magenta, and yellow dyes. As a result, the image-measurement method used typically results in three values: the density to red light (D_r), the density to green light (D_g), and the density to blue light (D_b); these are referred to as three-filter integral densities. The problems associated with obtaining and interpreting such data have previously been discussed.

The construction of the characteristic curve for a color film follows the convention discussed in Chapter 4 for black-and-white materials, except that three curves will be plotted for each image instead of one. In cases where a visual neutral has been formed in the image, it is possible to simply measure the visual density of each step and therefore plot a single curve that is representative of the image. The basic properties of speed and contrast are determined in the evaluation of the resulting curve. The sensitometric characteristics of a variety of color photographic materials follow.

REVERSAL COLOR FILMS

Examples of these films include Ektachrome, Kodachrome, Fujichrome, and Agfachrome. The visual density curve for a neutrally exposed and processed color reversal image is shown in Figure 15–11. Since this is a reversal material, increasing amounts of exposure produce decreasing amounts of density in the image. In order to obtain adequate contrast and color saturation in the transparency, the midtone slope is generally between 1.8 and 2.0, which is much higher than for negative materials. Far to the right on the curve is the minimum density, which

Reversal color films are capable of producing considerably higher densities than reflection print materials.

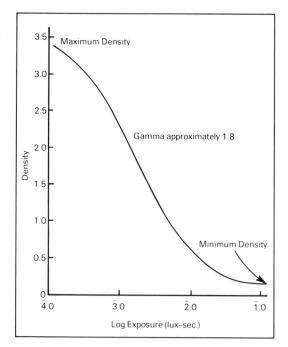

Figure 15–11   Visual density curve from a neutrally exposed and processed reversal color film.

is the combined result of residual color couplers, stain that was picked up during processing, and the density of the supporting base. The minimum density (visual density) of this area is seldom less than 0.20.

Far to the left of the plot is the shoulder of the curve, where the maximum density, representing no exposure, occurs. For most color reversal films, the maximum density obtainable is at least 3.0, and often greater. Consequently, color transparencies are capable of substantially greater output density ranges (near 3.0) than are reflection print materials. The contrast of reversal films is typically measured as the gamma, with these values usually between 1.80 and 2.0. Additionally, average gradient can be determined between any two points on the curve, and as a result will be different from (lower than) the gamma.

The useful log exposure range of the film is the range of log exposures contained between the minimum and maximum useful densities of the curve. Since the slopes of color reversal films are very steep, they are typically characterized by narrow useful log exposure ranges. Most such films do not exceed a useful log range of 1.90, indicating that an exposure ratio of at best 80:1 can be accommodated. As a result, the amount of exposure latitude (margin for exposure error) is usually quite small.

The speed of a color-reversal film is based on the location of two points on the curve that relate to shadow and highlight reproduction, as illustrated in Figure 15–12. Point A relates to highlight reproduction and is located at a density of 0.20 above the minimum density. From this point, a straight line is drawn so that it is tangent to the shoulder of the curve and locates the upper useful point. If this point of tangency falls at a density greater than 2.0, the upper useful point—labeled point B—is simply located at a density of 2.0 above the minimum density. The log exposures are determined for both positions and averaged by adding them and then dividing the sum by 2. The resulting value is termed the minimum log exposure (log H_m). The antilog is taken and used in the following speed formula:

$$ASA = \frac{I}{H_m} \times 10$$

Other sensitometric properties of color reversal films can be evaluated by using red, green, and blue density readings and constructing the corresponding curves, as illustrated in Figure 15–13. These curves are the result of reading a visually neutral color film image of a sensitometric strip on a densitometer equipped with the status A response functions. Note that the three curves are not exactly superimposed, indicating that although the image is a visual neutral, it is not a densitometric neutral. Since these curves were derived from a visually neutral strip, they can be used as a reference against which future strips can be compared in order to detect changes in image color.

For example, Figure 15–14 illustrates the red, green, and blue status A curves for a second strip. The shape of the red density curve is considerably different from that in Figure 15–13; the principal difference is an increase in the gamma. Such a condition is referred to as crossed curves and is generally related to problems in the film processing. In this image the shadows would probably ap-

The speed of reversal color films is calculated from the average of the log exposures required to produce two different densities.

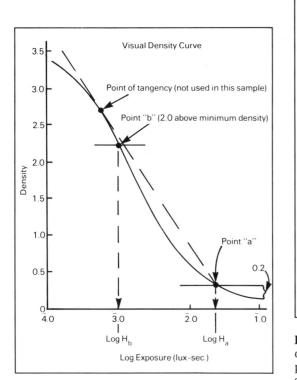

Figure 15–12 Location of the speed points in the ISO (ASA) speed method for a reversal color film.

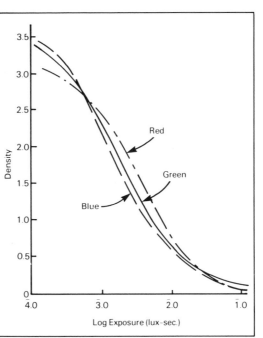

Figure 15–13 Red, green, and blue status A density curves from a neutrally exposed and processed reversal color film. This image is a visual neutral. Since the response of the densitometer is different from that of the human eye, the curves are not superimposed (i.e., the image is not a densitometric neutral).

The term "crossed curves" refers to a lack of parallelism of the red-green-blue characteristic curves.

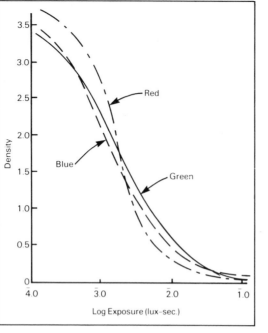

Figure 15–14   Red, green, and blue status A density curves from a non-neutral reversal color image. Since the red curve is significantly higher in the shoulder region, the shadows in the photograph would have a cyan cast. The highlights would show a reddish cast because the red curve is much lower in the toe region.

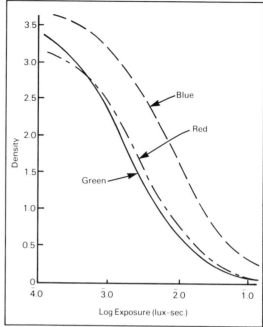

Figure 15–15   Red, green, and blue status A density curves from a non-neutral reversal color image. The image shows a high density to blue light in all regions and thus would appear yellow.

pear somewhat cyan because of the high red density, and the highlights would likely contain a red cast because of the lowered red density (indicating less cyan dye) in that region. The resulting color transparency would likely have unacceptable color quality.

Figure 15–15 illustrates the red, green, and blue status A curves for a third strip. In this case, the shapes of all three curves are approximately the same, except the blue density curve is displaced to the right. Thus there is an increase in blue density everywhere in the image. The image corresponding to the curves in Figure 15–15 would likely show an overall yellow-colored cast. Such an image could result from exposing a color reversal film balanced for daylight (5500 K) to a tungsten light source without the proper filtration.

NEGATIVE COLOR FILMS

Examples of this kind of film include Kodacolor, Vericolor, Fujicolor, and Agfacolor.

Color negative films, sometimes referred to as color print films because prints are made from them, employ the traditional integral tripack assembly. However, many modern-day films have double-coated emulsions in each of the three spectral bands to improve latitude and film speed. The spectral responses of the emulsions are selected principally for the ability to produce optimum color reproduction in the final color reflection print. Since the color negative represents only an intermediate step in the system, it makes no difference what the actual colors of the negative are, as long as they are matched to the properties of the color print material. This allows for the use of integral color masking, as discussed previously, which gives an overall orange cast to the negative.

The product of a set of neutral exposures and normal processing of the color negative material is a scale of non-neutral densities. The resulting sensitometric strip is evaluated by measuring the red, green, and blue densities on a densitometer equipped with the status M response functions. The red, green, and blue density plots for a sensitometric test

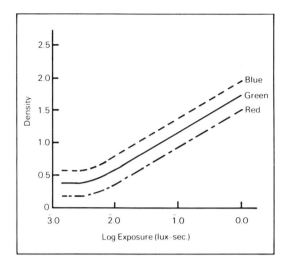

Figure 15–16   Red, green, and blue status M density curves from a neutrally exposed and processed negative color film. The integral color-correcting mask gives the image an overall orange cast, causing the curves to be vertically displaced.

of a typical negative color film are shown in Figure 15–16. The graph indicates that increasing amounts of exposure give increasing amounts of density, as is the case for negative-working materials.

Additionally, the slopes of the three curves are considerably less than those of a reversal color film, resulting in an increase in the exposure latitude. The three curves are separated vertically, due to the orange-colored integral mask in the image. The minimum densities for the curves are located at the far left of the graph and consist of the integral mask density, emulsion fog, residual color couplers, processing stain, and base density. The density to blue light is always the highest of the three because of the principal absorption of the yellow dye, the blue absorption of the integral mask, and the unwanted blue absorptions of the magenta and cyan dyes. Since the yellow and magenta dyes have only low absorptions to red light, the density to red light is generally the lowest of the three, resulting mostly from the cyan dye. It is sometimes assumed, erroneously, that the blue curve is the highest because it is the top emulsion layer, and the red curve is the lowest because it is the bottom layer of the three. The order in which the emulsion layers are coated has absolutely no effect upon the red,

green, and blue light-stopping ability of the image that is formed in the emulsion. This is a result of the nature of the dye-set and the integral color mask in the color negative.

The contrast of color negative films is generally determined by measuring the slope of the straight-line portion (gamma); slopes typically lie between 0.50 and 0.60. As with the black-and-white negative-positive process, the color negative has a relatively low slope, larger exposure latitude, allowing for a variety of subject contrast. The lowered slope of the negative is compensated for in the printing stage by using print materials with steep slopes. Theoretically, the slopes of the three curves should be equal over their entire length to avoid undesirable shifts in color balance between the shadows and the highlights. However, practical experience indicates that some divergence in slopes can be tolerated, depending to a great extent upon the nature of the scene.

The speed of a color negative film is derived from locating the minimum useful points (0.15 above the minimum density) on the green status M density curve and the slowest status M density curve of the three curves. Since the slowest curve inevitably is the red status M density curve, it is the green and red status M density curves that are used. The log exposure (in lux-seconds) at each speed point is determined and averaged. The antilog of the resulting log H value is then used in the following formula:

$$\text{Speed} = \frac{1}{H_\text{m}} \times 1.5 \,.$$

The blue status M density curve is not used to determine film speed, since it is the fastest of the three and would produce a slightly inflated speed.

Since three filter color densities do not give precise information about either visual appearance or printing characteristics, such data for color negatives must be interpreted with the same care as discussed previously with reversal color images. Nevertheless, it is possible to make meaningful inferences about the printing properties when there are large differences between the aim curve shapes and those actually encountered. For example, the red, green, and blue status M density curves in Figure 15–17 illustrate the sensitometric

The red-green-blue characteristic curves for typical color negative materials are displaced vertically due to the absorption of blue light and green light by the color-correcting mask.

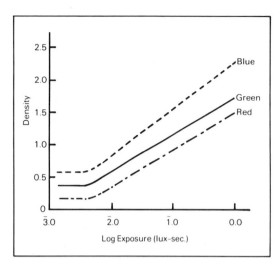

Figure 15–17   Red, green, and blue status M density curves from a neutrally exposed negative color film that was overdeveloped. Notice the increased slope of the blue density curve.

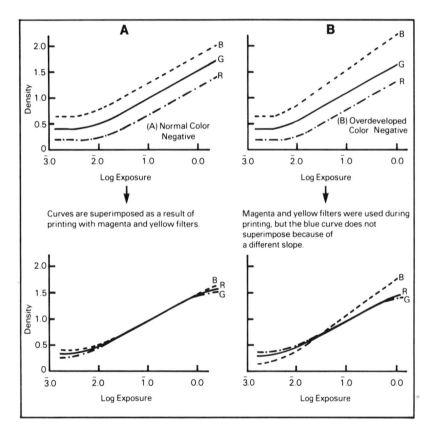

Figure 15–18   The sensitometric effects of printing color negatives in an enlarger using supplementary filtration. The curves in graph A are from a color negative with proper contrast in all three layers. The use of supplementary magenta and yellow filters in the printer brings the curves into nearly exact alignment, as shown in graph A bottom. The curves in graph B are from a negative with high blue contrast and normal green and red contrast. When this negative is printed, there is no combination of filters that can eliminate the mismatch seen in graph B, bottom.

results of overdevelopment of a color negative. Comparing these curves to those shown in Figure 15–16 (normal processing conditions) reveals that the extended development time produces a noticeably higher contrast in the blue density plot.

Such a change in curve shape would lead to the problem of crossed curves, with the negative producing a color print with a noticeable color shift between the shadows and the highlights (i.e., the shadows would have a yellowish cast, while the highlights would have a bluish cast). This problem can be illustrated graphically by superimposing the red, green, and blue density curves, which is essentially what happens when the negative is printed on conventional color print material using supplementary filtration in the enlarger.

Figure 15–18A illustrates how the red, green, and blue density curves exactly superimpose for a normally exposed and processed color negative. Figure 15–18B illustrates what happens when an attempt is made to balance the color negative shown in Figure 15–17. Notice that the blue density curve is steeper, and if a midtone is printed to a neutral, as shown here, the blue light density in the shadows will be low, producing larger amounts of yellow dye in the print. The blue density in the highlights will be higher, resulting in a lack of yellow dye (a bluish cast) in the highlights. Thus when the red, green, and blue density curves are severely crossed, excellent color prints cannot be obtained.

COLOR PAPER

In the other half of the negative-positive process a color reflection print is made from the negative. Therefore, the properties of color photographic papers depend primarily upon the characteristics of the negatives to be printed. For example, the spectral sensitivities of the paper emulsions are typically matched to the spectral absorptions of the dyes in a color negative, thus allowing for optimum color reproduction. The inherent color balance of the color paper emulsion is designed so that using a tungsten enlarging lamp in combination with the red, green, and blue light-stopping abilities of a typical color negative will result in a properly balanced color print with a minimum of supplementary filtration. Furthermore, by making the paper's red-sensitive layer the slowest, the use of supplemental cyan filtration with its unwanted absorptions can be avoided.

If a color reflection paper is exposed and processed to produce a scale of visual neutrals, the visual densities of the steps can be measured and a characteristic curve constructed as shown in Figure 15–19. Notice that the curve shape is very similar to that of black-and-white reflection print material. The steep slope in the curve's midsection is necessary to expand the midtones, which were compressed in the negative as a result of the lowered slope in the negative. The result is a print with the desired tone reproduction characteristics, as discussed in Chapter 10.

Notice that the maximum density of color paper is somewhat greater than that of a black-and-white reflection paper. This is principally due to multiple internal reflections between the emulsion layers in the color paper, which cause less light to be reflected from the print. Since there are no standard methods for determining the speed and contrast of color print materials, there exists no common method for determining these properties. However, since the characteristics of color and black-and-white reflection print materials are similar, the same methods may be used.

On this basis the useful log exposure range for a conventional color reflection print material turns out to be approximately 1.15 to 1.25 (current ANSI standard paper range of 120), indicating that it is similar to a contrast grade 2 black-and-white printing paper. Also, the ANSI standard paper speed is approximately 100, indicating that this is a medium-speed paper. Such information is of little practical use, since color print materials are invariably balanced individually when the negative is printed. In effect, individual tests are performed for each negative/print combination, minimizing the need for standardized values of speed and contrast.

Figure 15–20 illustrates the red, green, and blue status A density curves for the same scale of visual neutrals. Again, notice that the three curves are not exactly superimposed, indicating that a set of visual neutrals is not the same as a set of densitometric neutrals. The inherent sensitivity of the red, green, and blue layers is illustrated in Figure 15–21, where it can be seen that the blue- and green-sensitive layers are faster (displaced to the left) than the red-sensitive layer. This is in part to compensate for the integral mask incorporated in the negative, which has a high density to blue and green light. Also, since the Tungsten lamps used in color enlargers are rich in red light, the paper speed in this region can be reduced.

Thus by matching the speeds of the three layers of the color print material to the red-,

The color sensitivities of the three emulsion layers of color printing papers are controlled during manufacture so that it will not be necessary to use the inefficient cyan filters when printing.

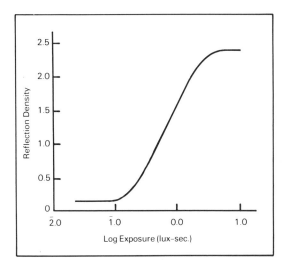

Figure 15–19 Visual density curve from a neutrally exposed and processed reflection color print material.

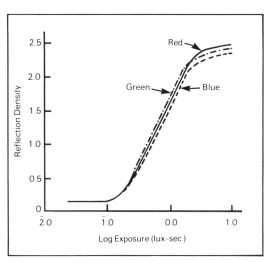

Figure 15–20 Red, green, and blue status A curves from a neutrally exposed and processed reflection color print material. The image is a visual neutral.

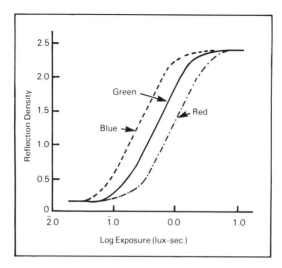

Figure 15–21   Red, green, and blue status A curves from reflection color print paper balanced for an orange mask but printed without it. Thus the actual speeds of the three layers are revealed. The blue- and green-sensitive layers are the fastest, which compensates for the high blue and green densities of the orange negative mask.

green-, and blue-light-stopping characteristics of the integral mask in the color negative, minimum amounts of supplementary filtration will be required to obtain optimum color reproduction. In essence, the three curves illustrated in Figure 15–21 will be superimposed as they appear in Figure 15–20, as a result of the filtration caused by the integral orange mask and the supplementary filtration in the printer.

REVIEW QUESTIONS

1. The major physical reason why color photographs cannot exactly duplicate subject colors is that . . . *(p. 334)*
   A. photographic lenses alter the color balance of the image-forming light
   B. there is variability in the processing of color films
   C. there is variability in the manufacture of color films
   D. existing dyes have limitations
2. The subtractive primary colors absorb . . . *(p. 335)*
   A. red, green, and blue light
   B. cyan, magenta, and yellow light

3. In comparison with daylight reversal color films, tungsten reversal color films have . . . *(p. 341)*
   A. relatively low red sensitivity
   B. relatively high red sensitivity
   C. about the same red sensitivity
4. The poorest of the three dyes, with respect to unwanted absorption, is the . . . *(p. 342)*
   A. yellow dye
   B. magenta dye
   C. cyan dye
5. A practical method of obtaining analytical densities with integral tripack color film such as Ektachrome is to . . . *(p. 344)*
   A. expose only a single emulsion layer
   B. develop only a single emulsion layer
   C. expose and develop all three emulsion layers and then bleach the dyes in two layers
   D. isolate one of the three dye layers by using a filter on the densitometer
6. The illumination system in reflection densitometers is based on the assumption that when prints are displayed they are illuminated . . . *(p. 346)*
   A. directly from the front
   B. at an angle of 45 degrees
   C. uniformly from all angles
7. The gamma of reversal color films is typically about . . . *(p. 348)*
   A. 0.7
   B. 1.0
   C. 1.2
   D. 1.4
   E. 1.9
8. Assuming that the red, green, and blue curves for a reversal color film should be superimposed for a neutral color balance, a transparency represented by a set of curves with the red curve higher than the other two would appear too . . . *(p. 350)*
   A. red
   B. green
   C. blue
   D. cyan
   E. magenta
9. The gammas of negative color film curves are . . . *(p. 351)*
   A. somewhat lower than the gammas of reversal color films
   B. about the same as the gammas of reversal color films
   C. somewhat higher than the gammas of reversal color films

10. The color balance of color printing papers is controlled during manufacture so that when exposing the paper it is usually not necessary to use a . . . (p. 352)

A. cyan filter
B. magenta filter
C. yellow filter

Appendixes

Barry Myers. *Victoria Station, London. 1972.*

A Conversion Units and Scales

CONVERSION UNITS

Length

Feet	Inches	Millimeters	Meters
1	12	305	0.305
0.0833	1	25.4	0.0254
0.00328	0.0394	1	0.001
3.28	39.37	1000	1

Miles	Kilometers
1	1.61
0.621	1

Weight

Pounds	Ounces	Grams	Kilograms
1	16	454	0.454
0.0625	1	28.3	0.0283
—	0.0353	1	0.001
2.20	35.3	1000	1

Liquid

Quarts	Fluid Ounces	Cubic Centimeters	Liters
1	32	3785	3.78
0.0312	1	29.6	0.0296
—	0.0338	1	0.001
1.06	33.8	1000	1

Illuminance

Footcandle	Metercandle (Lux)
1	10.76
0.0929	1

Luminance

Cd/square foot	Cd/square meter
1	10.76
0.0929	1

Fahrenheit, Celsius, and Kelvin Temperature Scales

F	C	K	
9440	5230	5500	Photographic daylight
5300	2930	3200	Photographic studio lamps
2780	1530	1800	Candle light
1340	727	1000	Min. for Blackbody emission of light
212	100	373	Boiling point of water
98.6	37	310	Body temperature
68	20	293	Typical film development
32	0	273	Freezing point of water
0	−17.8	255	Zero fahrenheit
−459	−273	0	Absolute zero

To change F to C:

$$C = \frac{5(F - 32)}{9}$$

To change C to F:

$$F = \frac{9C}{5} + 32$$

B Exposures, Depth of Field, Zone System, and Film Speed Information

Existing-Light Photography: Suggested Trial Exposures Based on ISO/ASA 400 Speed film

Subject-Lighting Condition	Camera Settings	
Photograph of a full moon	1/500 sec.	f/16
Photograph of a half moon	1/125 sec.	f/16
Subject illuminated by light of full moon	8 sec.	f/2
Skyline at sunset	1/125 sec.	f/11
Skyline just after sunset	1/125 sec.	f/8–5.6
Star trails	Time	f/2.8
Aurora borealis (Bright)	10–20 sec.	f/2
Aurora borealis (Medium)	40–80 sec.	f/2
Lightning bolts (Night)	Time	f/16
Underwater vs. above water, 5–15 feet	Open 1 stop	
Underwater vs. above water, 20–30 feet	Open 2 stops	
Bright signs at night	1/60 sec.	f/5.6
Fires	1/60 sec.	f/5.6–4
Fireworks	1/125 sec.	f/5.6–2
Night baseball, football, etc.	1/60 sec.	f/5.6–4
Night city streets and store windows	1/60 sec.	f/4–2.8
Moving traffic light streaks	1 sec.	f/11
Floodlit buildings	1/15 sec.	f/2.8–2
Niagara Falls lit by floodlights	1 sec.	f/2.8–1.4
Galleries	1/60 sec.	f/5.6–2
Offices, work areas	1/60 sec.	f/2.8–2
Television screen	1/8 sec.	f/11

Note: Other equivalent combinations of exposure time and f-number may be used. Extra exposure may be required to compensate for reciprocity effects with longer exposure times. (No compensation has been included in these numbers.)

Reciprocity Exposure and Development Compensation for General Purpose Black-and-White Films

Indicated Exposure Time	Adjusted Exposure Time	Time Exposure Factor	Developing Time Adjustment
1/10 sec.	1/10 sec.	1.0	None
1 sec.	2 sec.	2.0	−10%
2 sec.	5 sec.	2.5	
3 sec.	9 sec.	3.0	
4 sec.	14 sec.	3.5	
5 sec.	20 sec.	4.0	
6 sec.	25 sec.	4.2	
7 sec.	31 sec.	4.4	
8 sec.	37 sec.	4.6	
9 sec.	43 sec.	4.8	
10 sec.	50 sec.	5.0	−20%
15 sec.	80 sec.	5.3	
20 sec.	120 sec.	6.0	
30 sec.	200 sec.	6.7	
40 sec.	300 sec.	7.5	
50 sec.	420 sec.	8.4	
60 sec.	560 sec.	9.3	−25%

F-Numbers and Shutter Speeds

F-Numbers	Shutter Speeds
f/64	1/4000 second
f/45	1/2000 second
f/32	1/1000 second
f/22	1/500 second
f/16	1/250 second
f/11	1/125 second
f/8	1/60 second
f/5.6	1/30 second
f/4	1/15 second
f/2.8	1/8 second
f/2	1/4 second
f/1.4	1/2 second
f/1	1 second

Note: The f-numbers and shutter speeds above represent an increase in exposure of one stop (or 2×) for each change from top to bottom.

Relative Depth of Field for Different Focal Length Lenses on 35 mm Cameras, with the same object distance and f-number

Lens Focal Length	Relative Depth of Field
16 mm	9.8
20 mm	6.2
24 mm	4.3
28 mm	3.2
35 mm	2.0
50 mm	1.0
75 mm	1/2.2
100 mm	1/4
135 mm	1/7.3
200 mm	1/16
400 mm	1/64

ISO Arithmetic and Logarithmic Film Speeds (1/3 stop increments)

Arithmetic	Logarithmic
6	9
8	10
10	11
12	12
16	13
20	14
24	15
32	16
40	17
50	18
64	19
80	20
100	21
125	22
160	23
200	24
250	25
320	26
400	27
500	28
650	29
800	30
1000	31
1250	32
1600	33
2000	34
2500	35
3200	36
4000	37
5000	38
6400	39
8000	40

Zone System Scene and Print Values

Scene	Value	Print
Dark No detail	0	Dmax No detail
Dark Just detectable detail	I	90% of Dmax Just detectable detail
Dark Good detail	II	Dark Good detail
	III	
Dark midtone	IV	
Midtone, 18% reflectance	V	Facsimile copying: D = 0.75 Pictorial: Slightly lighter
Light midtone	VI	
Light Good detail	VII	Light Good detail
Light Just detectable detail	VIII	D = 0.04 above paper white Just detectable detail
Specular highlights No detail	IX	Paper white No detail
Light sources No detail	X	Paper white No detail

SZ-100 RULE

The speed relationship between ISO 100 film and typical grade 2 enlarging paper is approximately 100 to 1. To expose enlarging paper in a view camera, as illustrated at right, find the recommended exposure time for an ISO 100 film and multiply it by 100. To make a transparency or diapositive on ISO 100 film from a negative, first make a good print on an enlarging paper and then divide the exposure by 100 when exposing the film. A convenient way to do this is to add a 2.0 neutral density filter under the enlarger lens and use the same exposure time.

F/16 RULE

The camera exposure settings for an average scene in direct sunlight can be determined without an exposure meter by setting the lens at f/16 and using a shutter speed equal to the reciprocal of the ISO film speed. Thus for an ISO 125 speed film, the recommended settings would be 1/125 second at f/16, or any comparable combination.

C Calculations and Basic Logarithms

CALCULATION OF THE MEAN*

To determine the mean of a population or sample, all of the data are simply added together and divided by the number of pieces of data. For example, suppose that three tests of the resolving power of a film produced the following data: 59 lines/mm, 55 lines/mm, and 51 lines/mm. Using the formula $\overline{X} = \Sigma X/n$ (where the symbol Σ means "sum of") the steps are:

1. Add up all of the X values $(59 + 55 + 51 = 165)$.
2. Divide the total by the number of X values (n) $(165 \div 3 = 55.0)$.

Notice that the mean is carried to one more decimal place than the original data. Thus, the mean resolving power for this experiment is 55.0 lines/mm.

CALCULATION OF THE STANDARD DEVIATION

Although the formulas for determining the standard deviation contain more terms, the calculations are relatively easy to do if taken step by step. To continue the example given above with resolving powers of 59, 55, and 51 lines/mm requires the use of one of the formulas for the sample standard deviation.

Formula I:
1. Determine the mean. Here $\overline{X}$ was found to be 55.0.
2. Determine the absolute difference (ignoring the sign) between each X value and the mean. In order, the differences are 4, 0, and 4.
3. Square each of the differences. The squares are 16, 0, and 16, respectively.

---

*Calculations should be carried out to one more significant figure than in the original data.

4. Add the squares. The total is 32. This is the numerator of the fraction under the radical sign.
5. Divide the total by the sample size minus 1 $(n - 1 = 2; 32 \div 2 = 16)$.
6. Finally, find the square root of this value. $\sqrt{16} = 4.0$. The standard deviation is therefore 4.0 lines/mm correctly taken to the same number of decimal places as the mean.

It is often convenient to set up the calculations in tabular form as shown:

Formula I

$$s = \sqrt{\frac{\Sigma(X - \overline{X})^2}{n - 1}}$$

X	(Step 2) $X - \overline{X}$	(Step 3) $(X - \overline{X})^2$
59	4	16
55	0	0
51	4	16
$\Sigma X = 165$		$\Sigma = 32$
$165 \div 3 = 55.0$		(Step 4)
(Step 1)		

$32 \div 2 = 16$; $\sqrt{16} = 4.0 = s$.
(Step 5)   (Step 6)

BASIC LOGARITHMS

Both *interval* scales of numbers (1-2-3-4-5-6-etc.) and *ratio* scales of numbers (1-2-4-8-16-32-etc.) are used in photography. Interval scales, for example, are found in thermometers and rulers. Ratio scales, for example, are used for f-numbers (f/2-f/2.8-f/4-f/5.6-f/8) where every second number is doubled, and film speeds (100-125-160-200-250-320) where every third number is doubled. Even though certain quantitative relationships in photography conform naturally to the ratio-type scale, there are disadvantages in using ratio scales of numbers in calculating and graphing. The disadvantages include the difficulty of determining the midpoint or other sub-division between consecutive numbers, and the inconvenience of dealing with very large numbers.

Using the *logarithms* of numbers in place of the numbers eliminates disadvantages of ratio scales by converting the ratio scales to interval scales. Mathematical operations are reduced to a lower level when working with logs, so that multiplication is reduced to ad-dition, division is reduced to subtraction, raising a number to a power is reduced to multiplication, and extracting a root is re-duced to division.

Logarithms, or logs, are derived from a basic ratio series of numbers: 10, 100, 1000, 10,000, 100,000, etc. Since $100 = 10^2$, $1000 = 10^3$, etc., the above series of numbers can be writ-ten as: 10^1, 10^2, 10^3, 10^4, 10^5, etc. The super-scripts 1, 2, 3, 4, 5, etc. are called powers, or exponents, of the *base* 10. If we always use 10 as the base, the exponents are called logs to the base 10. By pairing the original num-bers with the exponents, we have the begin-ning of a table of logs (see Table C–1). This table can be extended indefinitely down-ward. Since one million contains six tens as factors ($10 \times 10 \times 10 \times 10 \times 10 \times 10$), the log of 1,000,000 is 6. For any number con-taining only the digit 1 (other than zero), the log can be found by counting the number of decimal places between the position to the right of the 1 and the position of the decimal point. The log of 1,000,000,000 is therefore 9. The number whose log is 5 is 100,000, and 100,000 is said to be the *antilog* of 5. If the columns in Table C–1 are extended upward one step, the number 1 is added to the first column and the log 0 is added to the second. Thus the log of 1 is 0 and the antilog of 0 is 1.

We also need the logs of numbers between 1 and 10. Since the log of 1 is 0 and the log of 10 is 1, the logs of numbers between 1 and 10 must be decimal fractions between 0 and 1. The logs of numbers from 1 to 10 are listed in Table C–2. It is customary to write the log

Table C–1

Number	Logarithm
10	1
100	2
1,000	3
10,000	4

Table C–2

Number	Logarithm
1	0.00
2	0.30
3	0.48
4	0.60
5	0.70
6	0.78
7	0.85
8	0.90
9	0.95
10	1.00

of a number to one more signficant figure than the number itself. Thus the number 8 has one significant figure, and 0.90 (the log of 8) has two significant figures.

Table C–2 is used to illustrate some basic relationships between numbers and their logs:

1. $6 = 2 \times 3$. Note that the log of 6 is the sum of the logs of 2 and 3. When numbers are *multiplied*, their logs are *added*. The log of 10 equals the sum of the logs of 2 and 5.
2. $4 = 8 \div 2$. Note that the log of 4 is the difference of the logs of 8 and 2. When one number is *divided* by another, the log of the second is *subtracted* from that of the first. The log of 3 equals the difference of the logs of 6 and 2.
3. $4 = 2^2$. Note that the log of 4 is twice the log of 2. When a number is *squared*, the log is *doubled*. Similarly, $8 = 2^3$, and the log of 8 is three times the log of 2. When a number is raised to a *power*, the log of that number is *multiplied* by the power.
4. $\sqrt{9} = 3$. Note that the log of 3 is half the log of 9. Similarly, 2 is the cube root of 8, and the log of 2 is one-third the log of 8. When a *root* is taken of a number, the log of the number is *divided* by the root.

By the multiplication rule, the log of 50 is the log of 5 + the log of 10, or 0.70 + 1.00, or 1.70. Similarly, the log of 500 is 0.70 + 2.00, or 2.70; the log of 5000 is 3.70, etc. Also the log of 200 is 2.30, the log of 300 is 2.48, etc. All numbers in the hundreds have logs beginning with 2, all numbers in the thou-sands have logs beginning with 3, etc.

From this comes the concept that the log of a number consists of two parts: a whole number, determined by whether the original number is in the tens, hundreds, thousands, etc.; a decimal part determined by the digits. The whole number part of a logarithm is called the *characteristic*; it is solely determined by the location of the decimal point in the original number. The decimal part of a log is called the *mantissa*; it is found from a table of logs. To find the characteristic, count the number of places from a spot just to the right of the first digit of the number to the decimal point. For 500,000, the count is 5, and therefore the log begins with 5. The decimal part of the log, found in Table C–2, is 0.70, and the entire log is thus 5.70.

If a log is given, the antilog (the number corresponding to the log) is found by the reverse process. That is, use of the decimal part of the log to find the digits in the number and use the characteristic of the log to place the decimal point in the number. For example, to find the antilog of 3.78, note that 0.78 is the log of the number 6 in Table C–2. Therefore, 6 is the antilog of 0.78. The 3 in the log 3.78 indicates that the decimal point is moved three places to the right, changing the number from 6 to 6000.

We will now extend the basic log table to include decimal numbers between 0 and 1. In Table C–3, each number has one-tenth the value of the number just below it, whereas each log is reduced by a value of 1 for each step upward. Thus the next number above 1 will be one-tenth of 1, or 0.1, and the log will be 1 less than 0, or −1, and so on. The basic log table can now be considered to extend indefinitely in both directions. Note the symmetry in Table C–3 about the number 1: for example, all numbers in the thousands have logs containing 3, and all numbers in the

thousandths have logs containing −3. (Note that the number column is identified as a ratio scale, which will never reach zero, and that the log column is identified as an interval scale.)

The logs of decimal fractions are found by the same procedure as that used for larger numbers. To find the characteristic (the whole number part of the log), count the number of places from the position to the right of the first nonzero digit to the decimal point. To find the mantissa (the decimal part of the log), locate the nonzero digit in the number column and note the corresponding value in the log column. For example, to find the log of 0.0003, the decimal point is 4 places to the left of the digit 3, so the log has a characteristic of −4. The mantissa is 0.48, opposite 3 in the number column in Table C–2.

The most awkward thing about logs is that there are several ways of writing the logs of numbers between 0 and 1. Most often the mantissa (a positive value) and the characteristic (a negative value) are kept separate and are written as an indicated unfinished computation. For example, three ways of writing the log of the number 0.004 are:

1. 0.60 − 3, where 0.60 is the mantissa and −3 is the characteristic. This is the most direct form, but it is not often used.
2. 7.60 − 10, where 0.60 is the mantissa and the combination 7 − 10 (which is equal to −3) is the characteristic. This form, which is always written as some number minus 10, is the form most commonly used outside photography.
3. $\bar{3}.60$, where 0.60 is the mantissa and $\bar{3}$ is the characteristic. The minus sign is placed over the 3 to emphasize that only the 3 is negative, not the 0.60. This form is referred to as *bar notation*, and the log above is read "bar three point six zero." This is the form most commonly used in photography.

The procedure for finding the antilog of a log having a negative characteristic is the same as described above for positive logs, except that the decimal will be moved to the left in the number rather than to the right. With $\bar{3}.60$ as the log, for example, the antilog of 0.60 is 4, and moving the decimal three places to the

Table C–3

Number	Logarithm
—	—
0.0001	−4
0.001	−3
0.01	−2
0.1	−1
1.0	0
10.0	1
100.0	2
1000.0	3
10000.0	4
—	—

left produces 0.004. Thus, the antilog of $\overline{3}.60$ is 0.004.

An expanded table of logs for numbers from 1.0 to 10.0 to one decimal place, with logs to three decimal places, is provided in Table C–4.

In addition to the three forms of writing negative logs for numbers between 0 and 1 is a fourth, in which both the characteristic and the mantissa are negative numbers. This form is obtained by adding the negative characteristic and the positive mantissa in bar notation logs. To illustrate, the bar notation log of the number 0.2 is $\overline{1}.3$. Adding -1 and $+0.3$ equals -0.7, a totally negative log. Totally negative logs are used with pocket calculators, which are discussed in the next section. Table C–5 shows a comparison of bar notation logs and totally negative logs of the number 2 in multiples of 10 from 0.0002 to 2000.

LOGARITHMS AND THE USE OF A CALCULATOR

The use of scientific hand calculators greatly eases the tasks of determining logs and anti- (or inverse) logs. To find the log of a number greater than 1, simply enter the number and then press the key labeled "LOG." The answer will appear on the display. To find the antilog of a positive log, simply enter the log value and press the keys labeled "INV" and "LOG" in that order. The answer will appear on the display.

To find the log of a number less than one, the procedure is the same as for numbers greater than one, except that the resulting value will be a totally negative log, since the calculator automatically combines the characteristic and the mantissa.

Example
Problem: Find the log of 0.004.
Solution: Enter the number 0.004 into the calculator and press the key labeled "LOG"; the answer of -2.39794 rounded to -2.40 is shown in the display.
Answer: The log of 0.004 is -2.40.

The answer is a totally negative number, and in a sense is the "true" log of 0.004; it must

be carefully distinguished from $\overline{3}.60$, which is part negative and part positive.

To find the antilog of a totally negative log, the above procedure is reversed. The totally negative log is entered into the calculator and the keys labeled "INV" and "LOG" are pushed, in that order.

Example
Problem: Find the antilog of -2.40.
Solution: Enter the value -2.40 into the calculator and press the keys labeled "INV" and "LOG" in that order; the number 0.003981 appears in the display and is rounded to 0.004.
Answer: The antilog of -2.40 is 0.004.

Note: On some calculators, there is no key labeled "INV." For these calculators, the key labeled "10^x" should be substituted for the entire "INV" and "LOG" key sequence.

Since the bar notation system is often encountered in photographic publications, a procedure for using a calculator with this system is given below.

To find the log of a number less than 1.0 expressed in bar notation using a calculator:

1. Move the decimal point to the right of the first nonzero digit and count the number of places it was moved. The number of spaces moved is the characteristic it will be, but assigned a negative value.
2. Enter the number (with the decimal moved) in the calculator and press the button labeled "LOG"; the value that appears in the display is the mantissa of the number and will be a positive value.

Example Number 1
Problem: Find the log of 0.2.
Solution: 1. The decimal point is moved one place to locate it to the right of the first nonzero digit (2); thus the characteristic is bar one ($\overline{1}$).
    2. Enter the number 2.0 in the calculator and press the button labeled "LOG"; the value of 0.30 appears and is the mantissa.
Answer: The log of 0.2 is $\overline{1}.30$.

Table C–4   Abbreviated Table of Logarithms

Number	Logarithm	Number	Logarithm	Number	Logarithm
1.0	0.000	5.5	0.740	0.0001	−4
1.1	0.042	5.6	0.748	0.001	−3
1.2	0.080	5.7	0.756	0.01	−2
		5.8	0.763	0.1	−1
1.26	0.100	5.9	0.771	1	0
				10	1
1.3	0.114	6.0	0.778	100	2
1.4	0.147	6.1	0.785	1000	3
		6.2	0.792	10,000	4
1.414	0.150	6.3	0.799	10,0000	5
		6.4	0.806		
1.5	0.175			2	0.301
		6.5	0.813	20	1.301
1.6	0.204	6.6	0.819	200	2.301
1.7	0.230	6.7	0.826	2000	3.301
1.8	0.255	6.8	0.832		
1.9	0.278	6.9	0.839		
2.0	0.301	7.0	0.845		$0.301 - 1$
		7.1	0.851		or
2.1	0.322	7.2	0.857	0.2	$9.301 - 10$
2.2	0.342	7.3	0.863		or
2.3	0.361	7.4	0.869		$\bar{1}.301$
2.4	0.390				
		7.5	0.875		$0.301 - 2$
2.5	0.398	7.6	0.881		or
2.6	0.415	7.7	0.886	0.02	$8.301 - 10$
2.7	0.431	7.8	0.892		or
2.8	0.447	7.9	0.898		$\bar{2}.301$
2.9	0.462				
		8.0	0.903		
3.0	0.477	8.1	0.908		
3.1	0.491	8.2	0.914		
3.2	0.505	8.3	0.919		
3.3	0.518	8.4	0.924		
3.4	0.532				
		8.5	0.929		
3.5	0.544	8.6	0.934		
3.6	0.556	8.7	0.940		
3.7	0.568	8.8	0.944		
3.8	0.580	8.9	0.949		
3.9	0.591				
		9.0	0.954		
4.0	0.602	9.1	0.959		
4.1	0.613	9.2	0.964		
4.2	0.623	9.3	0.968		
4.3	0.634	9.4	0.973		
4.4	0.644				
		9.5	0.978		
4.5	0.653	9.6	0.982		
4.6	0.663	9.7	0.987		
4.7	0.672	9.8	0.991		
4.8	0.681	9.9	0.996		
4.9	0.690				
		10.0	1.000		
5.0	0.699				
5.1	0.708				
5.2	0.716				
5.3	0.724				
5.4	0.732				

Table C–5

Number	Bar Log	Negative Log
0.0002	$\bar{4}.3$	− 3.7
0.002	$\bar{3}.3$	− 2.7
0.02	$\bar{2}.3$	− 1.7
0.2	$\bar{1}.3$	− 0.7
2	0.3	0.3
20	1.3	1.3
200	2.3	2.3
2000	3.3	3.3

Example Number 2
Problem: Find the log of 0.004.
Solution: 1. The decimal point is moved three places to locate it to the right of the first nonzero digit (4); thus the characteristic is bar three ($\bar{3}$).
2. Enter the number 4.0 in the calculator and press the button labeled "LOG"; the value of 0.60 appears and is the mantissa.
Answer: The log of 0.004 is $\bar{3}.60$.

To find the antilog (inverse log) of a log expressed in bar notation using a calculator:

1. Enter the mantissa of the log into the calculator and press the buttons labeled "INV" and "LOG," in that order; the value that appears on the display will contain all of the digits of the antilog with the decimal point immediately to the right of the first digit.
2. Locate the decimal point at the correct position by moving it to the left the number of places indicated in the characteristic.

Example Number 3
Problem: Find the antilog of $\bar{1}.3$.
Solution: 1. The mantissa of 0.3 is entered in the calculator and the keys labeled "INV" and "LOG" are pushed and the value 1.9952623 appears, which is rounded to 2.0.
2. The characteristic is bar one, indicating that the decimal should be moved to the left one position, giving 0.2.
Answer: The antilog of $\bar{1}.3$ is 0.2.

Example Number 4
Problem: Find the antilog of $\bar{3}.6$.
Solution: 1. The mantissa of 0.6 is entered in the calculator, the keys labeled "INV" and "LOG" are pushed, and the value 3.9810717 appears, which is rounded to 4.0.
2. The characteristic is bar three, indicating that the decimal should be moved three positions to the left, giving 0.004.
Answer: The antilog of $\bar{3}.6$ is 0.004.

Answer Key

Question Number	Chapter Number														
	1	2	3	4	5	6	7	8	9	10	11	12	13	14	15
1	A	D	B	A	A	C	D	D	E	B	C	B	D	D	D
2	B	A	C	E	D	C	B	B	B	B	B	A	E	B	A
3	C	C	D	D	C	C	A	D	B	C	C	B	D	C	A
4	A	C	C	A	B	D	C	B	C	D	C	B	C	C	C
5	C	A	C	E	D	A	E	B	C	D	B	C	E	C	A
6	A	D	C	D	A	B	E	C	C	C	A	B	A	D	B
7	B	C	D	C	B	D	D	D	A	C	B	C	C	B	E
8	A	D	B	B	B	A	A	B		D	C	D	B	D	D
9	D	C	D	B	E	B	D				B	D	C	A	A
10	C	A	B	E	B		C					D	D		A
11	D	C		D	E		B					A			
12	B	C		B	D		A					C			
13	B	A		C	B		B					B			
14	C	A		D	A		C					E			
15	E	B		C	B		D								
16		A		B	C		D								
17		A		D	B		C								
18		C		D	B		B								
19		C		C	A		E								
20		C		B	C		D								
21				B	C		B								
22				D	A		E								
23				A	A										
24				D	D										
25				A	C										
26				C	B										
27				B	D										
28				D	B										
29				A	B										
30				D	C										
31				B											

Index

Douglas Ford Rea. Untitled Still Life. 1988. Still video photograph printed on an ink-jet printer.